Post-earthquake Fire Analysis in Urban Structures

Risk Management Strategies

Post-earthquake Fire Analysis in Urban Structures

Risk Management Strategies

Behrouz Behnam
Department of Civil & Environmental Engineering
Amirkabir University of Technology (Tehran Polytechnic)
Tehran, Iran

CRC Press is an imprint of the
Taylor & Francis Group, an **informa** business
A SCIENCE PUBLISHERS BOOK

CRC Press
Taylor & Francis Group
6000 Broken Sound Parkway NW, Suite 300
Boca Raton, FL 33487-2742

First issued in paperback 2021

ISBN-13: 978-0-367-78271-9 (pbk)
ISBN-13: 978-1-4987-4391-4 (hbk)

Visit the Taylor & Francis Web site at
http://www.taylorandfrancis.com

and the CRC Press Web site at
http://www.crcpress.com

Preface

Earthquakes are generally considered among the most destructive natural disasters; when an earthquake occurs, it leaves in its wake massive debris, destroyed buildings, and overwhelmed infrastructure. This, however, is not the end of the damage, as the situation can worsen if followed by new destructive events, such as post-earthquake fires (PEF). The history of PEF events shows that they have been much more destructive than the earthquakes that preceded them. Burned-out cities, dead and injured people, homelessness on a vast scale, and damaged infrastructure are only some of the consequences of PEFs.

It is thus of paramount importance to investigate PEF from different aspects. From the urban planning point of view, the social, environmental, and economic aspects need to be highlighted. Safety engineering has to address the strategies that can be employed to provide adequate safety to inhabitants under such conditions. Structural engineering should first scrutinize how structures would behave under earthquake loads alone, and then when combined with fire loads. These all point to the need for urban regions to be constructed to an integrated plan, where multi-hazard conditions are concurrently taken into consideration. This integrated planning is not an easy task, since it depends on very good quantification and estimation of all the mentioned aspects, something that is almost impossible. With regard to urban planning, for example, the environmental, social, and economic effects are often long-term and might be experienced over a period of decades after a disaster. Quantification of such effects is thus difficult. Providing a high level of safety is, of course, a necessity, but the difficulty is in identifying the desired level of safety. As there are different defined levels of safety, it can be difficult to reach a consensus. Structural engineering is also faced with some difficulties in quantifying the applied loads to a building structure over its lifetime. As for earthquake loads, they have a dynamic inherent, and thus they should be dynamically applied to building structures. The requisite of such application is further underlined when it is understood that the nonlinearity of the stiffness and damping are two very important factors in the structural materials, and the neglect of these may reduce the accuracy of the load application. Performing such a dynamic analysis is a very complex and time-consuming process, meaning that, if alternative methods were used to make some simplifications, the results would not be as accurate as when using dynamic methods.

Regarding the application of fire loads, the complexities are even greater. When an earthquake is considered, it is generally understood whether it could potentially occur or not. For instance, if there is no tectonic fault, there will be no fault-induced earthquake. Fire, however, is a highly random phenomenon, and its severity is highly dependent on numerous factors. The structural occupancy, the residents' behavior, timing, the season, etc., are only some of the factors that, although they have no meaningful correlation to one another, can all influence the occurrence of fire. Also, the existence (or absence) of fire fighting facilities inside and outside the structural buildings can highly influence a fire's severity. These all point to numerous difficulties in quantifying a fire load. On the other hand, when dealing with earthquake loads, it is fair to say that no changes to the material characteristics would occur during the loading process. This might be correlated with the small duration of earthquake loading. Therefore, ignoring some of the inherent structural behaviors (such as creep) might be justified. When dealing with a fire, however, the loads are often applied over a considerable duration of time, where it has a dominant role in changing the materials' characteristics. The changes might even continue long after the fire has been extinguished; thus, more complexity is involved in the quantification processes.

Although earthquake and fire codes have each been developed to provide minimum safety to inhabitants, neither currently consider a situation in which earthquake and fire loads are applied in sequence. Therefore, the structural components can be expected to have been weakened when subjected to PEF loads. The relevant safety provisions, which are often utilized in responding to one disaster, cannot also meet the need of a second disaster, in this case, a PEF event. For example, building structures might be equipped with sprinklers to battle against a possible fire, but sprinkler systems have been reported to incur considerable damage when faced with an earthquake. Therefore, there is often almost no measure in place to protect against a PEF event, and structural robustness is the last line of defense.

This book aims to quantify the resistance of moment-resisting reinforced concrete and steel structures under combined earthquake and fire loads. As there are no PEF codes currently available, the book aims to investigate how seismically-designed structures would be resistive when exposed to PEF loads. The book will be particularly useful for structural engineers, who often do not receive formal training concerning fire engineering; a fire engineering course is rarely included as part of systematic engineering curricula. In addition, although there are a number of useful books for addressing PEF events at a regional scale, few address such an event at a building scale. On the other hand, while most fire engineers would obviously possess in-depth information about fire engineering, they are not generally expected to be familiar with structural seismic engineering. Therefore, this book will also be useful for fire engineers whose intent is to become familiar with the behavior of seismically-designed structures under PEF loading.

It is evident that PEF creates a high-risk situation, particularly in urban regions. Based on conventional risk management procedures, where the potential risks in a system are first identified and then analyzed, some solutions can be proposed in order to respond to these risks. These solutions can be at both regional and building scale. In this book, more stress is placed on the building-scale solutions, and the main part of the text is dedicated to building structures. The risk management strategies in the book can particularly be useful for urban decision-makers who would like to improve their knowledge about multi-hazard conditions, and learn how to adopt some pre-disaster strategies to deal with such conditions. The book thus endeavors to fill the current gap that lies between different areas of research: fire engineering, structural engineering, and urban management. To meet this objective, the book delivers at least the minimum information pertinent to the mentioned research areas, presented through an engineering approach, in seven chapters. Each chapter commences with a short introduction (as headlines) followed by the required information. Examples are provided where applicable.

Chapter 1 provides historical information about past PEF events, from the beginning of the last century until recent years. The history covers not only general information, such as the number of fatalities, it also points to the reasons for the events. The chapter also provides a review of PEF modeling in urban regions. Modes of fire spread, as very important stages of the development of conflagration, are also discussed in this chapter.

Chapter 2 provides a review of the history of building codes relating to fire and seismic events. It is well-documented that the history of building codes falls into two separate eras, in which prescriptive-based codes and performance-based codes were developed. The approach behind these codes is first explained, and then examined, by revealing the differences between the prescriptive-based fire and seismic codes and the performance-based codes. Since damage to either the fire protection layers of steel structures or the cover of reinforced concrete structures can directly affect a building's PEF resistance, these forms of damage are also examined. Using the information about performance-based earthquake and fire codes, the main approach for performance-based PEF analysis is then introduced.

Chapter 3 provides detailed information regarding fire definition. As the application of design fire has a very important role in accounting for the fire resistance of structural components, the majority of this chapter is dedicated to introducing various fire curves. This chapter covers the history of fire design from the early 1800s, when time-temperature curves were developed, to more recent years, when advanced traveling fire simulations were developed. In addition, along with some of the best-known time-temperature curves, parametric curves, zone fires, localized fires, and traveling fires are also introduced and examples provided.

Heat transfer and materials behavior under elevated temperature are two important elements of fire engineering, and these are discussed in **Chapter 4**. First, the fundamental information regarding the mechanism of heat transfer is given, along with the equations that are required for computing that mechanism in the structural materials. Then, there is a discussion regarding the ways materials would behave under different rates of temperature increase. Only concrete and steel materials are discussed here, because the focus of this book is on steel and reinforced concrete structures; other materials are beyond the scope of the book.

Chapter 5 gives a brief summary of seismic analysis. Only the key points of various seismic analyses are presented, since it is believed that most earthquake engineers possess adequate information about this. Thus, to avoid redundancy and repetition of already-revealed documents, no detailed information is given. The fundamentals of plastic hinges and acceptance criteria are also pointed out in the chapter.

Chapter 6 explains the simulation of PEF loads in steel and reinforced concrete structures. In order to do that, sequential analysis is introduced, whereby the results of the nonlinear seismic analysis (performed by structural seismic software) are stored in a file and then transferred to subsequent software to perform the structural fire analysis. As any damage to cross-sections of reinforced concrete and fireproofed steel components can alter the frontiers of fire, this chapter provides adequate information to address this point. The chapter also provides numerous case study structures, of both low- and high-rise reinforced concrete and steel structures, under diverse fire and PEF scenarios. Most fire curves (as explained in Chapter 3), such as standard fires, parametric fires, and vertically and horizontally traveling fires, are employed in these case studies. A sensitivity analysis is also performed to understand which structural element, beams or columns, is more vulnerable under PEF conditions. Furthermore, the response of irregular steel structures under various fire and PEF scenarios is investigated, in order to discover whether they are more vulnerable to fire than regular structures.

Chapter 7 proposes some practical solutions as risk management strategies for PEF. The strategies are designed to comply with the conventional stages of natural disaster management, i.e. pre- and post-disaster strategies. The pre-disaster strategies include preparedness and prevention, while post-disaster strategies address response and recovery. This chapter, however, deals only with the pre-disaster strategies. To do that, separate investigations at regional (macro) scale, mesoscopic scale, and finally, building scale are performed. At the regional scale, the massive effects of PEF on urbanism, e.g. environmental and social, are investigated. The mesoscopic investigation is performed to find how some fire fighting facilities established inside and outside buildings can influence the PEF risk. Strategies at the building scale are mostly proposed to improve the PEF resistance of structural components;

existing structures and those yet to be designed are addressed separately. This chapter ends with some final remarks.

Finally, to familiarize the readers with input thermal and structural files, **Appendices A** and **B** detail the input files of the RC portal frame explained in section 6.4, and the five-story steel structure explained in section 6.8. The input files are for post-earthquake fire analyses.

This book aims to serve as a reference to deepen the knowledge of structural engineers regarding the fundamentals of fire engineering, to provide fire engineers with the essentials of structural engineering, and to give urban disaster managers some engineering solutions to reduce PEF risks. The review of collected literature does cite key contributions to the course of development, but it is not a comprehensive representation of all sources. The content is in the line of essential information for the abovementioned research areas, and is not meant to constitute a design guide, either for structural engineers or for fire engineers. It is also worth noting that timber structures remain popular in some seismic regions; they have been excluded from discussion in this edition of the book. Pre- and post-tensioned concrete structures are also used in the building of urban structures, but they are not discussed in the book, as it is believed that they require a wider and separate investigation. Lastly, it is hoped that this book can become a reliable resource for all concerned with PEF safety.

Behrouz Behnam
behrouz.behnam@uqconnect.edu.au

Acknowledgements

Writing a book about post-earthquake fires has been a goal for the author for some time. In 2014, the Taylor and Francis Group offered the opportunity to write a book about this important subject. Completion of this text would not have been possible without the direct and indirect support of numerous people. The author has endeavored to acknowledge the information collected from work conducted by other scholars; however, he believes that acknowledgement alone cannot tell the whole story. The author has been privileged to work with some highly competent scholars from different fields: structural engineering, fire engineering, and risk management; and at different universities. The author's most profound thanks go to Prof. Peter Dux and Prof. Jose Torero Cullen from the University of Queensland, Prof. Hamid R. Ronagh from Western Sydney University, and Prof. Martin Skitmore from Queensland University of Technology for their constructive support during his academic life to date. The author also wishes to express his deep appreciation to A/Prof. Ron Blackwell from the University of Queensland, Dr Chrysanthos Maraveas from the University of Manchester, Dr Mahmud Ashraf from the University of New South Wales, and A/Prof. Ahmad Tahershamsi from Amirkabir University of Technology (Tehran Polytechnic) for their helpful comments on different chapters of the book. The author would like to thank Prof. Jean-Marc Franssen and Dr Thomas Gernay from the University of Liege, Belgium, for their support while the SAFIR program was employed for solving the case studies presented in the book. The author most sincerely thanks Ms Debra Hamilton for her comments in better organizing the book chapters. She reviewed the entire manuscript, and not only assisted in straightening out the author's English but also made many constructive comments on the contents. The author also would like to thank his students at Amirkabir University of Technology (Tehran Polytechnic), most notably Mr Ali R. Razzazi, Mr Shahhid Abolghasemi, and Mr Mazdak Hatamianfar, who were involved in some of the studies reported here.

Completion of this book took a long time, and while the author was reading, writing or lost in thought, his family was not only patient, but supportive. The author's heartfelt thanks go to his wife Maryam and his son Amir for their patience and support during the work on this book.

Contents

Notations

Symbol	Description
K_i	initial elastic stiffness
rK_i	post-yield stiffness
K_e	secant stiffness
Δ_d	maximum displacement
Δ_y	yield displacement
F_u	design base-shear
θ	rotation
Δ	deflection
L	length of the member
θy	yield rotation
Δy	yield deflection
Z	plastic section modulus
E	modulus of elasticity
F_{ye}	expected yield strength of materials
l_b	beam length
l_c	column length
$I_{b,c}$	moment of inertia
P	axial force
P_{ye}	expected axial yield force
A_g	cross-section of the member
P_{CL}	axial compression capacity
P_{UF}	axial force in the member
M_x	bending moment in the member for the x-axis

M_y	bending moment in the member for the y-axis
M_{CEx}	expected bending strength of the column for the x-axis
M_{CEy}	expected bending strength of the column for the y-axis
m_x	value of m for the column bending about the x-axis
m_y	value of m for the column bending about the y-axis
M_{UFx}	bending moment in the member about the x-axis
M_{UFy}	bending moment in the member about the y-axis
M_{CLx}	lower-bound flexural strength of the member about the x-axis
M_{CLy}	lower-bound flexural strength of the member about the y-axis
t^2	t-squared
Q	heat release rate
Q_0	reference heat release rate
t_g	fire growth rate
R	risk
P	probability of exceedance
L	probability of loss
$P(S)$	success probability
$P(F)$	failure probability
t_u	untenable conditions
t_{oe}	occupant evacuation conditions
t_i	ignition times
T (°C)	temperature
t (hr)	time
T_0	ambient temperature
L''	fire load (wood)
A_t	area of the internal envelope
A_v	ventilation area
h	height of the ventilation opening
A_F	floor area of the compartment
q_t	fuel load per unit
k_f	factor
H_N	normalized heat load
K	thermal conductivity

ρ	density
δ	proportion of heat evolution in the compartment
c	specific heat
$(K\rho c)^{0.5}$	thermal inertia of the compartment boundaries
q_{fd}	design fire load density per unit floor area
c'	conversion factor
w_f	ventilation factor
y_{n1}, y_{n2}	safety factors
b	wall factor
λ	thermal conductivity
O	opening factor
h_{eq}	average height of the openings
Γ	expansion factor
$q_{t,d}$	fire load density of the area of enclosure
$q_{f,d}$	occupancy of the fire compartment
A_{roof}	floor area
A_t	total area of the enclosure
δ_{q1}	risk of fire activation
δ_{q2}	risk of fire activation owing to the type of occupancy
δ_n	fire fighting measure factor
m	combustion factor
$q_{f,k}$	characteristic fire load density
t_{lim}	the shortest possible duration of the heating phase
t_{max}	the longest duration of the heating phase
$\dot{Q}$	heat release rate
$\dot{m}$	mass flow rate
γ	combustion coefficient
H_{net}	net calorific value
A_w	area of the openings
h_w	height of openings
$\dot{Q}_{max}$	maximum heat release rate
Z	vertical axis
Δt	time interval

ΔZ	a part of the wall height
D	diameter of the fire
L_f	flame length
$\dot{h}$	heat flux received by the surface area at the ceiling level
y	a parameter
L_h	horizontal flame length
z'	vertical position of the virtual heat source
Q^*_H	dimensionless heat release rate
Q^*_D	Froude number
$\dot{h}_{net}$	net heat flux
α_c	convection heat transfer coefficient
T_g	gas temperature
T_s	surface temperature
ε	surface emissivity of the element
σ	Stephan Boltzmann constant
t_b	characteristic burning time
t_{otal}	theoretical maximum time
T_{max}	maximum temperature
r	distance from the center of the fire
s	spread rate
L_f	length of the fire
Δx	grid size
T_{nf}	near field temperature
T_{ff}	far field temperature
$\dot{q}''_{tot}$	total energy balance
$\dot{q}''_{rad}$	radiation heat flux
$\dot{q}''_{conv}$	convection heat flux
$\dot{q}''_{cond}$	conduction heat flux
K_i	conductivity coefficient
h_c	convection coefficient
B	vertical surface configuration
g	the ground acceleration
P_r	Prandtl number

v	air viscosity
k	air conductivity
ΔT	temperature difference between the air and the surface
$\dot{q}''_{rad, g}$	radiation for the gas
$\dot{q}''_{rad, w}$	radiation between a structural element and wall
Φ	configuration factor of the element
ε_g	gas emissivity
T_w	wall's temperature
A_1	areas of the solid and furnace
A_2	view factor
$F_{1\text{-}2}$	distance between the walls and the solid
κ	extinction coefficient of the gas between the wall and the solid
α	thermal expansion
ν	Poisson's ratio
$\Delta\varepsilon$	total strain
ε_σ	mechanical strain
ε_{Th}	thermal strain
ε_{cr}	creep strain
ε_{tr}	transient strain
C_a	specific heat of steel
c_p	specific heat of concrete
K	nonlinear stiffness matrix
U	displacement vector
P	predefined load vector
P_e	reaction vector
ΔU	calculated displacement increment within an iteration
K_T	current nonlinear stiffness matrix
λ	load factor within the corresponding load increment
P_0	initial load
B	strain-displacement matrix of each element
σ_{NL}	vector of the element's nonlinear stress
K_i	elastic lateral stiffness

K_e	effective lateral stiffness
V_y	effective yield strength
K_s	post-yield slope
V_y	yield base-shear
V_T	total base-shear
Δ_y	yield displacement
Δ_T	target displacement
T_e	effective fundamental period
S_a	spectral response acceleration
$S_a (T_e/2\pi)^2 g$	spectral displacement
V_1	first increment of the lateral load
δ_1	first increment of the displacement
T_1	first increment of the period
C_0	coefficient to modify the roof displacement
w_i	portion of the seismic weight
φ_i	amplitude of the shape vector
C_1	coefficient to modify the expected maximum inelastic displacement
C_2	coefficient to modify the energy absorption and dissipation capacities
C_3	coefficient to modify the P-Δ effects
M_{cr}	cracking moment
M_y	yield moment
M_u	ultimate moment
l_p	plastic hinge length
Φ_{cr}	cracking curvature
Φ_y	yield curvature
Φ_p	plastic curvature
Φ_u	ultimate curvature
C_{PEF}	post-earthquake fire coefficient

Abbreviations

PEF	Post-earthquake fire
GNP	Gross national product
PGA	Peak ground acceleration
AHP	Analytical hierarchy process
NBC	National Building Code
UL	Underwriters Laboratories
NFPA	National Fire Protection Association
UBC	Uniform Building Code
NEHRP	National Earthquake Hazards Reduction Program
FEMA	Federal Emergency Management Agency
MCE	Maximum Considered Earthquake
O	Operational
IO	Immediate Occupancy
LS	Life Safety
CP	Collapse Prevention
DDBP	Direct displacement-based approach
MDOF	Multi-degree-of-freedom
SDOF	Single-degree-of-freedom
MRRCF	Moment-resisting reinforced concrete frames
MRSF	Moment-resisting steel frames
BSF	Braced steel frames
RCSW	Reinforced concrete shear walls
UMIW	Unreinforced masonry infill walls
UMNW	Unreinforced masonry non-infill walls

DCR	Demand over capacity
RC	Reinforced concrete
JBDPA	Japan Building Disaster Prevention Association
SFRM	Sprayed fire-resistant materials
WTCT	World Trade Center Towers
GDP	Gross Domestic Product
HRR	Heat release rate
ASTM	American Society for Testing and Materials
NBS	National Bureau of Standards
ISO	International Organization for Standardization
CFD	Computational fluid dynamic
NRC	National Research Council
BJFRO	British Joint Fire Research Organization
HERA	Heavy Engineering Research Association
FDS	Fire Dynamic Simulator
HSC	High-strength concrete
LWC	Lightweight concrete
NC	Normal concrete
LSA	Linear static analysis
LDA	Linear dynamic analysis
SRSS	Square root of the sum of squares
CQC	Complete quadratic combination
CERT	Community Emergency Response Teams
MCDM	Multi-criteria decision-making
SGTT	Safety Guaranteed Time Target
FRP	Fiber-reinforced polymers
GFRP	Glass-based fiber-reinforced polymers
CFRP	Carbon-based fiber-reinforced polymers

Chapter 1

Introduction to Post-earthquake Fire

1.1 Introduction

The Earth is always faced with various disasters, both natural and man-made. It is well-accepted that, second to devastating wars, natural disasters are among the greatest enemy of humankind, with earthquakes being amongst the most severely destructive. The history of earthquakes that occurred in the last century shows that millions of people have lost their lives or been injured, while many more have become homeless. The combined death toll from the earthquakes that took place in 1908 in Messina, Italy, in 1920 in Haiyuan, China, in 1923 in Kanto, Japan, in 1976 in Tangshan, China, in 2004 in the Indian Ocean and in 2010 in Haiti, was greater than two million. It is evident that if it was possible to collect the data from all earthquake casualties during earlier centuries, the results would be much more horrific.

Most earthquakes are followed by post-earthquake events, e.g. liquefaction, aftershocks, landslides, etc., and the damage from these events may even be worse than that from the earthquake itself. Among these, post-earthquake fire (PEF) is considered one of the most destructive earthquake-based events to be widely experienced in urbanized regions. The scale of PEFs has rarely been vast and they have often occurred as scattered fires, but they have sometimes been of catastrophic proportions. Recorded information has proven that damage from PEF has, on occasion, been four times that from the earthquake itself. PEF can cause damage not just to buildings, but also to the environment, society, economy, and industry. While investigation of these forms of damage in urban regions deserves attention, this book places more emphasis on the damage to buildings. The damage from PEF to building structures is two-fold: the damage due to the burning of combustible materials, which possess substantial value; and the damage caused by excessive structural load on the structural members.

As urban structures are often designed to withstand only one extreme load, such as wind, fire or earthquake, they are thus too weak when encountering two subsequent loads, such as earthquake followed by fire. In that case, the structures would collapse much more quickly than in a single-load situation. As well, when a building is exposed to fire alone, that fire can normally be extinguished through facilities established in the building, such as sprinklers and vertical pipes. Urban facilities, such as fire fighting and rescue teams, can also be used to extinguish the fire before it leads to a catastrophe. For these reasons, most fire events in ordinary situations do not get out of control and become a conflagration, i.e. where the fire spreads to surrounding structures and grows to a vast scale. After an earthquake, however, there is no guarantee that the facilities established in the buildings will work properly. Rather, there is ample evidence showing that internal fire extinguishing facilities commonly sustain severe damage after an earthquake. There is also a strong possibility that buildings will collapse, often blocking roads, thus increasing the difficulty experienced by rescue teams in reaching and helping people trapped by the fire. Therefore, the number of casualties of a fire following an earthquake will probably be much higher than for other fire situations.

1.2 Historical information about post-earthquake fires

PEFs are extremely variable events, and damage from these fires may range considerably from very minor to extremely major. The variability of PEFs pertains to two factors: variability in the number of ignitions; and variability in the extent of spread of the fire. Both of these factors can be influenced by a building's materials, its density, the wind speed and direction, etc. While PEF events have been observed in almost all large earthquakes, some of the most significant of the recorded events are reviewed here.

1.2.1 The 1906 San Francisco, California PEF

On 18 April 1906, the city of San Francisco was extensively damaged by a powerful earthquake with a magnitude of 8.3 on the Richter scale [1]. The earthquake's origin was the San Andreas Fault, one of the longest faults on the Earth, with a length over 960 km. The earthquake itself caused significant damage to infrastructure and buildings. However, the quake did not account for the total loss. Instead, over 50 ignitions that occurred just a short while after the shock caused a conflagration over four successive days and nights. The fire itself produced around 80% of the total damage, and the economic loss alone was calculated at more than $350 million (in 1906 dollars). More than 3,000 people died and over 28,000 buildings (mainly timber structures) were destroyed. As recorded, the fire's causes were ruptured gas pipes, broken chimneys and short circuits in internal electrical wiring. Very quickly, the fire went out of control and developed into a conflagration. In addition to a failure of the water supply [2], there were other reasons for the spread of the fire, the most important of which was the incorrect use of dynamite by some inexpert firefighters (when they endeavored to destroy a line of buildings to create a fire-break). Those buildings caught fire as well, causing the fire to further intensify. Additionally, it is said that some owners had intentionally set fire to their partially-damaged buildings, in order to claim insurance, fire being the only event for which they had been able to get insurance cover. It has also been documented that while most fire departments sustained no significant earthquake damage, the lack of communication systems and fire alarms (due to the quake damage), caused major failures in getting information to the fire services. Climate conditions at the time, such as low humidity and changing winds, also had a significant role in spreading the fire.

1.2.2 The 1923 Kanto, Japan PEF

On 1 September 1923, the Kanto region of Japan, including the Tokyo/Yokohama conurbation, was hit by a strong earthquake with a magnitude of 8.3 on the Richter scale. This shock was considered the worst natural disaster ever to hit the seismic region of Japan. The earthquake, however, was not the only catastrophe. It was followed by a two-day PEF, resulting in over 140,000 deaths and many more people injured. It has been reported that more than 694,000 houses were partially damaged or completely destroyed, mainly due to the fire. The total economic damage was estimated around $1 billion (in 1923 dollars), with the PEF itself contributing about 78% of the overall loss. Thousands of people were reported trapped by the fire, since destroyed bridges and railroad tracks and traffic congestion prevented their escape. Meanwhile, hundreds of people had found refuge in places safe from the conflagration. However, a sudden change in the wind direction redirected the fire toward these people, almost all of whom perished.

Regarding causes of the fire, since the earthquake occurred around midday, many people were cooking over coal- or charcoal-based stoves at the time. This in turn was one of the main reasons for the fire spreading throughout the city just a few minutes after the shock. As most urban facilities (including water supply systems) had sustained extensive earthquake damage, there was a serious difficulty in extinguishing the fire. Thus, people were besieged by the fire but had no way to extinguish it. Also, as there was significant earthquake damage to the communication systems, it was almost impossible to inform people outside the city and to ask them for help. At the time, no one even knew whether this disaster has occurred in the Kanto region alone, or if it was a countrywide catastrophe. Inappropriate storage of flammable materials also intensified the fire's spread. Additionally, another horrific event also occurred on the first night after the earthquake. The harbor had been occupied by evacuees, who were embarking ships to leave the city on the next day. Sadly, flammable materials, such as oil, had seeped into the water and caught fire. Panicked people tried to flee and many of them were killed or wounded. In addition to the reasons listed above for the rapid spread of the fire, the high density of mainly wooden buildings in the city was also an important factor, as were the weather conditions [4].

1.2.3 The 1931 Hawkes Bay, New Zealand PEF

On 3 February 1931, the Hawkes Bay area of New Zealand, including the towns of Napier and Hastings, was hit by an earthquake with a magnitude of 7.8 on the Richter scale. While the primary shock brought about significant damage to the bay area, including buildings, infrastructure and public health facilities, it was additionally followed by numerous aftershocks, around 550 in two weeks. Napier's hospitals were severely damaged, such that patients had to be dispatched to nearby regions that had adequate facilities or to emergency hospitals that had been set up outside the earthquake-affected regions. Shortly after the main earthquake, some PEFs occurred. Although Napier's gas supply was wisely shut off immediately after the earthquake, and the risk of electrical short circuits was reduced (because of properly working fuses on street power poles), three separate fires broke out from three chemist shops, where quantities of flammable materials had been stored. At first, it was presumed that the fires would not spread to other buildings, due to the very calm conditions at the time. Soon after, nevertheless, fast winds started to blow up and caused the fire to spread to the main parts of the central business district. As recorded, more than a hundred fires igniting inside half a day were the reason for a major conflagration, resulting in the destruction of over 400 buildings in an area of 10 acres and the deaths of more than 260 people. While there is evidence that firefighters used dynamite to destroy a number of buildings located in the fire's path as a way to create a fire-break, it is believed that if there had not been wide open spaces in the fire-affected regions, the fire would not have been brought under control. As documented, the central fire station in the earthquake area had been severely damaged just after the main shock and, as a result, fire engines had been covered in debris. These engines could not be deployed when the fires started. The earthquake had also caused the water supply system to sustain substantial damage. Even in the places where it had sustained only minor damage, the water pressure was quite weak, and after an hour it failed completely. Therefore, those fire services that had remained operational after the shock did not have adequate water to fight the fire. For those areas that were accessible, however, the fire brigades utilized some seawater and wastewater as a means of controlling and extinguishing the fire. It is also worth mentioning that while the main earthquake had damaged the communications systems, such as telephone lines and the telegraph, requirements for help were transferred by ships near the bay using wireless systems. The Hawkes Bay's PEF is considered the deadliest fire-based disaster in the history of New Zealand [6].

1.2.4 The 1933 Long Beach, California PEF

On 10 March 1933, Long Beach in California experienced an earthquake with a magnitude of 6.3 on the Richter scale. As reported, due to considerable damage to weak buildings, more than 140 people died, and 2000 were injured. It is believed that if not for the time of the shock (around 6.00 pm, when schools had finished their daily activities), the number of casualties would have been significantly higher, since the schools' structures were also weak. The earthquake caused the main water supply to sustain damage and, as a result, there was some decline in the water pressure. On the other hand, while around 15 PEFs occurred shortly after the earthquake, none of them resulted in a conflagration. This is mainly related to the well-engineered automatic cut-offs for the gas and electricity supplies. There are also reports showing that while more than 60% of the sprinkler systems inside homes had sustained damage and were thus out of commission, those that remained operational had worked properly [8].

1.2.5 The 1964 Niigata, Japan PEF

On 16 June 1964, Niigata city was shaken hard by an earthquake with a magnitude of 7.5 on the Richter scale, resulting in severe damage to buildings and infrastructure, including water supply systems. Some factories and oil refineries were also severely damaged. The earthquake caused a tsunami to occur and, as a result, a large amount of floating oil leaked out from the damaged industrial facilities and was conveyed to some parts of the city by the wave of water. Soon after this, nine PEFs occurred in various urban and industrial areas, four of which were instantly extinguished by fire brigades, with the rest going quickly out of control and resulting in a conflagration [9]. It was said that road closures and access conditions had prevented fire fighters from attending all of the fires sufficiently quickly. Some fires lasted for about two weeks, by which time there was nothing remaining to burn. While there is no information documented about the weather conditions when the fires occurred, it is believed that the oil conveyed by the tsunami played a significant role in spreading the fire. Overall, more than 2200 buildings were destroyed, about 350 of which were due to the PEFs. Niigata's PEF is considered one of the most complicated industrial fire events in the history of Japan.

1.2.6 The 1971 San Fernando Valley, California PEF

On 9 February 1971, the San Fernando Valley was shaken by an earthquake with a magnitude of 6.5 on the Richter scale, followed by more than 35 aftershocks. The earthquake caused significant destruction to the city and surroundings with about 20,000 buildings sustaining damage, approximately 830 of which were demolished [10]. It was reported that 64 people lost their lives and more than 2500 were injured. Infrastructure, including power supplies and gas pipes, was severely damaged. Leaking gas caused some fires to develop; others were caused by electrical short circuits. As the telephone lines had also been disconnected, there were communication difficulties for fire brigades and rescue teams trying to reach the scene. Overall, there were 116 fire events across the earthquake-affected region, mostly occurring in residential buildings, though none of them resulted in a conflagration. That result can largely be attributed to the absence of conflagration-inducing factors, such as extreme weather conditions or dense fire load. As well, although the earthquake caused water lines to sustain damage and thus some decline in water pressure, it was reported that the fire brigades used swimming pools as an available water source.

1.2.7 The 1972 Managua, Nicaragua PEF

On 9 December 1972, Managua in Nicaragua was shaken around midnight by three consecutive earthquakes, with magnitudes of 5.5 to 6.5 on the Richter scale. While earthquakes of these magnitudes would not usually cause serious damage, in this case there was significant damage to buildings and infrastructure. It is believed that this damage was due to non-seismic-resistant structures and poor emergency planning [11]. As reported, over 6000 people died, more than 20,000 were injured and about 250,000 became homeless. Damage to both the electrical distribution systems and to the telephone lines was reported to be significant, which caused difficulties in getting the rescue teams to the damaged regions. Most fire departments collapsed, and their fire engines and equipment were buried under the debris. After the earthquakes, the water supply system was impaired, resulting in a disruption to the water distribution and hydrants, causing a grave impediment for fire fighting operations. Shortly after the shocks, five fires broke out, resulting in a conflagration that lasted over three days. Very dense areas with high fire loads per home caused vast areas of the city to burn completely, including numerous multi-story buildings. As reported, the building of the First National City Bank (a six-story reinforced concrete building) was among those that completely collapsed, even though it had not sustained significant earthquake damage. Its failure was related to the inadequate ability of the plaster-reinforced concrete to withstand the fire. It has also been reported that some areas were still burning, even ten days after the shocks, but it is believed that at least some of these fires were not due to the earthquakes, but had been deliberately lit by criminals.

1.2.8 The 1984 Morgan Hill, Northern California PEF

On 24 April 1984, Morgan Hill in the north of California was hit by an earthquake with a magnitude of 6.2 on the Richter scale. The earthquake was felt over a large part of the San Francisco Bay area, but it did not cause significant damage to buildings or infrastructure. There was, however, some minor damage to telephone lines, gas line systems, electricity power lines, and water supply systems, none of which were severe enough to cause major disruption to the rescue teams accessing the shocked areas [12]. Minor damage to non-structural components was also reported in some buildings. The earthquake caused six residential and commercial fires to break out, though none of them developed into a conflagration. It was believed that the fire sources were damaged gas pipes inside homes, and that a prompt response from the fire brigades prevented the fires from getting out of control.

1.2.9 The 1985 Mexico City PEF

On 19 September 1985, Mexico City was severely shaken by an earthquake with a magnitude of 8.1 on the Richter scale, lasting for about one minute. This shock is generally considered to be the most devastating earthquake to have occurred in Mexico (and even in the whole of North America) in the last century. A large aftershock then occurred on the following day, with a magnitude of 7.5 on the Richter scale, which intensified the damage from the primary earthquake. Combined, these shocks left more than 10,000 fatalities, 30,000 injured and beyond 50,000 homeless [13]. Hundreds of buildings (both low- and high-rise) collapsed, resulting in many streets in the city becoming blocked. The water distribution systems were greatly damaged, resulting in a complete loss of water supply to many districts. The earthquake also caused most electricity distribution networks to fail, leading to a severe outage in a central part of the city. As well, the telephone lines in most of the affected areas were severely damaged, causing communications to be interrupted and preventing updates about the city's condition from reaching the rescue teams.

Importantly, this earthquake triggered about 200 PEFs, although none of these resulted in a major conflagration. That fact was mainly attributed to the absence of potential fuel for the fires – most buildings in Mexico City had not been made of combustible materials, i.e. they were not wooden structures. As well, there were no gas supply systems and, as such, no gas leaks could occur after the shock.

1.2.10 The 1987 Whittier Narrows, Los Angeles PEF

On 1 October 1987, Whittier Narrows in Southern California experienced an earthquake with a magnitude of 5.7 on the Richter scale. Some aftershocks were also felt during the next days; the largest of these had a magnitude of 5.5 on the Richter scale. As reported, the earthquake and its aftershocks caused significant damage to buildings (around 10,000) and infrastructure, and as a result, a few people were killed and hundreds were injured. The earthquake caused some disruption to the electricity systems and the gas pipes. Shortly after the main shock, the fire departments in Los Angeles were informed of 112 fire events in residential and commercial buildings, although none led to a conflagration [15]. In addition, the fire departments responded to around 55 gas leaks, but it is not reported how many of these led to fire. It is believed that the major reason for the successful extinguishment of almost all of the fires was the good performance of both the water and communication systems, which had sustained only minor damage after the earthquake.

1.2.11 The 1989 Loma Prieta, Northern California PEF

On 17 October, 1989, Northern California was hit by a powerful earthquake with a magnitude of 7.1 on the Richter scale, centered on the Loma Prieta peak in the Santa Cruz Mountains. This was the largest earthquake to occur in Northern California since the San Francisco earthquake of 1906. The earthquake caused over 60 deaths, around 3500 people were injured and more than 12,000 were made homeless. It also was the cause of electricity outages, gas leaks, and water supply damage in the Bay area. Most buildings designed based on modern seismic codes sustained only minor damage, but those not designed to meet the latest seismic codes showed weak performance during the shock [16]. The earthquake caused 27 PEFs to break out, mostly due to gas leaks from damaged pipes. While there were some massive fires, no conflagration was reported, mostly due to the weather conditions at the time. For instance, there was no wind after the earthquake and fires. Rain just prior to the shock had also led to high moisture levels, which naturally hampered the spread of the fires. If these circumstances had not existed, a very different situation might have been observed.

1.2.12 The 1993 Hokkaido Nansei-oki, Japan PEF

On 12 July 1993, Hokkaido Nansei-oki was intensely shaken by a powerful earthquake with a magnitude of 7.8 on the Richter scale. The earthquake lasted for about a minute and caused a destructive tsunami to occur. The death toll was reportedly about 240, mostly due to the tsunami. More than 540 buildings, both residential and public, were severely damaged. Infrastructure also sustained significant damage, particularly railways, highways, bridges, and ports. Shortly after the shock, some fire events caused a conflagration to develop, which caused the destruction of about 190 buildings in a major part of the city [18]. The conflagration lasted more than 11 hours. The sources of the fires were never understood, but it was guessed that cooking appliances may have been the reason. Although firefighters had tried to stop the progression of the fires, they were not successful. This failure was attributed to the presence of outdoor fuel tanks, which were commonly installed on the front of houses. The fire's spread also pertained to the closeness of most buildings and the narrow streets, with a width around 3 m. Debris from the destroyed buildings also hampered the firefighters in their attempts to combat the fires. Although there is no documented report regarding the water supply situation after the earthquake, it is known that the water stored in the city was not adequate for extinguishing such large-scale fires.

1.2.13 The 1994 Northridge, Los Angeles PEF

On 17 January 1994, Northridge in the San Fernando Valley experienced an earthquake with a magnitude of 6.7 on the Richter scale, which resulted in one of the most disastrous events in the history of Southern California. The earthquake was large enough to be felt even hundreds of kilometers away from its origin. The earthquake was also unique in that it occurred on a relatively small local fault, instead of along a major boundary. In spite of the widespread physical damage from the earthquake, the number of casualties reported was less than could be expected – 58 people died, and about 1500 were injured. The earthquake caused the region's infrastructure to sustain severe damage, including electricity power lines, and gas and oil pipelines. Damage to the water supply system was severe, with more than 3000 leaks in the pipes and six breakages in the pumping stations and storage tanks reported. This damage caused weak sequential water pressure over a central part of the San Fernando Valley. More than 800 buildings sustained damage, including both seismic-designed and non-seismic-designed structures, although only non-earthquake-resistant structures were damaged to the degree that they collapsed.

Soon after the shock, more than 110 fire events were reported, which mostly started from residential buildings and which were attributed to short circuits or gas leaks [20]. However, a considerable number of the fires were extinguished due to the rapid response of the fire brigades. Thus, no conflagration took place. It is believed that the well-equipped fire services and the high quality of their training had significant roles in overcoming the fires. This achievement is highlighted by the point that the fire brigades had to respond to more than 2200 requests for help in the days following the earthquake. There were other reasons as well that hampered the spread of fire, including weak wind, humidity, the nature of the building materials and the separation between buildings. It is also worth mentioning that many well-established sprinkler systems in the seismic-designed buildings remained serviceable, while no documented report was found to describe the condition of fire detection systems after the earthquake.

1.2.14 The 1995 Kobe, Japan PEF

On 17 January 1995, Kobe (a long, narrow city with a width of around 3 km) was shaken by a powerful earthquake with a magnitude of 7.2 on the Richter scale. The earthquake was said to be the second most devastating earthquake in Japan (after the 1923 Kanto earthquake) and it left around 6000 dead and 33000 injured. More than 100,000 buildings, including residential, commercial and industrial structures, collapsed due to the quake and its following fire [22]. The city's infrastructure also sustained severe damage, including to pipes and telephone lines. Although Kobe had an advanced control center to respond to events such as fire, it was not able to work properly due to interruption of the telephone lines. Therefore, every fire station in every suburb had to act separately. This caused an irregularity in the distribution of available teams in responding to fires. It is reported that around 15 minutes after the earthquake, 53 fires broke out, a number which rapidly tripled [23]. Figure 1.1 shows the PEF's distribution during the period 17 to 19 January in the city. As the firefighters could not control the fires, they developed into a number of conflagrations. Most fires lasted longer than 24 hours, although some of them burned for more than two days. The fires destroyed about 5000 buildings over an area of approximately 660,000 m^2. Almost all types of buildings suffered from the earthquake and its following fires, including steel, timber and old concrete frames. As reported, there were several factors in the fires' development and spread; the most important were broken gas pipes, short circuits, and demolished buildings. Combustible debris had also provided a path for the fire to spread. As well, some fires that had been lit to provide heating for survivors also went out of control (though some of these were suspected to be due to arson). It is also said that some of the fires occurred as a result of 'rapid recovery' actions, e.g. reconnecting the electricity for home customers without adequate fire safety prerequisites in place. According to some reports, the fires traveled across narrow alleys due to the position of burning vehicles or to thermal radiation.

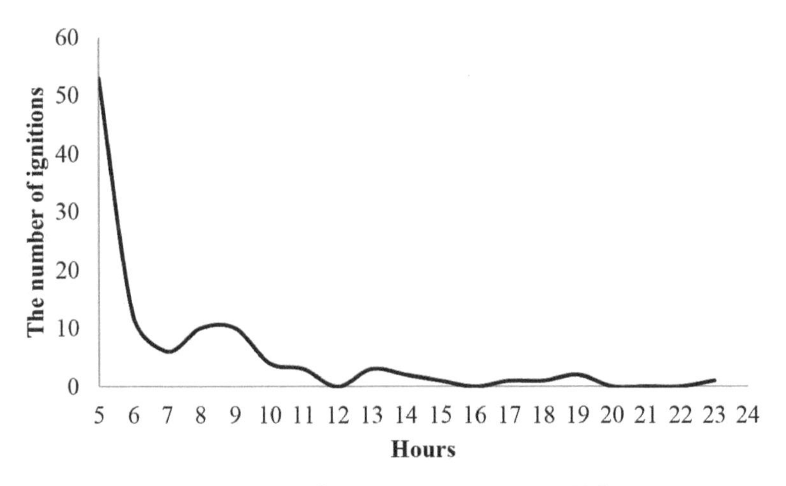

Fig. 1.1: PEF events from 17 to 19 January in Kobe city.

1.2.15 The 1999 İzmit, Turkey PEF

On 17 August 1995, the İzmit region in northwest Turkey was severely shaken by a great earthquake with a magnitude of 7.6 on the Richter scale (this is also known as the Kocaeli earthquake). The earthquake lasted for about 37 seconds and caused more than 17,100 deaths, while 44,000 people were injured. Some reports put the number of people killed at almost equal to the number injured [25]. The earthquake affected about 15 million people in a large area of Turkey and left at least 500,000 homeless. The total direct damage was estimated to be more than 23 billion dollars. Reports also mentioned that more than 120,000 buildings sustained severe damage, and more than 2,000 buildings collapsed (these reports implicitly showed that most of the buildings had not been designed based on seismic codes). In addition, the infrastructure was damaged extensively, including the water and electricity systems, bridges and roads. The industrial sector of İzmit, which was one of the more heavily industrialized parts of Turkey, was badly damaged. As reported, this part of Turkey generated around 35% of the gross national product (GNP) [26]. The earthquake also brought about two remarkable events: a PEF at TUPRAS İzmit refinery in Korfez, Kocaeli, which lasted for more than five days and was suppressed only with international support; and a release of toxic substances from the AKSA production plant in Ciftlikkoy, Yalova, which resulted in environmental contamination affecting domestic animals and agriculture. It took more than five years to clean up the effects of this contamination. Both of these post-earthquake disasters led to the mandatory evacuation of people from the affected regions. The evacuation process in turn hampered the rescue operations.

Regarding the PEF, it began within a few minutes after the shock, from three different locations in the refinery. The first fire started from the chemical warehouse but was successfully brought under control in less than one hour. The second fire started from one of the oil processing plants, which had flammable substances throughout the pipes. This was also extinguished in a few hours. The third fire started successively from four tanks located in the refinery center. While fire fighters were struggling to extinguish these fires, one of the tanks failed suddenly, resulting in a jet of fuel spreading the fire, which quickly became a conflagration. The fire fighters were consequently unsuccessful in extinguishing the conflagration and they had no choice but to call on the national services. However, even the national services could not put the fires out and the government asked for the international help. The fires were finally extinguished after five days, with huge damage left to the refinery and neighborhood. As reported, the main reasons for the failure to quickly extinguish the fires related to the damaged electricity and water systems after the shock. An inadequate sprinkler system in the refinery was also said to be an important factor in spreading the fires.

1.2.16 The 2003 Tokachi-oki, Japan PEF

On 26 September 2003, the Tokachi-oki region was hit by an earthquake with a magnitude of 8.0 on the Richter scale. It is said that the earthquake was ten times larger than the Kobe earthquake in 1995. The earthquake was classified as a shallow earthquake with a depth at epicenter of about 42 km. Some aftershocks also followed the main earthquake, the biggest of which had a magnitude of 7.0 on the Richter scale. The earthquake left over 280 people injured, but no deaths were reported. In addition, thousands of people were forced to evacuate the affected regions, mostly to provide adequate safety against possible tsunami. Overall, buildings in the affected areas did not sustain significant damage, except for some damage in foundations, mostly due to liquefaction [27]. While there are reports mentioning that some areas were affected by a tsunami with very small waves, e.g. in the Toyokoro-cho region, almost no serious tsunami occurred after the shocks. Immediately after the earthquake, the electricity and gas supplies to more than 370,000 houses were shut down, as there were established automatic shutdown systems in place. However, these outages were brief, and the power and gas were reconnected within a few hours. The water supply to around 16,000 houses was cut off, due either to the electricity outage or to damage to the piping. There are also some reports of water contamination after the earthquake. Other infrastructure, such as roads, ports, and phones, sustained some damage, but this was not significant.

Shortly after the main earthquake, a fire started from a tank containing raw oil at the Hokkaido Refinery in Tomakomai. This PEF then spread through the piping systems in the refinery and caused a large fire to develop. As documented, the fire's cause was most likely due to the roof type of the oil tank. The roof was a kind of floating cover moving up or down as the volume of reservoir oil rose or fell. During the earthquake, there was possibly some friction between the roof cover and the sidewall of the tank, resulting in the fire igniting [28]. It took about eight hours to control and extinguish the fire. Importantly, another fire (possibly with the same cause) started from another tank on 28 September, two days after the main shock. This fire was also extinguished quickly and did not get out of control.

1.2.17 The 2011 Tōhoku, Japan PEF

On 11 March 2011, the Tōhoku region in Japan was severely shaken by a strong earthquake with a magnitude of 9.0 on the Richter scale. This earthquake is also known as the Great East Japan earthquake. It is believed to be not only one of the most devastating earthquakes in the history of Japan, but even worldwide. The earthquake's epicenter was underneath the sea, as a result of which, tsunami waves with heights as high as 40 m and lengths of over 180 km impacted on a long line of coastal cities. As reported, 15891 people were killed, 6152 injured, and 2584 people went missing. In addition, more than 23,000 people lost their homes [30]. The earthquake and the tsunami that followed brought about very severe damage to buildings and infrastructure, including factories, roads, bridges, and ports. Most urban regions in northeast Japan were affected by this mega-disaster, resulting in about 4.4 million homes losing electricity and 1.5 million being without water. The tsunami also caused the Fukushima Nuclear Plant to sustain severe damage, resulting in leaks in three reactors. As the result of this, residents within an area of 20 km around the plant were forced to evacuate.

The earthquake and tsunami then caused numerous fires to break out, rapidly reaching more than 293 separate fires, causing conflagrations [31]. It is important to note that there were differences between the fires caused by the earthquake and those caused by the tsunami. Most of the earthquake-based fires had a root cause in the buildings, e.g. overturning of fire appliances, short circuits in the electricity wires, etc. The tsunami-based fires seemed to be due to the release and spread of hazardous materials caused by the tsunami waves. Most houses in Japan are equipped with propane cylinders. When tsunami waves hit those cylinders, they became detached from their regulators, as the result of which, the gas started to leak. Some of the detached cylinders traveled a long distance on the tsunami waves, further increasing the fire risk. Many houses, particularly wooden houses, were ignited by these floating cylinders. In addition to the gas-based fires, another cause for the spread of the fire related to oil stored in the tanks of industrial plants and ships. Many industrial plants and ships had badly been damaged and, as a result, considerable amounts of oil were distributed across a wide area by the tsunami. The oil was then absorbed by timber materials, which became very flammable. The city of Kesennuma is one example of such an ignition, and almost the city entirely was engulfed by flames. It is also worth mentioning that, according to the report released by the World Bank, the total damage due the earthquake and its aftermath was above 235 billion dollars, certainly one of the most costly natural disasters in world history [32].

1.2.18 The 2014 northern coast of Chile PEF

On 1 April 2014, Iquique, in the north of Chile, was severely shaken by a powerful earthquake with a magnitude of 8.2 on the Richter scale. The earthquake caused about 200 buildings to incur severe damage or collapse, but only 6 people died and 9 were injured (it is reported that four of the victims died due to heart attacks). The earthquake and its following aftershocks also caused more than 90,000 residents to be evacuated from the affected regions. About 45 minutes after the main shock, tsunami waves of about 2 m height arrived at the coast, but as people from low-lying coastal regions had already been evacuated, no one was caught by the waves. The earthquake caused power outages affecting thousands of buildings, damage to the Iquique airport, and PEFs affecting several commercial buildings. The PEFs, however, were rapidly extinguished by fire fighters and were never out of control. No documents report any casualty due to the PEFs.

1.3 PEF simulation at regional scale

It can be understood from the history above that there are high degrees of randomness and complexity inherent to PEF events, which make it difficult to design an accurate model. Numerous researchers have tried to develop methods to simulate PEF at regional scale. Although these models are different in terms of the methodologies proposed, all point to the level of earthquake damage – the more buildings collapsed due to the earthquake, the higher the number of ignitions. From the earthquakes and PEFs described above, it can be seen that this assumption lacks the necessary complexity. Most PEF models contain two parts: ignition, which describes how the number, location and time of ignitions can be predicted; and modes of spread of fire after ignition has occurred.

1.3.1 Studies on PEF modeling

The background of studies conducted on PEF at regional scale goes back to the 1950s. It is believed that Hamada was the pioneer in addressing PEF [35]. He proposed an empirically-based model for estimating fire spread over urban areas using available fuel load, wind speed and direction, etc. The model proposed by Hamada became then a base for further research on PEF in Japan [36, 37]. Scawthorn also widely investigated PEF at regional scale [38]. In Scawthorn's models, the responses of fire departments were also considered, along with the number of ignitions. That work was considered to be of high enough quality to be used as a source by the insurance industry. HAZUS software [39], which is one of the most widely used computer packages, was then developed by making some modifications to the Scawthorn models. Scawthorn later modified his previous models by involving data collected from a number of earthquakes that had occurred in the past, and proposed his new equation based on peak ground acceleration (PGA) [40].

Cousins and Smith [41] developed a probabilistic-based model to estimate PEF losses in three cities of New Zealand. Using shaking intensity (MMI), and assuming that no PEF would occur for MMI6.0 and lower (by contrast, most PEFs would occur for MMI11.0), they proposed a linear equation for estimating the number of post-earthquake ignitions. Then, assuming that some ignitions would be suppressed by available resources, the overall PEF losses were accounted for. They also considered wind speed as an important factor in spreading fire. Ren and Xie [42] presented a GIS-based model for the prediction of PEF prone areas. As per their research, the PEF risk is highly dependent on the level of structural seismic damage. As well, the number and locations of fires, the situation of urban fire fighting facilities, weather conditions, location of temporary housing, road closures, types of materials used in buildings, etc., all are parameters influencing the spread of PEFs. They also introduced some fire-stopping factors, for example, the distance between buildings, wind direction, and rain. Davidson [43] used a statistical approach to develop a generalized linear PEF model. The model is mainly based on detailed data collection for the California region. The data includes building types (residential, commercial, industrial, etc.), percentage of built area, material used in construction (e.g. steel, concrete, wood), and the age of the buildings. The results show that Davidson's model can be used as a tool for estimating the number and location of future PEFs. Zolfaghari *et al.*

[44] have also developed a probabilistic-based model for PEFs. The model proposed is based on intra-structure ignition and thus more focus has been placed on building-related factors that could intensify an ignition after earthquake. They included earthquake intensity, building vulnerability, and ignitable non-structural components, such as stoves, etc., in developing their model. A GIS-based model was developed by Yildiz and Karaman [45] for PEF simulation. They involved three important factors for post-earthquake ignitions in buildings, which were gas and electrical distribution systems and overturning appliances. Yildiz and Karaman used an analytical hierarchy process (AHP) model as a semi-quantitative-based assessment method, in order to determine the relative weight of effective factors in post-earthquake ignitions. Himoto and Nakamura [46] developed a physics-based model for urban PEF spread in historic buildings in Kyoto, Japan. They considered fires in individual buildings and their effects on nearby ignited buildings. In that way, both regional and building scales were considered in the Himoto and Nakamura model. At building scale, structural behavior under fire was found according to fundamental equations of mass and energy. The fire spread from building to building was correlated to several factors, such as weather conditions. They also calibrated their model with data collected from previous PEF events, the Sakata fire in 1976 and the Kobe fire in 1995.

As this book focuses more on PEF modeling at building scale, no further information is provided regarding modeling at regional scale. To be more familiar with PEF modeling at regional scale, there are a number of reviews, such as [47].

1.3.2 Modes of fire spread

An ignition can spread from one cell to the next through four modes: *direct spread* to an adjacent point, *spontaneous ignition* of an adjacent cell due to radiation, *sparking*, and *branding*.

- Direct spread to an adjacent cell

For a fire to directly spread to the adjacent cell, it is assumed that the cladding materials in the adjacent cell are combustible. The adjacent cell can be the next room or the next building. The time taken for a fire to directly spread therefore varies, depending on the building type, but in some references the value of 2.5-5.0 minutes as an approximate is given [48].

- Spontaneous ignition (Auto-ignition)

For a fire to spread from one cell to the next through radiation, a rise in temperature in a compartment has the main role. The temperature itself depends on the fuel load density and the size of the compartment. In some references, it is mentioned that for this ignition to occur, the temperature value changes to be in the range of 800 to 1000 °C in small and medium compartments [49] and up to 1200 °C in large compartments [50]. However, there are some studies that have shown that the temperature in large compartments is lower than that in small compartments [51, 52]. Emissivity values can also significantly influence radiation. The emissivity values are dependent on the surfaces in the compartments, and vary from 0.5 to 1.12. In a simplified manner, spontaneous ignition would happen if the radiation received by the surface of the compartment went beyond a critical level. For example, the critical radiant heat flux for a wooden surface is 28 kW/m^2, and the equivalent temperature is 600 °C [49]. This means that when the surface temperature goes beyond 600 °C, there is a high possibility of auto-ignition. In some documents it is stated that 30 kW/m^2 is enough for a cell to ignite automatically, while the weather conditions have no effect. It is also believed that when the distance between two adjacent buildings (two adjacent cells) is less than 3.0 m, and the radiant heat flux is more than 30 kW/m^2, it is enough to cause spontaneous ignition. This, albeit, depends on the existence of combustible cladding.

- Sparking ignition

Burning cells can produce sparks from 7.0 to 22.0 minutes after ignition. For a fire to travel by sparking, it is necessary for the critical value of the radiant heat flux to reach the range of 10-18 kW/m^2. The sparks can then spread in all directions, dependent on wind speed and direction. Therefore, instead of spontaneous ignition that does not depend on the weather, sparking ignition is dependent on wind, in such a way that higher wind speeds will lead to sparks traveling greater distances. Table 1.1 shows the effect of downwind and upwind speed on spreading sparks [8]. The table clearly shows that the minimum width of passageways should be 12.0 m if there is a possibility of even a light wind.

Table 1.1: The effect of wind speed in spreading fire through sparking

Wind strength	*Calm*	*Moderate breeze (20 km/h)*	*Fresh breeze (30 km/h)*	*Near gale (50 km/h)*
Spread distance downwind (m)	12	15	21	45
Spread distance upwind (m)	12	12	12	12

- Branding ignition

As ignition progresses, it may produce firebrands, which quickly fly in the presence of strong winds (near gale). Branding would occur, for example, when the roof has ignited and broken into flaming particles. Branding can thus occur if the roof is made of ignitable materials, as the result of which fatalities can be intensified. During the Kobe PEF event, some fire was observed to have spread due to branding.

1.3.3 Accounting for PEF risk in urban regions: A preliminary example

Here, through a semi-quantitative-based approach, it is shown how to account for the risk of PEF in urban areas. To do that, it is necessary first to define the probability of occurrence (P) through a prescriptive-based scale, as shown in Table 1.2. The consequences (C) of these risks are shown in Table 1.3.

Table 1.2: Risk levels based on the probability of occurrence

Degree of occurrence	*Very rare*	*Almost impossible*	*Impossible*	*Moderate possibility*	*High possibility*	*Almost certain*
Quantitative scale	$P < 1\%$	$1\% \leq P < 5\%$	$5\% \leq P < 10\%$	$10\% \leq P < 30\%$	$30\% \leq P < 85\%$	$85\% \leq P$

Table 1.3: Intensity of consequences

Degree of consequences	*Description*	*Quantitative scale of consequences*
Very small	The PEF has not caused any casualties and the financial damages are less than 2% that of the direct earthquake damages	$C < 5\%$
Small	The PEF has not caused any casualties and the financial damages are less than 5% that of the direct earthquake damages	$5\% \leq C < 25\%$
Moderate	There are very limited casualties caused by PEF and the financial damages are more than 5% that of the direct earthquake damages	$25\% \leq C < 50\%$
Extensive	There are some casualties caused by PEF and the financial damages are more than 10% that of the direct earthquake damages	$50\% \leq C < 85\%$
Very destructive	There are extensive casualties caused by PEF and the financial damages are more than 10% that of the direct earthquake damages	$85\% \leq C$

In order to prepare a matrix of risks and then account for the PEF risks, a number of assumptions are made:

- The area of the considered region is 50 km^2 with a population density of 1 person per 100 m^2.
- PGA is 0.3 g.
- The earthquake occurs in summer, at 7 pm, while the temperature is 34 °C.
- There is a moderate breeze of 20 km/h speed.
- About 60% of buildings are non-seismic masonry buildings, 30% are seismically designed reinforced concrete (but not designed based on new seismic codes), and the remaining 10% are steel structures which are relatively new buildings.
- The region is very dense, with an average width between structures of 3 m.
- There are four fire stations available.

The possibility of PEF is a function of building type, earthquake damage density, and weather conditions. This can be shown using Equation 1.1.

$$P = P_1 \times P_2 \times P_3 \times P_4 \tag{1.1}$$

where:

- P_1 is the possibility that a building contains combustible materials or it has been made of combustible materials. Here it is assumed that P_1 is 1.0.
- P_2 shows the flammability rate of structural types. It might be assumed that when the building does not contain flammable materials or it is not made of flammable materials, there would not be any possibility of PEF occurrence. Although this assumption might be correct in an ideal situation, it is not a realistic assumption. Here, the data provided by Ren and Xie [42] is used as a reference to determine the values of flammability of different structural types, as shown in Table 1.4.

Table 1.4: The values of flammability of different structures

Class	*Characterization*	*Value of risk*
1	Flammable and explosive chemical materials	0.97
2	Flammable materials	0.89
3	Wood construction	0.795
4	Brick construction with wood window, and door,	0.675
5	Steel construction	0.50

As per the assumptions made for the considered region, the value of P_2 = 60% × 0.675 + 40% × 0.5 = 0.605

- P_3 represents the weather conditions. As it is here assumed that the earthquake occurs on a hot day in summer, the value of P_3 is given as 0.95.
- P_4 shows the effect of damage to seismic-resistant buildings on the probability of PEF occurrence.

This probability is a function of spectral acceleration (P_5), the probability of leakage of flammable materials (P_6), and the degree of damage as the result of the possible fire (P_7). Here, it is assumed that the non-seismic masonry buildings are extensively damaged (say 0.95), that the reinforced concrete buildings sustain considerable damage (say 0.75), and that the steel buildings sustain moderate damage (say 0.40). Thus:

$$P_5 = 60\% \times 0.95 + 30\% \times 0.75 + 10\% \times 0.40 = 0.835$$

Here, the probability of leakage of flammable materials (P_6) is assumed to be 0.97 when there is extensive damage, 0.89 when there is considerable damage, and 0.675 when there is moderate damage. Hence, for the example here, $P_6 = P_7 = 60\% \times 0.97 + 30\% \times 0.89 + 10\% \times 0.675 = 0.916$. Thus, based on values of P_5 to P_7, $P_4 = 0.835 \times 0.916 \times 0.916 = 0.7$.

Therefore, for the region considered, the PEF probability (P) = $1 \times 0.605 \times 0.95 \times 0.7 = 0.4$. This means, on average, the probability of PEF is about 40%. Using the method described in HAZUS software [39], the number of ignitions in the region considered is about 15 fires. It is assumed that 12 of these 15 fires would spread to other buildings. In addition, in the region considered, it is assumed that the piping systems are very old. Consequently, they would probably sustain considerable damage after earthquake, and perhaps only 20% would remain workable after the earthquake. It is also assumed that among the four fire stations in the region, one of them would collapse immediately after the earthquake, and two would not be ready to work on the first day after the earthquake. The last fire station, however, can be dispatched immediately after the earthquake, but its personnel are not totally ready. Therefore, the reliability of the last station is $0.5 \times 0.85 = 0.425$; and that of the two that will be available to work on the second day after the earthquake is $0.7 \times 0.5 = 0.35$. Thus, the total reliability is 0.28. This means that the possibility that the fire stations cannot work properly is 72%. As the region considered is under stress, and as per its very narrow thoroughfares, it is also assumed that the reliability of the communication systems is 60%. Using the information above, the probability of the abovementioned risks can be presented in Table 1.5. It is, however, significant that the data mentioned here are based on direct risks, while other indirect risks, such as the weather conditions, the population, and the building density, are not included.

Involving all of the possibilities requires a very complex simulation, which is beyond the scope of the example here. Here, in order to assess the values of consequences in the assumed region, the Kobe PEF information is used. As reported after the Kobe PEF, about 175 fires developed, resulting in around 820,000 m^2 and 7300 houses being burnt out. The earthquake left about 5500 fatalities, around 550 of which were due to the fires. For this example, it can be expected that about 70,000 m^2 including 625 houses will burn out and 50 people will lose their lives. To account for the quantitative value of consequences, and given that the importance of the ratio of fatalities is assumed to be 0.8, 0.2 is the value given to the economic loss. Table 1.6 shows the probability of consequences after the PEFs.

Table 1.5: The probability of occurrence

No.	*Description of direct risk*	*Probability of occurrence (%)*	*Local probability of occurrence (%)*	*Total probability of occurrence (%)*	*Quantitative description of risk*
			PEF reasons		
1	Short circuits	40	55	22	Moderate possibility
2	Gas leaking	40	25	10	Moderate possibility
3	Gas-based appliances	40	15	6	Impossible
4	Chemical release	40	5	2	Almost impossible
			PEF spread		
5	Pressure drop	78	80	62	High possibility
6	Road closure	78	50	39	High possibility
7	Damage to infrastructure	78	40	31	High possibility
8	Loss of reliability of the fire stations	78	72	56	High possibility

Table 1.6: The probability of consequence

No.	*Description of direct risk*	*Possible fatalities*	*Ratio of fire damage to the overall damage (%)*	*Quantitative value of consequences (%)*	*Quantitative description of risk*
			PEF reasons		
1	Short circuits	11	4	32	Moderate possibility
2	Gas leaking	5	1.8	15	Moderate possibility
3	Gas-based appliances	3	1.1	9	Impossible
4	Chemical release	1	0.4	3	Almost impossible
			PEF spread		
5	Pressure drop	31	11.5	91	High possibility
6	Road closure	20	7.2	57	High possibility
7	Damage to infrastructure	16	5.7	45	High possibility
8	Loss of reliability of the fire stations	28	10.3	82	High possibility

Table 1.7: Risk factor

Risks		*Consequences*					
		Negligible	*Trivial*	*Moderate*	*Considerable*	*Destructive*	*Almost certain*
Short circuits	Moderate possibility			RF=0.47			
Gas leaking	Moderate possibility		RF=0.23				
Gas-based appliances	Impossible		RF=0.14				
Chemical release	Almost impossible	RF=0.05					
Pressure drop	High possibility						RF=0.97
Road closure	High possibility					RF=0.74	
Damage to infrastructure	High possibility					RF=0.62	
Loss of reliability of the fire stations	High possibility						RF=0.92

Table 1.8: Descriptions of risk profile

Risk profile	*Actions required*	*Quantitative risk value*
Trivial	There is no need to adopt any instant strategy.	RF < 10%
Moderate	There is a need to mitigate the risks using limited sources.	30% > RF ≥ 10%
Considerable	There is a need to mitigate the risks using required sources.	60% > RF ≥ 30%
Critical	There is a vital need to mitigate the risks using required sources and some specific strategies.	100% > RF ≥ 60%

Using Equation 1.2, the risk factor is determined, where RF, P, and C stand for risk factor, probability of occurrence, and probability of consequence, respectively.

$$RF = P + C - P \times C \tag{1.2}$$

Results of the risk factor for the example here are shown in Tables 1.7.

Results from Table 1.7 show that some consequences of the PEF risks in the example here need to receive immediate remedies, e.g. piping systems and fire stations. Road closure is also a high-risk PEF consequence and needs to be treated with short time strategies.

Chapter 2

Performance-based Approach

2.1 Introduction

The main purposes of building codes are to provide minimum safety to inhabitants, and to protect public health. These are achieved by starting with predictions of future possibilities, i.e. when buildings may be faced with some particular conditions. Those conditions may be nature-based, such as wind and earthquake, or they may be rooted in human activities, such as fire or explosion. The second purpose of building codes is to provide uniformity in the construction industry, which aims to keep construction costs down. Building codes might also address the durability and quality of construction materials, thus securing the welfare of the community over a long period. This chapter presents the literature covering two different approaches in building codes, prescriptive-based and performance-based. As per the main focus of this book, seismic and fire codes are discussed in particular.

2.2 Performance-based versus prescriptive-based codes

The history of building codes is very long, going back even beyond 4000 years, when some rules and regulations were gathered by Hammurabi to protect public health. The first 'modern' fire building code seems to have been documented after the Great Fire of London in 1666, when more than 70,000 homes were destroyed. The code mostly contained some requirements in order to provide buildings with a minimum level of fire resistance.

In the United States, the first construction regulation was proposed in New York City in 1860. In 1872, Boston experienced a great fire event, and as a result of this, many insurance companies were bankrupted. That fire caused the remaining insurance companies to think seriously of establishing an organisation that would be responsible for developing fire safety regulations, resulting in the formation firstly of the National Board of Fire Underwriters (NBFU), and then of the National Building Code (NBC). The

NBC shortly became a reference for almost all cities. Cities needed favorable insurance rates if progress was to continue, and the NBC had a direct link with insurance companies. More people gradually became involved in founding Underwriters Laboratories (UL) in 1894. Shortly after, in 1896, the National Fire Protection Association (NFPA) was established. By the beginning of the twentieth century, most insurance companies and cities were using the regulations developed by the NFPA as a source of fire protection regulations. It was in 1927 that the Uniform Building Code (UBC) was developed as a unified code throughout the US.

Some other countries were pioneers in establishing their fire codes, such as Canada, Australia and New Zealand, whose codes were first released in the 1840s. The main difference between their codes and those of the US was that most of the national codes in the US were developed by the private sector and then put into effect by governments, whilst most codes in the other mentioned countries were established and enforced by the national governments.

In terms of seismic building codes, the first regulation was mandated in Lisbon, Portugal, after the devastating 1755 earthquake, which left extensive damage and a high number of fatalities. It was after the 1911 Messina, Italy, earthquake and the 1923 Kanto, Japan, earthquake that some regulations were released that considered only the horizontal effect of earthquake forces for seismic-resistant structures. Surprisingly, no earthquake code was developed after the horrific 1906 San Francisco earthquake. In 1927, the first seismic code, named the Uniform Building Code (UBC), was released in the US, although its use was not mandated. The 1933 Long Beach, California, earthquake and others in the three following decades caused most engineers to believe that there was a vital need to develop a seismic building code that could provide adequate stability to buildings in earthquake events. The 1971 earthquake in Sylmar, California, again highlighted that need, when it was observed that even the most recently built structures were vulnerable to earthquake loads. In 1977, the seismic building code was mandated throughout the US as a program named the National Earthquake Hazards Reduction Program (NEHRP) [53]. The Federal Emergency Management Agency (FEMA), now one of the most well-known authorized organizations, was then established, mainly in order to address effective management of earthquake risks.

Over the history of building codes, two different paradigms have emerged, signifying profoundly different approaches to form and function. Building codes are often classified as either prescriptive- or performance-based in nature. The history of building codes can be divided into two periods: the era when the codes were mainly prescriptive-based (also known as "force-based codes") ; and the era when they were designed as performance-based. The history of performance-based codes is rather short, as they have become accessible mainly in the last two decades, whereas the prescriptive-based codes have a long history. Most prescriptive-based codes mention solutions to an issue without indicating the intention of the requirement. In other words, these codes specify construction requirements by specific materials and construction methods. Thus, in prescriptive-based codes, the building elements are designed independently, and the objective (which is a set of attributes) is only indirectly reached. In prescriptive-based codes, a distinct and discrete action is mandated in order to arrive at a goal. These mostly present a menu describing minimum and maximum numbers for different elements in a structural system, in such a way that the designer can just choose what is needed for the assigned intention. Prescriptive-based codes are thus considered easy to use, since they explain in relatively simple words what is acceptable and what is not.

Nevertheless, prescriptive-based codes suffer from several deficiencies, some of the most important of which will be discussed. First, when using these codes, it is not required to consider the whole of a building – elements are considered separately, and what is achieved will thus not necessarily be the optimum. Second, when using these codes, there is no need to control whether all elements are working properly over time, as it is optimistically assumed that all elements are set up correctly. Third, as there is a steady state approach towards consuming less budget while furthering efficiency, more flexibilities are required to achieve those goals. As prescriptive-based codes seemingly do not have adequate flexibility, they regularly have to be updated, which is a time-consuming process. Fig. 2.1 shows prescriptive approach, where there is a linkage between attributes and parts.

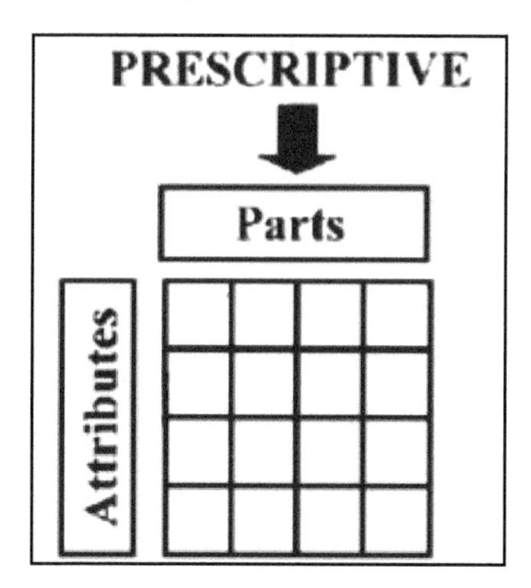

Fig. 2.1: Prescriptive-based approach [54].

On the other hand, in performance-based codes, the construction requirements are specified according to defined performance criteria in an explicit manner. In other words, in performance-based codes, the performance is targeted rather than the attributes and the methods. This approach can provide more opportunities for a higher quality of building, as there are various approved materials, building systems and advanced methods, while there are fewer limitations. When using performance-based codes, there will be more flexibility in design and most likely a reduction in unnecessary construction costs. This is because, in performance-based codes, building structures are seen as a whole, rather than separate parts, as shown in Fig. 2.2.

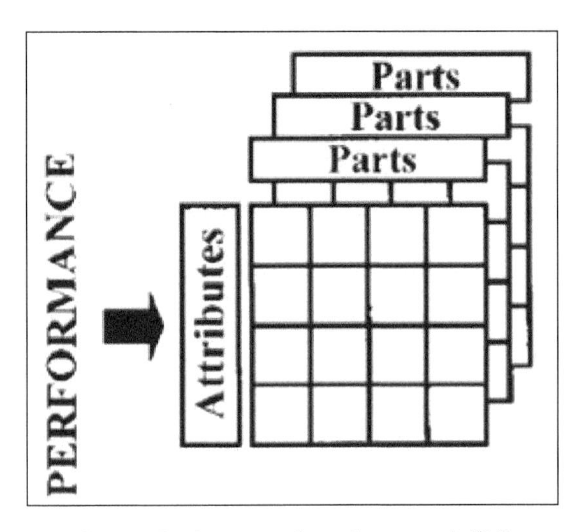

Fig. 2.2: Performance-based approach [54].

From a different point of view, literature reviews of performance-based codes often show an emphasis on making improvements through innovative approaches. However, the question arises, "What is innovation?" As there are different definitions of innovation with no clear consensus, confusion may arise. As a result, a designer should exceed code prescriptions and should predict the response of structural systems during extreme events [55]. For example, to design a floor system in a multi-story building, the first goal is to ensure that the story will withstand the loads applied, including gravity loads and lateral loads. However, the story shall also resist against deflection and vibration, both of which might cause some inconvenience to the occupants. Since these are mostly human-dependent parameters, they might vary from one to another. Thus, there would be a continuous process of trial and error for the designer to find the optimum solution. To achieve this, complex structural analyses, as well as advanced laboratory testing, might be required, both of which are time-consuming processes.

From another aspect, in performance-based codes, modeling results are highly dependent on data input. Hence, if the input is faulty, the results will be faulty too. While the penetration rate of performance-based codes in construction projects is generally low (regardless of what area of construction the codes concern), these codes require considerable time to review [56].

Performance-based codes also require special education for operators if they are to be appropriately enforced, that is, decisions must be made by expert people, not by those with limited training. This is another obstacle to the extensive use of performance-based codes [57]. Controlling and inspecting are also complicated processes in performance-based codes, since being sure about a system that is still being established is not an easy task. For example, being certain that a fire protection system set up in a home is working properly requires an owner's manual to be prepared, highlighting how the fire protection system works and how to maintain it. While this normally should be the inspector's responsibility, some liabilities are, in effect, conceded to the owners.

On the other hand, although performance-based codes are flexible in terms of their design process, this also means that making even a minor change to the arrangement of a building requires a total revision in the design, in order to ensure that the design objectives are still met. For example, if owners of a building change the occupancy type of their property, they need to be certain that those changes will not create an undue hazard [58]. Therefore, a great deal of attention needs to be placed on the maintenance of the building. Insurance companies may also be concerned about using performance-based codes and, as a result, they may increase their fees. A question can arise in the minds of the underwriters, "What would happen if the assumptions in the system are incorrect?" Because of this doubt, insurance companies would prefer to work with codes that may impose a lower risk on the project and which are more conservative, and it seems that prescriptive-based codes have been considered more conservative than performance-based codes [59]. This conclusion is mainly drawn from the fact that performance-based codes are often promoted for economic rather than safety reasons. Performance-based codes have been developed to consider a building as a whole while, as mentioned before, prescriptive-based codes focus on one single element at a time. This may cause a conflict between different codes when they are used concurrently in a particular building. For example, a building structure may be designed for earthquake loads using a performance-based code, and for fire loads using a prescriptive-based code. When the building is seen as a whole, this may be challenging.

While the points mentioned above are still in open discussion, performance-based codes are increasingly being used in building design. To address these points, there have been suggestions to combine performance- and prescriptive-based codes with each other as a practical compromise. It can be understood from Fig. 2.3 that a higher level of specification can provide a greater level of freedom to the designer. Either way, when there is a high degree of specification, more endeavors are made to find an accepted method for the verification of performance. Verification is an essential part of the performance-based approach that is implemented to ensure that a particular method or material will satisfy the considered performance criteria

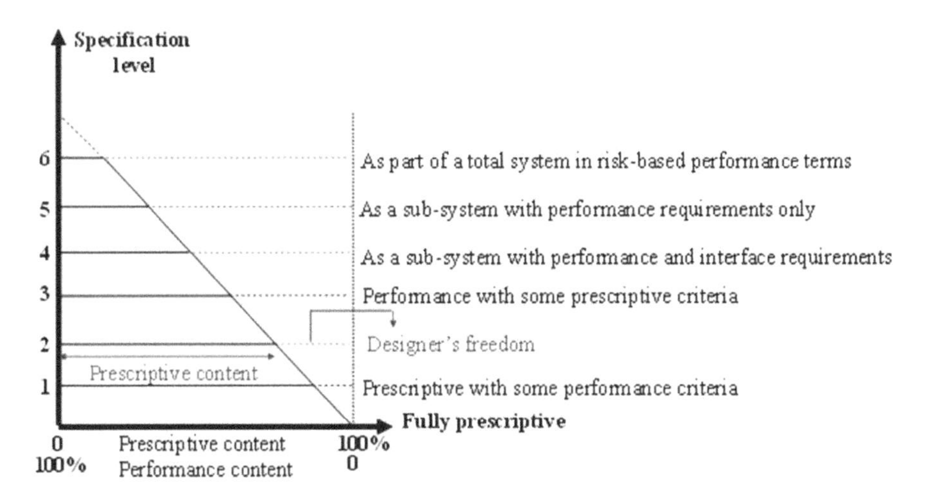

Fig. 2.3: Levels of specification with different performance-prescriptive mixes [54].

in a building. The verification can be performed by testing, by simulating, or by a combination of the both.

In addition to verification, there is an equally important step, which is the application of the performance-based approach to buildings. For doing this, building performance models are provided – to develop quantitative criteria, to design a building as per a performance target, and then to assess its response to service situations. For instance, in the floor system example mentioned earlier, a performance-based approach can be followed in: a) developing quantified criteria for acceptable floor deflection and vibration to meet the occupants' demands; b) designing innovative floor systems to meet the performance criteria; and c) evaluating the performance of an existing floor system as part of an audit process.

To provide an illustration of the above explanation, the simplest way is to use Fig. 2.4a, which illustrates a five-level hierarchical method suggested by the Nordic Committee on Building Regulations [60]. As is seen, a performance-based approach to a building is initiated by defining a goal, followed by functional requirements. Setting operative requirements and verifying are the next two steps. Finally, some acceptable solutions are suggested. While the illustration shown in Fig. 2.4a shows the minimum needed for the application of a performance-based approach, there might be situations when more details are required. For instance, there might be a need to develop some layers between the operative requirements and the verification in order to define the performance groups, the performance levels and the performance criteria. For achieving this, Meacham proposed an eight-layer performance-based hierarchical process, as shown in Fig. 2.4b. It is worth mentioning that the goal(s) may vary from case to case. They might be established in terms of health, safety and economy, while a particular user

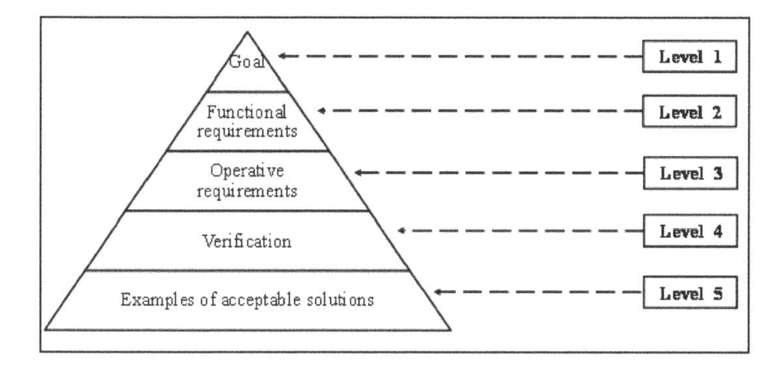

(a) A five-layers process

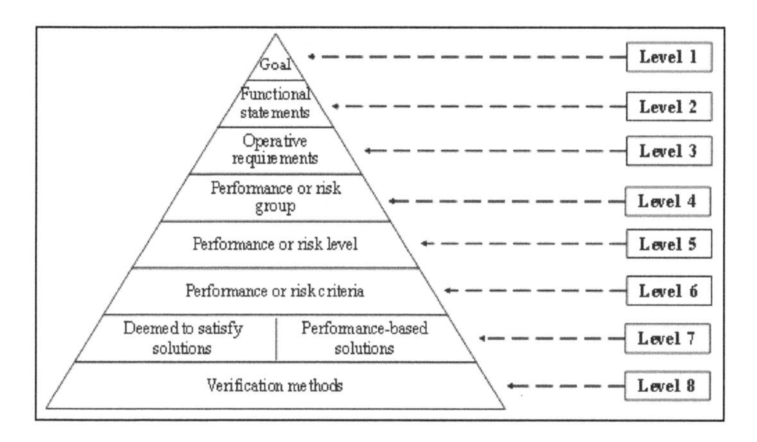

(b) An eight-layer process

Fig. 2.4: Performance-based hierarchy approaches [61].

may even be willing to further them. For instance, government occupancy may add more factors, such as national security, heritage, and public welfare to the factors already mentioned. Thus, different building groups, such as those that must remain under operational conditions during all types of loadings, e.g. hospitals, or those that can sustain significant damage but must still provide life safety to the occupants, can be defined. It is apparent that this type of classification can add to the level of required performance. After all levels of the hierarchical processes in Fig. 2.4b have been attentively defined, they can be developed by engineering tools to be used for design and evaluation. Importantly, most of this development is conducted on the performance requirements and the verification. It should also be pointed out that increasing the layers of levels may lead to a challenging situation, since it may increase the number of interrelations between requirement, criteria and verification, which in turn would create a complex process, as shown in the example in Fig. 2.5.

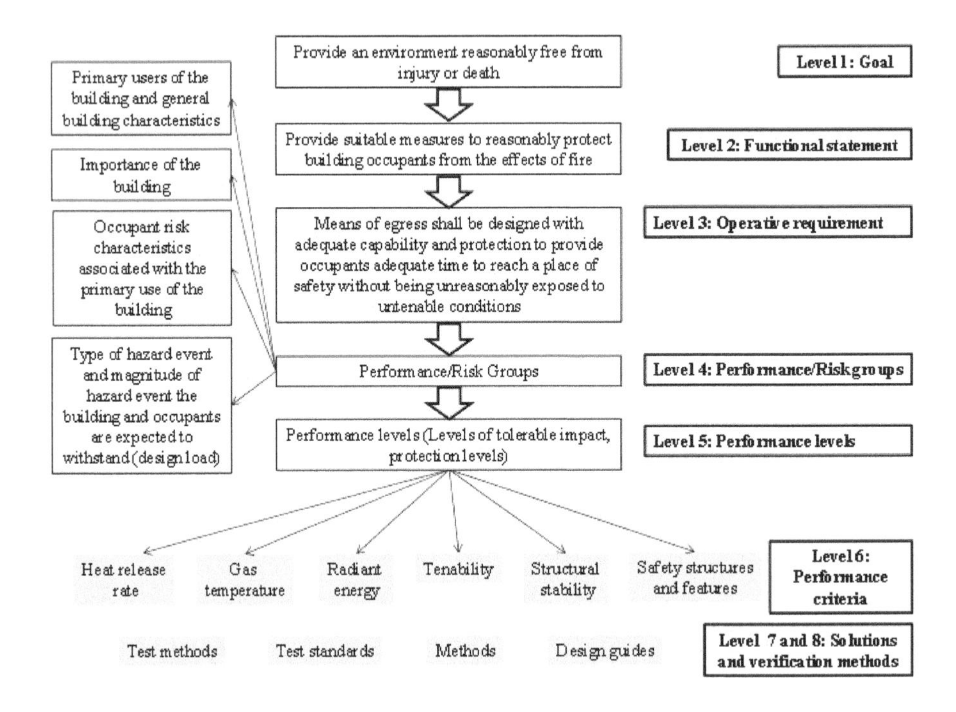

Fig. 2.5: Interrelations between different layers in a performance-based hierarchical system [61].

Figure 2.5 shows a sample use of the eight-layer performance-based approach for exit pathways in case of fire in a building. The doorways and thoroughfares must be designed to provide adequate time for occupants to safely leave the building. Therefore, both the occupancy type and the occupants' type are vital points to be carefully addressed. For example, a distinction must be made between the design of exit ways in a hospital and those in a residential complex, as they can change the thresholds and the structural response to fire. Hence, alongside increasing the differences between the performance levels, there would also be some improvements to the ways of assessing design methods. It is understandable from the above explanation that creating a well-defined linkage between different layers in a performance-based approach is an essential point that must be considered throughout the process, since otherwise the possibility of reaching a wrong assessment increases greatly.

When a performance-based approach is used in a building code, the hierarchical system mentioned above should be distinctively illustrated and explained. Starting with goals and objectives, they are defined in such a way that a building tolerates a specific level of an event within acceptable limits of damage. This means that, for a designated use and occupancy, the acceptable level damage may vary, from no damage to major. That "acceptable damage", however, must not lead to the life safety of occupants being put at risk. Use and occupancy are designated as per several risk factors. In this respect, the nature of the hazard – whether it has an external source such as earthquake or wind, or an internal source such as fire or explosion – plays an important role because it can have a direct effect on the occupants and on the building's serviceability. The number of the building's users, whether residents or visitors, is also of utmost importance in designating what level of damage is acceptable. Other factors, including sleeping characteristics, familiarity with the building and the times when people use the building, i.e. the occupancy type, are also active factors to be considered when assigning tolerable damage. In addition, the occupants' type shall be considered too. For example, a distinction must be made between the acceptable damage in a building to be occupied by disabled people and in one which is to be occupied by a wider cross-section of people, as the former group is obviously more vulnerable than the latter [62].

Overall, the performance of buildings in conjunction with their levels of acceptable damage is categorized in Table 2.1. As is seen, there are four performance groups named from I to IV. Group I consists of buildings with low importance levels, such as temporary structures, which will present a very low hazard to human life if an event causes them to fail. Most buildings are included in Group II – they may sustain significant damage, but that damage shall not threaten human life. Group III comprises buildings that will provide a substantial hazard to human life if they collapse from an event. For example, educational buildings, health care facilities, and high-rise buildings are included in Group III. Lastly, Group IV consists of buildings with a very high importance level, such that their failure may significantly threaten human life or national security. Some urban facilities, such as hospitals, police and fire stations, government buildings, and industrial facilities (such as power plants) are included in this group.

All of the building groups explained in Table 2.1 can be summarized in Table 2.2, which illustrates the magnitude of a possible event on the one hand, and the importance level of the buildings on the other hand. The table explains that a minimum performance level for all types of buildings must be considered, even those that are considered of very low importance, such as temporary buildings. The performance level then increases as the buildings' importance increases, mandating higher levels of performance for those buildings that have more vital roles in urbanized and industrialized areas. It can also be seen that the use of the information provided in Table 2.2 is an iterative process, which means that the performance level of a building category is evaluated in a way that it can tolerate the effect of an event of particular magnitude. It deserves mention that, under some particular circumstances, the performance level of some specific buildings may be re-designated; thus they might be classified at a new performance level.

Table 2.1: Buildings' and facilities' performance in conjunction with acceptable damage [63]

Performance Group	*Use and occupancy **classifications for specific** buildings or facilities*
I	Buildings and facilities that represent a low hazard to human life in the event of failure, including, but not limited to: 1. Agricultural facilities 2. Certain temporary facilities 3. Minor storage facilities
II	All buildings and facilities except those listed in Performance Groups I, III, and IV.
III	Buildings and facilities that represent a substantial hazard to human life in the event of failure, including, but not limited to: 1. Buildings and facilities where more than 300 people congregate in one area. 2. Buildings and facilities with elementary school, secondary school or day-care facilities with capacity greater than 250. 3. Buildings and facilities with a capacity greater than 500 per colleges or adult educational facilities. 4. Health-care facilities with a capacity of 50 or more residents but not having surgery or emergency treatment facilities. 5. Jails and detection facilities. 6. Any other occupancy with an occupant load greater than 5,000. 7. Power-generating facilities, water treatment for portable water, wastewater treatment facilities and other public utility facilities not included in Performance Group IV. 8. Buildings and facilities not included in Performance Group IV containing sufficient quantities of highly toxic gas or explosive materials capable of causing acutely hazardous conditions that do not extend beyond property boundaries.
IV	Buildings and facilities designated as essential facilities, including, but not limited to: 1. Hospitals and other essential health care facilities having surgery or emergency treatment facilities. 2. Fire, rescue, and police stations and emergency vehicle garages. 3. Designated earthquake, hurricane or other emergency shelters. 4. Designated emergency preparedness, communication, and operation centers and other facilities required for emergency response. 5. Power-generating stations and other utilities required as emergency backup facilities for Performance Group IV buildings or facilities. 6. Buildings and facilities containing highly toxic gas or explosive materials capable of causing acutely hazardous conditions beyond the property boundaries.

(Contd.)

Table 2.1: (*Contd.*)

Performance Group	*Use and occupancy classifications for specific buildings or facilities*
	7. Aviation control towers, air traffic control centers and emergency aircraft hangars. 8. Buildings and facilities having critical national defense functions. 9. Water treatment facilities required to maintain water pressure for fire suppression. 10. Ancillary structures (including, but not limited to, communication towers, fuel storage tanks or other structures housing or supporting water or other fire suppression materials or equipment) required for operation of Performance Group IV structures during an emergency.

Table 2.2: Correlation between severity of event and the performance groups [61]

↑		*Increasing level of building performance Performance Groups* →			
		Performance Group I	*Performance Group II*	*Performance Group III*	*Performance Group IV*
Increasing magnitude of event	Vast (very rare)	Severe	Severe	High	Moderate
	Large (rare)	Severe	High	Moderate	Mild
	Medium (less frequent)	High	Moderate	Mild	Mild
	Small (frequent)	Moderate	Mild	Mild	Mild

Whether a building would meet its performance objective is controlled by some fundamental structural performance criteria, such as safety, repairability, and serviceability. These criteria refer to life safety, protection of assets, and the functionality of a building, respectively, which are then combined with five evaluation objects: structural frames, building materials, equipment, furniture and ground. These five evaluation objects are separately connected with one of the basic structural performance criteria as pointed out above and termed evaluation items, as shown in Table 2.3.

The last point that should be mentioned is that while it is difficult to find a general method for performance evaluation in buildings, the approach explained above can provide a base for the users, i.e. owners, designers, and innovators, such that it can lead to improved performance and quality of buildings. However, this performance evaluation method may change from situation to situation as the goals and objectives change.

Table 2.3: Performance control criteria [64]

Primary structural performance	*Life safety*	*Repairability*	*Serviceability*
Type of limit state/ Evaluation object	Safety limit	Repairability limit	Serviceability limit
Structural frames	Never lose vertical bearing capacity	Never suffer damage exceeding the established range	Never cause malfunction or sensory disorder
Building materials (structural members and interior and exterior materials)	Never fall out or be scattered	Never suffer damage exceeding the predetermined range	Never cause malfunction or sensory disorder
Equipment	Never tumble, fall over or move	Never suffer damage exceeding the predetermined range	Never cause malfunction or sensory disorder
Furniture	Never tumble, fall over or be scattered	Never suffer damage exceeding the predetermined range	Never cause malfunction or sensory disorder
The ground	Never collapse or severely deform	Never suffer damage exceeding the predetermined range	Never cause malfunction or sensory disorder

2.3 Performance-based seismic design

The history of seismic codes is distinguished into two eras: an extended period of using force-based codes where the design for earthquake resistance emphasizes the strength of the materials and employs one response parameter, i.e. base shear to attain a single performance objective, life safety [65]; and a short period during which the performance of structures was the focus. Using force-based codes, the design criteria are described by restraining the values of stresses and forces resulting from prescribed levels of considered earthquake loads. To design against earthquake loads, the elastic acceleration response spectra is reduced by response modification (R) coefficients corresponding to the capacity of the selected structural system, e.g. moment-resisting system, in which the system can provide adequate stability against structural failure or can guarantee life safety. To use specific values of R coefficients, the structural system should comply with the standard to which the building is designed. In most forced-based standards, the earthquake loading is considered for a single design-level earthquake that is characterized by an acceleration response spectrum having 2/3 the intensity of the Maximum Considered Earthquake (MCE) shaking. The MCE is probabilistically defined as per the 2%-50-year earthquake motion. However, the mentioned value may be intensified for near-fault sites. From a different point of view, damage to non-structural elements, such as internal partitions, often has a significant effect on life safety yet it is not included in the design process.

There had also been a moot question as to whether force-based design provisions could provide sufficient strength when applied in the design of buildings. As this question received no explicit response in the force-based seismic codes, proposed new design approaches were gradually propounded. There was also another reason to change the force-based codes in favor of the performance-based codes. Over late the 1970s to early 1990s, the US experienced numerous destructive earthquakes, such as the 1979 Imperial Valley, the 1987 Whittier-Narrows, and the 1994 Northridge events. It was not surprising to see that most older non-seismic-designed structures were destroyed in the mentioned earthquakes. In contrast, very few modern seismic-designed structures were destroyed. This showed the ability of contemporary code procedures to protect life safety as the principal objective of seismic building codes. Nevertheless, even some modern structures sustained considerable damage after those earthquakes, showing that some weaknesses remained in the codes.

While these weaknesses could be ignored for ordinary structures, it was a serious concern for important structures, as it is expected that important structures will remain serviceable after earthquake. This issue was due to the fact that, in force-based provisions, sustaining significant damage is allowed. Most engineers, however, had become accustomed to visualizing earthquake hazard on the basis of the quake intensity using a hazard map, rather than as presented by a physical earthquake event.

The concept of design based on performance was introduced in the late 1980s and early 1990s as an alternative to the available force-based design codes. Performance-based codes were able to respond to the need for designers and owners to ascertain the performance of structures prior to the application of the earthquake load, and consequently to decide if there was a need for earthquake retrofitting. FEMA provided strong support to researchers who were developing new approaches in seismic design, resulting in the development of a series of performance-based criteria and provisions, such as ATC40 and FEMA273/274, which rapidly generated great interest worldwide. The methodologies proposed in those provisions were then adapted into the ASCE31/41 standards as building codes. While the primary intention of developing such codes was to improve the performance of existing structures, i.e. adaption of a pre-earthquake mitigation strategy, they were employed more widely for the design of new buildings as the new century began. Such extensive use was not only due to these methods satisfying the intended performance of the building codes, but also that they did this at a lower cost. This latter factor also gave rise to great interest from building owners, particularly regarding the design of tall structures. In that new approach, detailed definitions for structural performance had to be stated, in a set of limit states and based on well-defined and predicted behaviors of building materials under gravity and earthquake loads [66]. The limit states include life safety and damage limitation [67]. To meet the life safety standard, a structure is designed to withstand the shock loads, either entirely or individually. To meet the damage limitation standard, the obvious intention is to prevent damage to the building. Some residual deformation is allowed, however, provided the life safety of the occupants is not threatened. In this way, the objectives of performance-based design are associated with optimizing the serviceability of seismic-resistant buildings, and are documented by selection of one of the four performance categories: Operational (O), Immediate Occupancy (IO), Life Safety (LS) and Collapse Prevention (CP). Referring back to Table 2.2, the correlations between seismic performance levels and earthquake severity are defined as shown in Fig. 2.6.

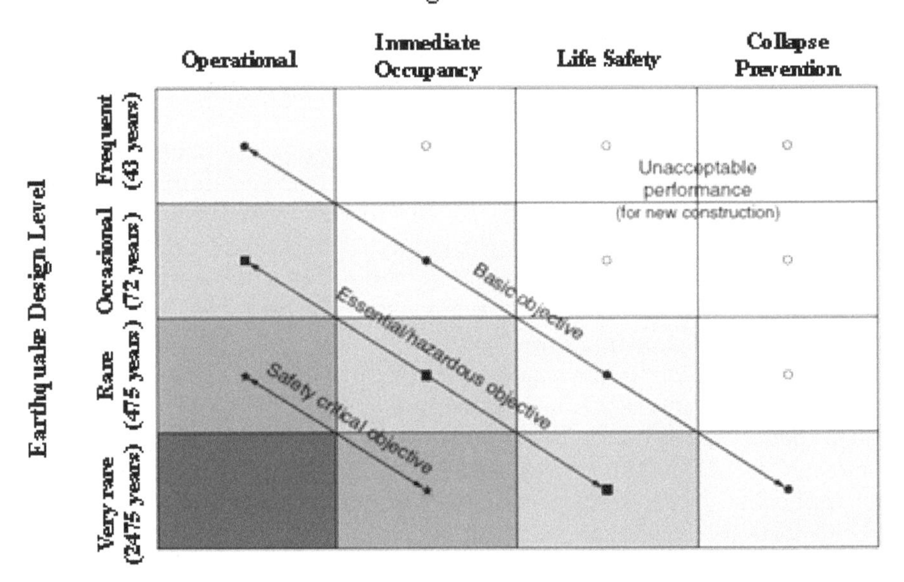

Fig. 2.6: Building performance levels versus earthquake severity [68].

Figure 2.6 shows the relationship between the four earthquake levels, and the annual probability of exceedance, which is not identical to the importance of a structure or an earthquake's severity. As an example, in frequent earthquakes, the probability in 50 years is 87%, while it is 33% for rare earthquakes. In addition, while the likelihood of earthquake exceedance for very rare earthquakes is about 2% in 50 years, it is 150% for rare earthquakes [69].

According to the design performance criteria, the expected performance of structures shall be controlled by the assignment of structures to one of the several seismic use groups. For example, based on the type of occupancy, there are three seismic use groups introduced in FEMA450 [70]. Structures that contain substances which, if released into the environment, are deemed to be hazardous to the public, are classified as Group I, or particular structures. Structures in this class are supposed to remain in the IO category when subjected to the design earthquake and remain operational under all frequent earthquakes. One class lower in performance requirement is the Group II structures, which contain a large number of occupants, or have occupants whose ability to exit is restrained. Schools, day-care centers, and suburban medical facilities are examples of this category. Group III, then, are ordinary structures, such as residential buildings, which shall remain in the LS level if subjected to the design earthquake.

The various performance levels required for buildings of different categories can implicitly be met by increasing the design earthquake by a factor called the importance factor. The importance factor basically adjusts the intensity of the earthquake in the design, so that the required performance level under the design earthquake is met. As is seen in Fig. 2.6, performance-based design codes specify progressively more conservative strength, serviceability and detailing requirements, so that structures can attain the minimum levels of earthquake performance suitable to their individual occupancies. Structures contained in these groups are not unique to a particular seismic zone; rather, they are spread across all zones from high to low hazard and, as such, the groups do not relate to hazard. In fact, the groups (categorized by occupancy or use) are employed to establish design criteria, which are intended to produce particular types of performance in design earthquake events, based on the importance of reducing structural damage and improving life safety. Most codes advise considering two performance objectives for the structural design:

1. The structure shall keep its serviceability when it is subjected to frequent earthquakes, with a 50% probability in 30 years and a 43-year return period.
2. There will be a slight likelihood of collapse if the structure is subjected to a very rare earthquake, with a 2% probability in 50 years and a 2475 return period, i.e. the MCE as defined earlier.

While performance assessment of structures under frequent earthquakes can be performed using both linear and nonlinear structural analyses, this evaluation under MCE is only performed using nonlinear time history analysis, which is an exhausting process. In Chapter 5, conventional structural seismic analyses will be explained.

On the other hand, there are numerous studies that have highlighted the importance of employing displacement as a performance quantifier. It is believed that structural damage depends heavily on inelastic displacement while subjected to earthquake loads. As for the three performance levels, i.e. serviceability, life safety, and collapse prevention, three corresponding structural characteristics, i.e. stiffness, strength, and deformation capacity, govern the performance (as shown in Fig. 2.7). Reinhorn [71] showed that defining more performance levels can cause more difficulties in determining which structural characteristics dictate the performance. It can also impose contradictory demands on strength and stiffness. It is worth mentioning that the structural systems can directly influence the ability to deform, i.e. the ductility. This means there will be various displacement controls for different structural systems. For instance, a limited displacement defined for collapse prevention in a moment-resisting structure will differ from that in a braced structure. Additionally, quantifying the structural repair cost and the earthquake hazard level is of importance, as the damage level can directly influence the decision as to whether a building is repairable.

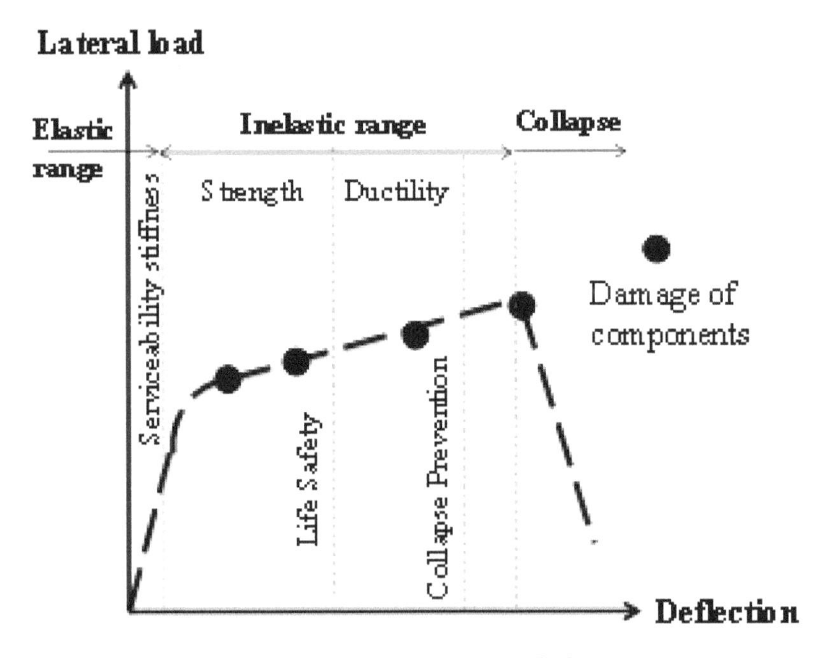

Fig. 2.7: Performance curve [72].

Using a direct displacement-based approach (DDBP), developed by Priestley *et al.* [73], displacement-dependent parameters, such as the overall displacement, the story drift and the plastic hinge rotation, can describe the intended performance. There are four fundamental steps in DDBP, as shown in Fig. 2.8. The first step is to convert a multi-degree-of-freedom

(MDOF) structure to a single-degree-of-freedom (SDOF) structure via the fundamentals mode as shown in Fig. 2.8a, where m_e is the effective mass of the structure that participates in the fundamental mode, and H_e is the effective height of the structure. The bilinear lateral force-displacement response of the assumed SDOF structure is then plotted as shown in Fig. 2.8b. As shown in this figure, K_i (corresponding to an initial elastic stiffness) is followed by rK_i (corresponding to a post-yield stiffness). Thus, secant stiffness (K_e) at the maximum displacement (Δ_d) is determined. The yield displacement (Δ_y) can be determined by the yield curvature or yield drift, both of which are independent of strength. Based on the results of a nonlinear time history analysis, the relationships between displacement ductility demand and the equivalent viscous damping are then found, as shown in Fig. 2.8c. The displacement spectra for various levels of equivalent viscous damping and the effective period are found via Fig. 2.8d. The design base shear (F_u) is calculated through K_e Δ_d. Thus there is no need to iterate the procedure. This is the reason the procedure is named DDBP.

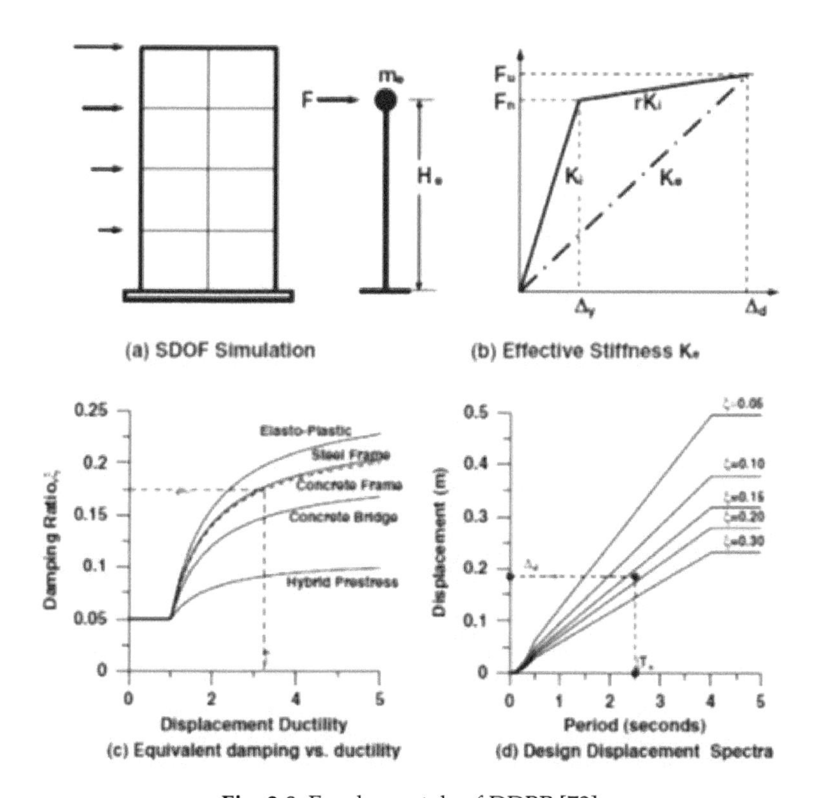

Fig. 2.8: Fundamentals of DDBP [73].

To meet the intended structural performance, most performance-based codes limit the transient and residual lateral displacement, or story drift, of buildings as for their limit state. The P-Δ effects, particularly, must be taken into account as they have vital impact on the stability of structures. This consideration is of particular importance for tall structures. A detailed review of the performance-based seismic design of tall buildings has been provided in [69]. FEMA356 [74], as one of the most widely used performance-based codes, has provided some limitations to different structural performance levels, as illustrated in Table 2.4. Both transient and permanent drifts are underlined in the table, since both can alter the intended performance level.

Table 2.4: Structural performance levels and drift limits [74]

Structural type	*Drift limit*	*Immediate occupancy*	*Life safety*	*Collapse prevention*
MRRCF	Transient	1%	2%	4%
	Residual	Negligible	1%	4%
MRSF	Transient	0.7%	2.5%	5%
	Residual	Negligible	1%	5%
BSF	Transient	0.5%	1.5%	2%
	Residual	Negligible	0.5%	2%
RCSW	Transient	0.5%	1%	2%
	Residual	Negligible	0.5%	2%
UMIW	Transient	0.1%	0.5%	0.6%
	Residual	Negligible	0.3%	0.6%
UMNW	Transient	0.3%	0.6%	1%
	Residual	0.3%	0.6%	1%

MRRCF: Moment-resisting reinforced concrete frames
MRSF: Moment-resisting steel frames
BSF: Braced steel frames
RCSW: Reinforced concrete shear walls
UMIW: Unreinforced masonry infill walls
UMNW: Unreinforced masonry non-infill walls

Inspired by Fig. 2.7, FEMA356 has introduced a normalized force/moment - displacement/drift ratio diagram for different performance levels (as shown in Fig. 2.9) and for two different primary (P) and secondary (S) structural members. The primary elements are those elements that, should they fail, can considerably reduce the vertical load-carrying capacity of the structure and impose a risk to occupants. The secondary elements, on the other hand, are those that, should they fail, can redistribute their assigned loads to other elements. Columns and load-bearing walls are often considered primary elements, and beams are assumed to be secondary elements. In Fig. 2.9, the first and second parts, i.e. AB and BC, correspond to the elastic behavior and the ductility of a structural component, respectively.

Point C corresponds to the beginning of the loss of capacity in such a way that, at point D, only some residual capacity has remained. Point E acts as a reference to the ultimate inelastic deformation/rotation capacity of the element. The normalized force in the vertical axis means the ratio of actual force or moment to the yield force or moment. Referring to Fig. 2.9, there is also another control in the rotation of beams and columns. It is important to ensure that the rotations of the structural elements will not exceed the limits that are introduced based on their performance levels, such as IO, LS, and CP. The rotation (θ) is defined as the ratio of deflection (Δ) experienced by the structural member to the length of the member (L). The yield rotation (θy) is then defined as the ratio of yield deflection (Δy) to the member length (L). The yield rotations of beams and columns are determined using Equations 2.1 to 2.3, respectively.

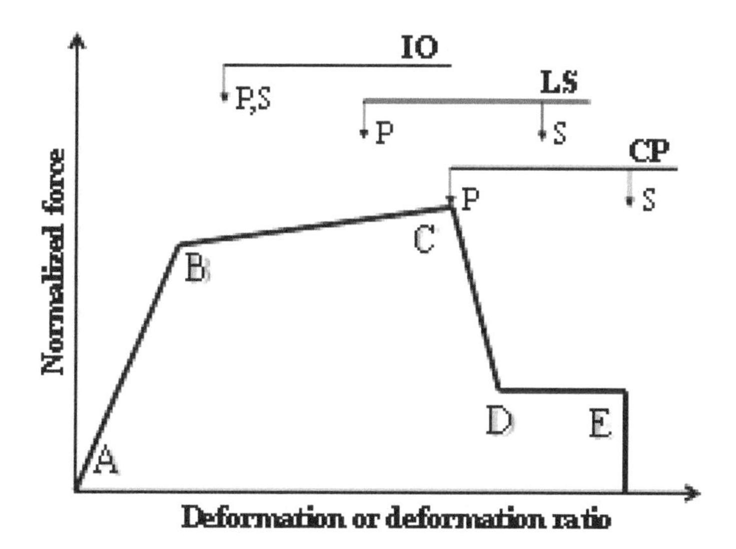

Fig. 2.9: Typical force/moment – displacement/drift diagram for various performance levels.

$$\theta_y = \frac{ZF_{ye}l_b}{6EI_b} \tag{2.1}$$

$$\theta_y = \frac{ZF_{ye}l_c}{6EI_c}\left(1 - \frac{P}{P_{ye}}\right) \tag{2.2}$$

$$P_{ye} = A_g\, F_{ye} \tag{2.3}$$

Controlling the actions of the columns depends on the axial load level applied to them, via which a deformation-controlled action, or a force-controlled action, should be considered in order to check the potential occurrence of failure. In this respect, when the compression axial loads are less than 50% of P_{CL}, i.e. the axial compression capacity, the column is considered deformation-controlled, while if this ratio goes beyond 50%, the column is considered force-controlled. The moment-force interaction of the column elements is determined using Equations 2.4 to 2.6. A column is called failed if the demand over capacity ratio (DCR), based on Equations 2.4 to 2.6, exceeds unity. For the beams, on the other hand, only the deformation-controlled action is required. This action checks rotations of the beams at every single step of the analysis and determines the performance level accordingly. A beam is considered failed if its performance level exceeds LS.

$$\text{For} \quad 0.2 \leq \frac{P_{UF}}{P_{CL}} \leq 0.5 : \frac{P_{UF}}{P_{CL}} + \frac{8}{9}\left[\frac{M_x}{m_x M_{CEx}} + \frac{M_y}{m_y M_{CEy}}\right] \leq 1.0 \tag{2.4}$$

$$\text{For} \quad \frac{P_{UF}}{P_{CL}} < 0.2 : \frac{P_{UF}}{2P_{CL}} + \frac{M_x}{m_x M_{CEx}} + \frac{M_y}{m_y M_{CEy}} \leq 1.0 \tag{2.5}$$

$$\text{For} \quad \frac{P_{UF}}{P_{CL}} > 0.5 : \frac{P_{UF}}{P_{CL}} + \frac{M_{uFx}}{M_{CLx}} + \frac{M_{uFy}}{M_{CLy}} \leq 1.0 \tag{2.6}$$

On the other hand, based on the theory behind performance-based approaches, as explained in section 2.1 (see Fig. 2.2) the performance of structural and non-structural components together constitutes a building's performance. Thus, along with evaluating the structural damage (as discussed above), a damage evaluation of non-structural components must also be performed. These assessments are of particular importance for the application of PEF loads to buildings, as the earthquake damage (both structural and non-structural) acts as an input for the subsequent fire. It is evident that the structural damage of steel structures differs from that of reinforced concrete (RC) and masonry structures. Most studies conducted have classified the structural damage into several classes, which are implicitly associated with different seismic performance levels.

2.3.1 RC structures

For RC structures, damage classifications mostly describe crack widths, concrete spalling, and rebar buckling. For example, using the Kobe 1995 earthquake experience, the Japan Building Disaster Prevention Association (JBDPA) categorized the RC seismic damage into five classes, as shown in Table 2.5 [75]. The damage classes and the load-carrying capacity, corresponding to different performance levels, are shown in Fig. 2.10.

Table 2.5: Damage classes of reinforced concrete structures [75]

Damage class	*Description of damage*
I	Visible narrow cracks on concrete surface (crack width is less than 0.2 mm)
II	Visible clear cracks on concrete surface (crack width is about 0.2.1.0 mm)
III	Local crush of concrete cover Remarkable wide cracks (crack width is about 1.0-2.0 mm)
IV	Remarkable crush of concrete with exposed reinforced bars Spalling off of concrete cover (cover width is more than 2.0 mm)
V	Buckling of reinforced bars Cracks in core concrete Visible vertical and/or lateral deformation in columns and/or walls Visible settlement and/or leaning of the building

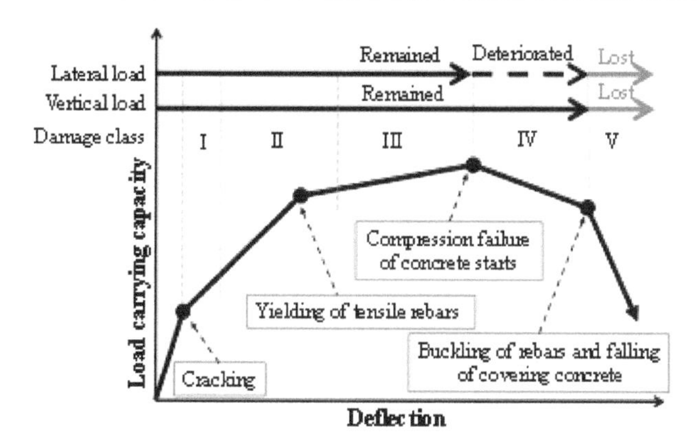

(a) Ductile member

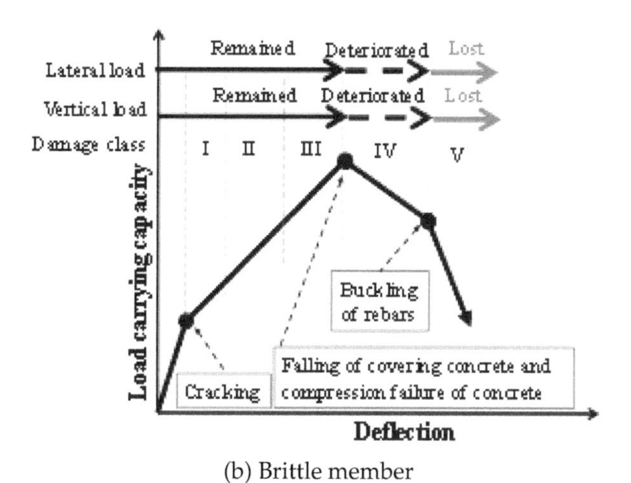

(b) Brittle member

Fig. 2.10: Damage classes versus load-carrying capacity [75].

For example, the damaged column shown in Fig. 2.11a is classified as Class III, with the assumption that the cracks are around 2 mm wide, but the damaged column shown in Fig. 2.11b is classified as Class IV because the rebars are directly exposed to the environment.

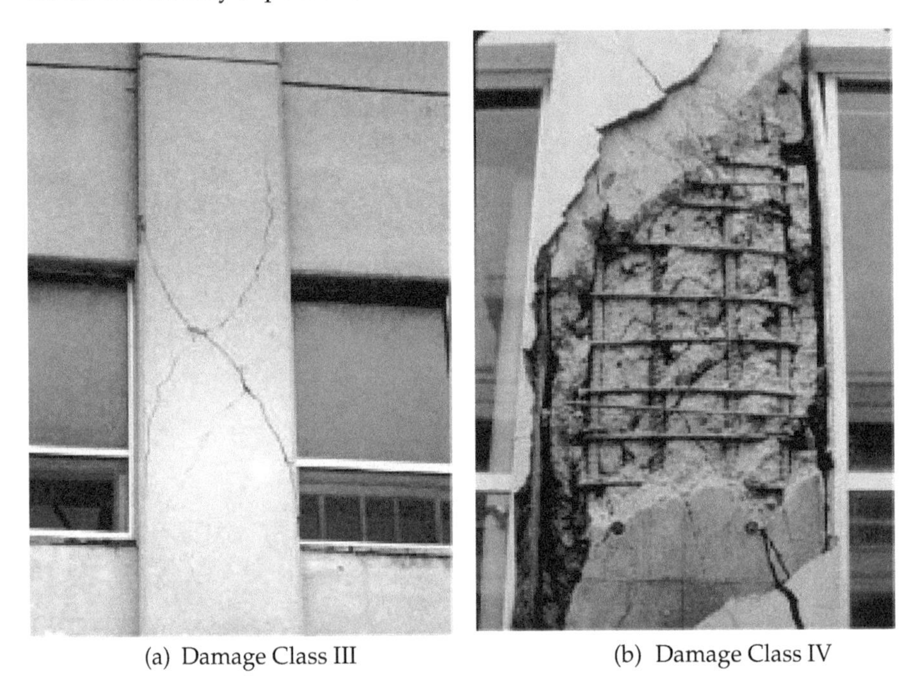

(a) Damage Class III (b) Damage Class IV

Fig. 2.11: Examples of damage categories, based on explanations in Table 2.5 [76].

In the same vein, FEMA356 has also provided guidance for evaluating damage to reinforced concrete structures at each seismic performance level and for primary and secondary elements. For the primary elements at the IO performance level, it is expected that only some minor hairline cracking would occur with some degree of yielding; no crushing would occur. For the same performance level but for the secondary elements, some minor cover spalling, and shear cracking with a width less than 1.5 mm in the vicinity of joints would occur. At the LS performance level and for the primary elements, widespread damage, cover spalling, and shear cracking with a width less than 3 mm in ductile elements would be expected. For the secondary components, on the other hand, the damage would be much more particularly in short columns, where some splice failure might be also observed. The earthquake damage at the CP performance level is very severe – both primary and secondary elements would sustain concrete cover spalling with core splicing and rebar buckling (Fig. 2.12).

Fig. 2.12: Severe concrete splicing after the 1985 Mexico City earthquake [77].

In order to relate the physical damage of RC members to various structural components, results were collected from a number of tests conducted in the laboratory at the University of Queensland, Australia. These tests were conducted on some RC joints, in order to scrutinize the extent of the simulated earthquake damage at various performance levels, such as IO, LS, and CP [78]. Figure 2.13 provides a schematic side-view of a joint sub-assembly under load in the testing apparatus. The specimen represented an external joint in a scaled-down (1:1.70) RC prototype. The prototype was a three-story RC frame, designed for peak ground acceleration (PGA) of 0.35g, and based on the ACI-318 code with the configuration shown in Fig. 2.14. The simulation of the scaled-down specimen was performed based on the Buckingham theorem [79]. The joints were continued to the mid-height of the column and mid-span of the beam, corresponding to the inflection points of the bending moment diagram under lateral loads. Details of the testing apparatus are shown in Fig. 2.15a. The cyclic loads were applied to the free end of the beam, and for drift values of around 1.0%, 2.0% and 4.0%, corresponding to the allowable drift values of IO, LS and CP levels of performance, respectively. The state of the damage at the mentioned performance levels was accordingly captured over the conduct of the test.

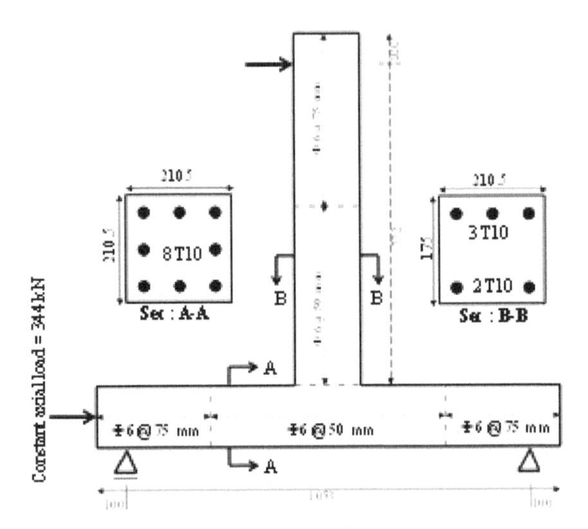

(a) Configuration of the specimen

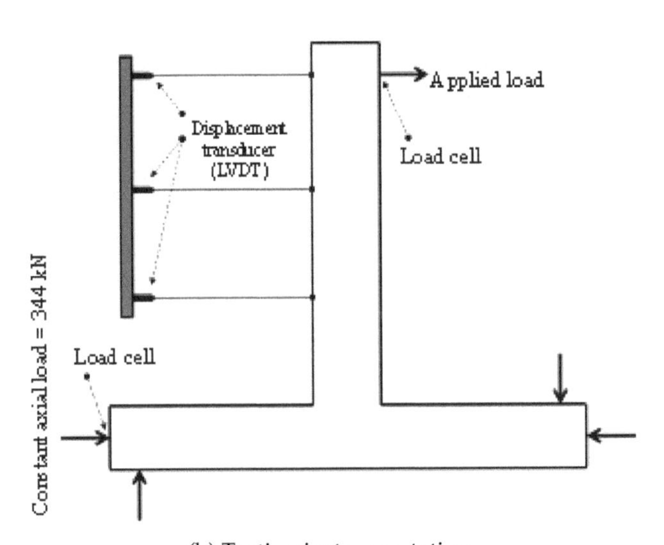

(b) Testing instrumentation

Fig. 2.13: Specimen (dimensions in mm) [78].

Figure 2.15b shows the load-displacement curve of the specimen, based on the cyclic loading. The envelope curve is also shown in the figure, to account for the drift values. Fig. 2.16 displays the specimen's damage state at different levels of performance. It was observed during the test that the damage was a localized geometrical phenomenon, even when the loading was continued up to the CP level of performance. For every performance level, the widths of the cracks were observed. At a drift value of around 1%, which corresponded to the IO level of performance, only minor cracking

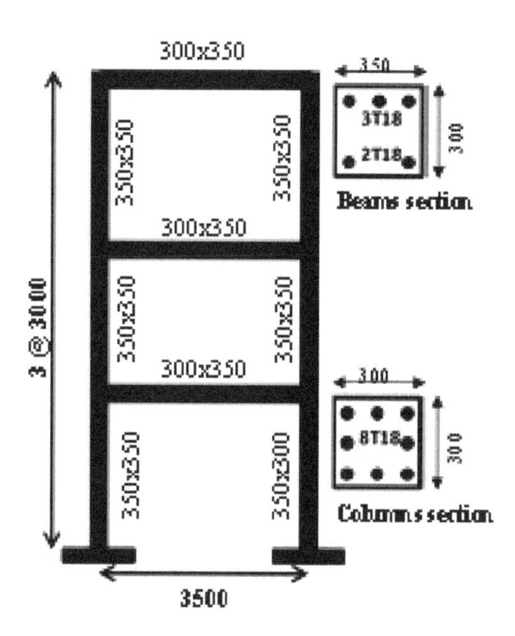

Fig. 2.14: The original prototype [78].

with a width of roughly 0.30 mm was observed, showing a good conformity with the prescriptive definition of damage mentioned by FEMA356. At a drift value of around 2%, i.e. at the LS level of performance, spalling of the cover, resulting in exposing of the rebars, was observed. Analogously, at the CP level of performance, alongside spalling of the cover, some parts of the concrete core were also spalled off. The lengths of severe damage (say at LS and CP) were similar to the plastic hinge length accounted for by empirical equations, such as Park and Paulay's equation [80].

While the results of the tests explained above show a close conformity with those presented by FEMA356, another test conducted by Kamath *et al.* [81] showed different results for the damage states at different performance levels. They constructed a single-story RC structure with a slab loaded by dead and live loads according to the Indian Concrete Code IS456. The structure was then subjected to cyclic loading in a quasi-static fashion. Target drift values of 2% and 4% were selected to push the structure to arrive at the LS and CP levels of performance, respectively. The first cracks appeared at a drift ratio of 0.92%, approximately corresponding to the IO level of performance. Some spalling in the concrete cover of beams and some widening cracks in columns were observed at a drift ratio of around 2%. The frame was then pushed further to arrive at 4% drift, to see whether the structural damage would be similar to that explained by FEMA356 at the CP level. While the test confirmed that at 4% drift there were extensive cracking and plastic hinges in beams and columns, the overall damage was not as severe as that mentioned by FEMA. They concluded that the damage expected by FEMA

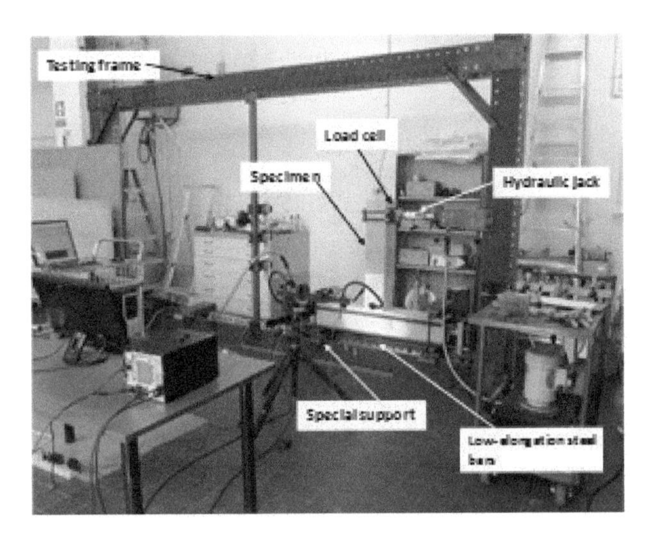

(a) The specimen under cyclic loading

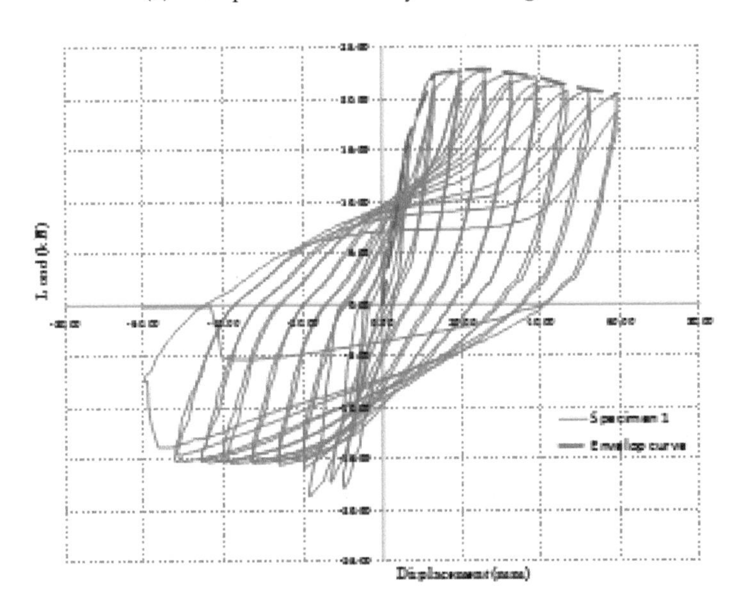

(b) Lateral load-displacement curve

Fig. 2.15: The test conducted in the laboratory at the University of Queensland [78].

was conservative. As well, there was no severe degradation of the strength and stiffness. Interestingly, their results showed minor residual drift, even after application of 4% drift to the structure, with no extensive spalling in the cover and core, and no rebar buckling. These results were plainly in contradiction to FEMA356. The differences between the Kamath *et al.* work and that of other researchers cited here can be attributed to their particular

(a) IO level of performance (b) LS level of performance (c) CP level of performance

Fig. 2.16: Damage state of the specimen at different performance levels [78].

test configuration; thus, their results cannot easily be generalized to a range of building structures. The results of the test conducted by Kamath *et al.* also contradict the author's personal observations from a number of earthquakes occurring in seismic regions.

2.3.2 Steel moment-resisting structures

Structural damage in steel moment-resisting structures includes yielding, buckling and fracturing in beams, columns, beam-to-column connections, and column panel zones, as schematically shown in Fig. 2.17. These forms of damage can be classified into several groups, from minor to major, in which they can influence the welding and the girder-to-column connections. Table 2.6 shows some of the most important types of damage that might occur in beam-to-column connections.

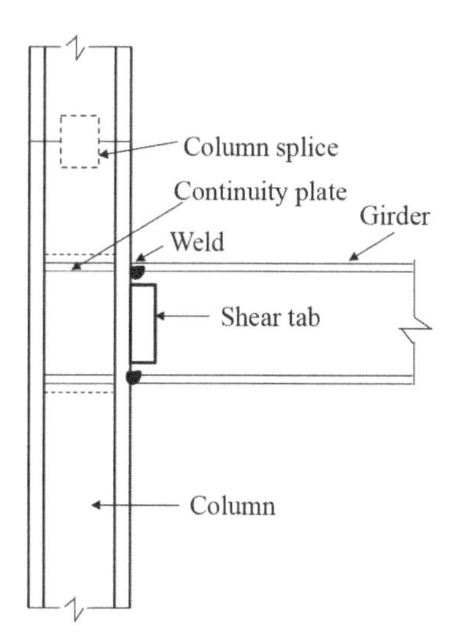

Fig. 2.17: Schematic side-view of beam-to-column moment connection.

Table 2.6: Damage classifications [82]

	Type	*Description*	
Girder	Group 1	There might be local buckling in the flange	 Type of column damage
	Group 2	The flange might yield	
	Group 3	There might be some fracture in the flange	
	Group 4	The web might buckle or yield	
	Group 5	There might be some fracture in the web	
	Group 6	There might be lateral torsion buckling	
Column	Group 1	There might be minor cracking on the surface flange	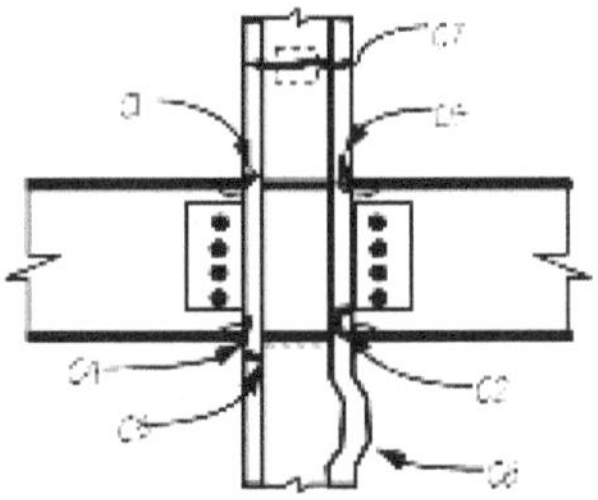 Type of column damage
	Group 2	There might be minor tearing to the flange	
	Group 3	There might be significant flange cracking	
	Group 4	There might be lamellar flange tearing	
	Group 5	The flange might buckle	
	Group 6	There might be column splice failure	
Panel zone	Group 1	The continuity plate might yield or buckle	Type of panel zone damage
	Group 2	There might be yielding in the web	
	Group 3	There might be partial fracture in the doubler plate	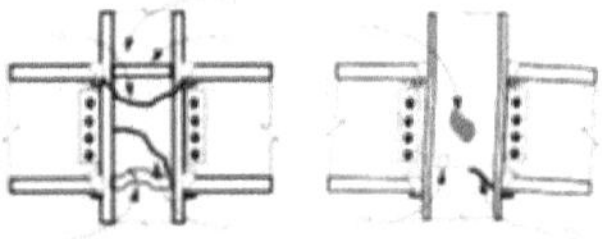
	Group 4	There might be fracture in the web	
	Group 5	There might be major fracture in the web or doubler plate	
	Group 6	The web might buckle	
	Group 7	The column might cut off	

The stability of a structure can be affected by one or more of the damage types mentioned above, in such a way that it can be considered to have sustained minor damage, or major damage. A classification as "minor damage" would mean that the structure could maintain its serviceability and defined performance level because its structural and non-structural components will have sustained almost no damage. This condition might be more relevant to buildings designed for the operational performance level, as introduced earlier. If any damage was seen, it would be limited to the non-structural elements in a way that the overall building still continues its serviceability. Some parts of the building, such as elevators, might be out of service due to a limited safety hazard. Buildings designed for the IO performance level can be assumed to fall into this categorization.

When there is significant (or "major") damage to both structural and non-structural components, it can be understood that the structure is at risk, although the structure will not necessarily fail. Most buildings designed for the LS level of performance would be able to sustain significant damage. Buildings that appear to have sustained major damage not only to non-structural but also to structural components to the extent that they should no longer be occupied are at severe risk. This classification is mainly attributed to buildings designed for the CP performance level, which are obviously unsafe to occupy.

The abovementioned classifications do not relate to whether or not the structure has been fireproofed. In line with the objectives of this book, more details need to be provided on this matter. There are documents, which are mentioned hereafter, that have directly addressed fireproofed structures, as their post-earthquake situation is of importance to the PEF analysis. Referring to Fig. 2.17, various zones of a fireproofed beam-to-column connection may be affected by earthquake loads. Overall, it is hard to guarantee that fireproofing materials will remain intact after an earthquake and, therefore, that the fireproofed structure will work at the same level as before the earthquake. This difficulty exists mainly because most fireproofing materials are weak under tension and compression forces – they are applied on the surface of the structural elements to withstand static loads and not necessarily dynamic loads. Some tests and numerical investigations have mentioned that the degree of damage sustained by fireproofing materials that have been bonded to bare steel profiles depends highly on the drift value sustained after earthquake. Figure 2.18 shows an example using fireproofing materials for protecting steel structures.

It is believed that even minor cracking to fireproofing materials can cause deterioration in their performance. Therefore, most investigations have scrutinized the displacement of the structural components under various lateral displacements or story drift. Investigations have included a portion of the beam within a distance of the column's face, the portion of column under the connection and within a distance of the bottom flange of beam, and the panel zone. These zones are supposed to be the regions most

Fig. 2.18: An example of fireproofed steel structures (photo courtesy of H. Salehi).

susceptible to damage, because most plastic formations would occur in the vicinity of joints. FEMA352 [82] mentions that any permanent drift of more than 1% should be seen as a situation in which the fire protection layer has been debonded and is thus no longer effective. Accepting this and referring to Table 2.4, most structures designed for the LS performance level can be assumed to have lost their fire protection layer, mostly in regions around the joints.

Some two-dimensional studies were conducted by Tomecek and Milke [83] to evaluate the effect of partial loss of fireproofing materials spread on steel columns. To investigate this, they assembled three steel columns fireproofed by cementitious material in a way that it could provide different fire resistance ratings, from one to three hours. Different protection losses on the flange or web, from 1% to 14% loss, were then considered. The fire resistances of the columns based on the new conditions were numerically accounted for. The results showed that at about 8% loss protection, the fire resistance of the columns (depending on the column's size) declined dramatically from around 40 to 30 minutes, even in those that were rated at a three-hour fire resistance.

A similar study but in three dimensions, was performed by Ryder *et al.* [84] to show the effect of loss of fire protection materials on the fire resistance of steel columns. Firstly, only fully protected columns were exposed to 90 minutes of fire, in order to find the temperature rise at the exposed surfaces. The minimum thickness of the protection layer was accounted for, to meet the thermal endpoint criteria (538°C) stipulated by ASTM E119 [85] for a one- or two-hour fire exposure. Various areas of the protection layers were then removed from the columns. The partially unprotected columns were then subjected to fire. The results showed that even in cases in which only a small area of fire protection materials had been removed, the fire resistance declined significantly. For example, while the average temperature at the fully protected column exposed to fire for one hour was about 400°C, it reached about 600°C when a 3.9 cm^2 area of protection was removed. This violated the thermal endpoint criteria.

Kwon *et al.* [86] considered various sizes and locations of damage to sprayed fireproofing materials and found that when there is an area of fireproofing removed from the surface of a structural member, the fire resistance drops significantly, even by more than 70%. This reduction was more sensitive to flange removal scenarios than to web removal scenarios. A similar study was also performed by Wang and Li [87], which showed that the fire resistance of steel columns with partial damage to their fire protection cover is much less than in those that are fully protected. Braxtan and Pessiki [88, 89] also investigated the situation of fireproofed steel structures after they had been subjected to earthquake loads. To comply with the seismic codes' regulations, they designed a weak beam, strong column moment-resisting steel frame (MRSF) in such a way that the first hinge formation occurred in the beams after the frame had been subjected to a quasi-static cyclic loading. The results of their work showed that when the drift value goes beyond around 1%, the beams yield and as a result, the fireproofed materials become debonded from the steel profile surfaces. Debonding of the fireproofed materials continued to the point of tearing when the frame experienced more drift, around 3-4%.

Arablouei and Kodur [90] numerically investigated the behavior of fire insulation materials sprayed onto the surface of MRSFs under earthquake loading, and found that there is a close similarity with the tests and numerical investigations conducted by Braxton and Pessiki. Also, the study performed by Arablouei and Kodur showed that even local buckling of the flange can lead to a substantial delamination of the insulation materials over a large area. In another work, Keller and Pessiki [91] conducted some tests in order to find a pattern for spalling of sprayed fire-resistant materials (SFRMs) on steel special moment-frame structures subjected to seismic loads. They showed that plastic hinge regions are the most vulnerable to spall. They then showed that, if the damaged frames were exposed to PEF loads, there would be a significant increase in sideways motion as well as lateral drift. Recently, Behnam and Abolghasemi [92] investigated a fully detailed seven-story MRSF (including details of beam-to-column connection, and beams and columns splice plates) covered with SFRM under earthquake-induced drifts. The required thickness of SFRM was calculated based on the structure's required fire resistance rating. Considering the height and performance of the frame, SFRMs of various densities were applied. The results showed that SFRM delamination occurred, even in small drifts of less than 1.15%, where plastic hinges were about to form. SFRM delamination in drifts of more than ≈ 1.70% was dramatically large in the plastic hinge locations. The results also showed that SFRMs on beams were more vulnerable than on columns, as shown in Fig. 2.19, the reason for which was attributed to the model's assumption (where beams were designed to be weaker than columns).

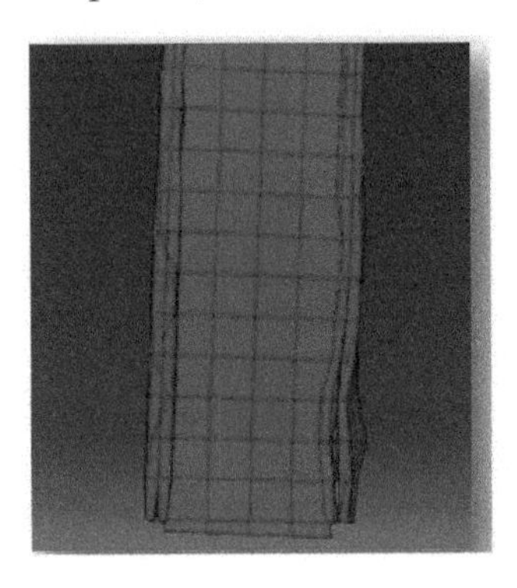

(a) Delamination at columns' base (b) Delamination in beams

Fig. 2.19: Response of SFRMs under seismic drift [92].

The abovementioned studies show that there is a vital need to focus on the behavior of SFRMs, in order to recognize the weak points in their performance under seismic loading. However, even with the great efforts made to this point, these studies have been neither comprehensive nor practical enough to be useful in the reliability assessment of the real fire ratings of steel structures exposed to PEF. The above results can nevertheless be correlated to the seismic performance levels, e.g. IO and LS. As buildings designed for the IO level should sustain a drift value less than 1%, it can thus be assumed that no debonding or tearing would occur and, as such, the fireproofing materials should remain serviceable. At the LS level, on the other hand, debonding (and possibly tearing) would occur. Thus, over the yielded regions, the steel profile should be considered as if there was no effective fire protection layer.

2.3.3 Irregular structures

While irregular structures are widely constructed in urban regions, recorded statistics have shown that these structures, on average, are more susceptible to earthquake damage than are regular structures. These irregularities take five main forms: vertical geometric irregularity, diaphragm discontinuity, stiffness irregularity, column stiffness irregularity (e.g. short columns), and mass irregularity. Figure 2.20 shows some conceptual and real examples of irregular structures seriously damaged after an earthquake. In terms of vertical irregularity, most codes mention that a structure is considered irregular if the ratio of one of the structural quantities (such as stiffness or mass between two subsequent stories) goes beyond a minimum value. For example, according to FEMA310 [93], a story is considered to be vertically

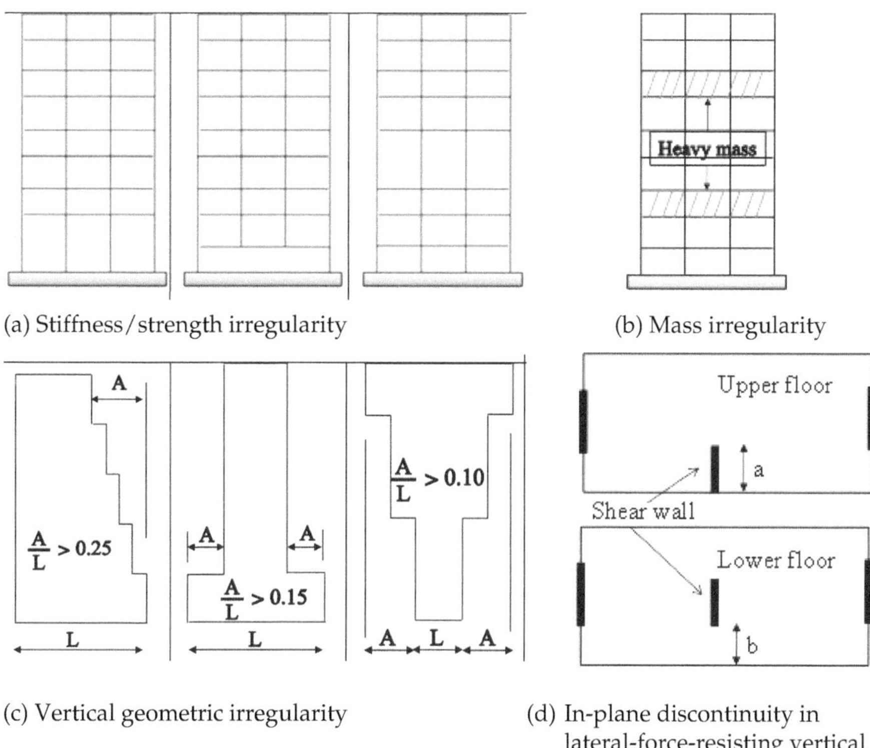

(a) Stiffness/strength irregularity

(b) Mass irregularity

(c) Vertical geometric irregularity

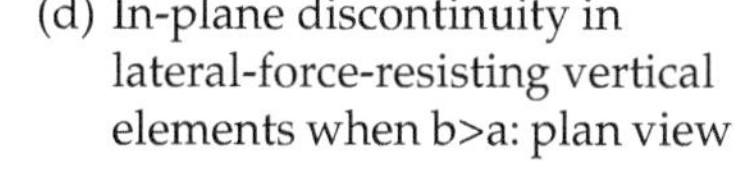

(d) In-plane discontinuity in lateral-force-resisting vertical elements when b>a: plan view

(e) A real example of a building that collapsed due to stiffness irregularity after the 2003 Bam earthquake (photo courtesy of H. Vosough)

(f) Collapse of an irregular structure with reduced plan dimensions after the 2003 Bam earthquake (photo courtesy of H. Vosough)

Fig. 2.20: Some types of structural irregularities [95].

irregular if the horizontal dimension of the lateral force resisting systems in a story is more than 130% of that in an adjacent story.

IBC 2000 [94] states that irregularities to structures shall be considered as follows:

Mass irregularity: When the effective mass of a story is more than 150% that of the adjacent story, the structure shall be considered as mass irregular. The mass irregularity, however, does not need to be checked for the last story, as the roof story is often much lighter than the story below.

Stiffness irregularity: When the lateral stiffness of a story is less than 70% that of the adjacent story or of the average of the three adjacent stories, the structure shall be considered to have a soft story.

Strength irregularity: When the lateral strength of a story is less than 80% that of an adjacent story, the structure shall be considered to have a weak story. The strength of the story is determined by the total strength of the earthquake-resisting components for a direction considered.

Irregular structures can be analyzed or designed for gravity loads with confidence because it is believed that gravity loads are definable. There are, however, numerous studies that have shown that the response of irregular structures under seismic loads is not as predictable as that of regular structures. The different response is correlated to the torsional component of the response, which is greater in irregular structures than in regular structures. There have been studies of the effects of the abovementioned irregularities on the performance of RC frames when they are subjected to seismic loads, to support the belief that irregular structures are more vulnerable to damage than regular structures [96, 97]. Figure 2.21 shows a comparison made between the linearized fragility curves of two regular and irregular structures designed for various performance levels. As is seen, the irregular structures are more susceptible to exceeding the considered limit states than the regular structures.

The figure also shows that irregularity causes a considerable increase in the probability of exceedance. There is, however, no clear consensus in the graphs as per the different performance levels. At the IO level of performance (Fig. 2.21a), the structures show divergence only over a weak earthquake (with a PGA lower than 0.2g), while during strong earthquakes they show a close convergence. The structures designed for the LS level of performance (Fig. 2.21b), by contrast, start to diverge during average and strong earthquakes. As for structures designed for the CP levels (Fig. 2.21c), the divergence interestingly starts over strong earthquakes, with PGAs greater than 0.3g. These graphs, meantime, confirm the very complex behaviors of irregular structures under seismic loading. That study also showed that the damage index versus PGA in irregular structures is considerably larger than in regular structures. The damage index mainly increases after a PGA of 0.2g, which implicitly confirms that irregular structures will experience more inelastic involvement from stronger earthquakes and, as such, are more sensitive. Behnam [97] also investigated the response of regular and irregular tall steel structures subjected to a PGA of 0.3g, and found that irregular structures sustained more damage after earthquake than regular structures. Behnam controlled the drift variations of irregular and regular

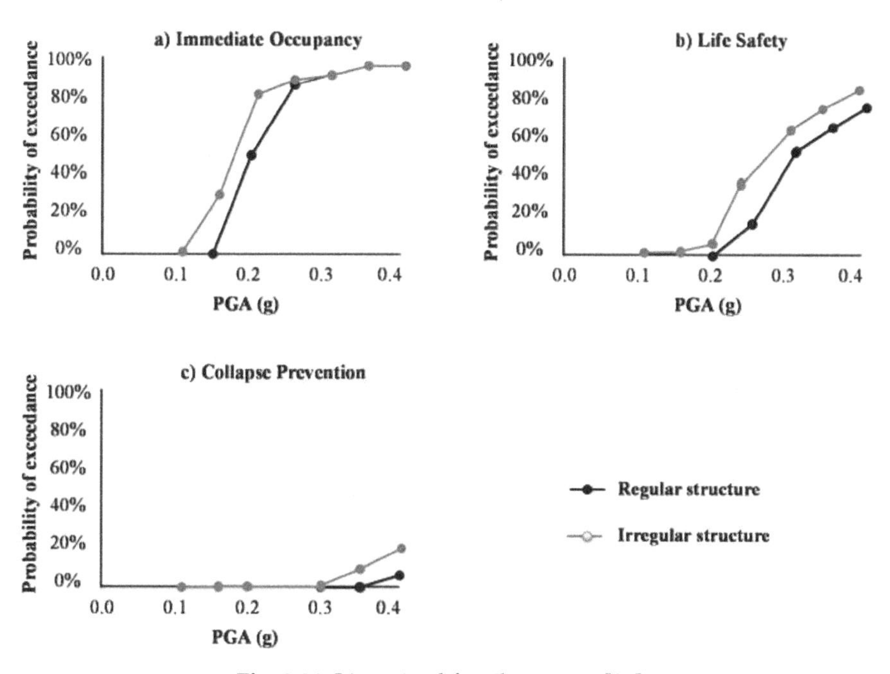

Fig. 2.21: Linearized fragility curve [96].

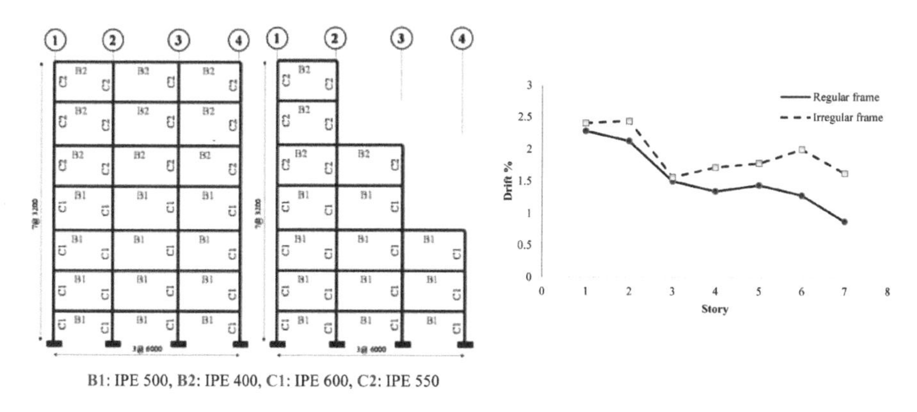

Fig. 2.22: Drift values of regular and irregular structures [97].

structures in different stories and found that irregular structures experienced more drift than regular structures, as shown in Fig. 2.22.

While the studies mentioned above focused on RC structures, the research conducted on steel structures shows similar results [98]. It can thus be concluded that more attention at the design stage must be paid to irregular structures than to regular structures. To address this concern, most codes and provisions provide stringent analyses and procedures for the design of irregular structures, such that the more irregular a structure is, the further is that structure from the restrictions of applicability of analysis and design methods.

One of the most commonly observed column stiffness irregularities is short column. The term 'short column' is often used as the opposite of slender column, and is determined by controlling shear span to depth ratio, $\alpha = M/Vh \leq 2.5$, in which M, V and h are the moment, shear force and column's depth, respectively. There is a correlation between the shear ratio and the brittle behavior of a column, in such a way that columns with $\alpha \leq 1.5$ behave in a very fragile manner. Indeed, when α becomes smaller, M decreases, while V remains the same. Thus, the column would most likely fail in a shear fashion. Given that for boundary conditions, M becomes minimal when a lateral load is applied to the middle height of the column, the point of inflation is also in the midst of the column. The synchronization of these two happenings would lead to the displacement being doubled (which is known as double curvature) with a very destructive outcome. Figure 2.23a shows double curvature and Figure 2.23b displays an example of a short column's behavior after an earthquake.

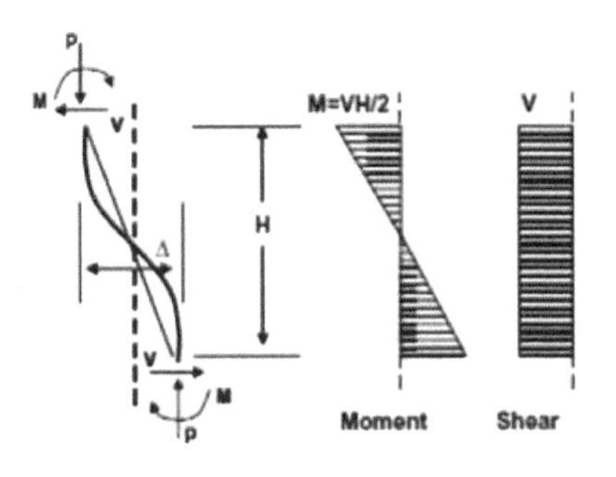

(a) Double curvature loading

(b) Short column's behavior after an earthquake

Fig. 2.23: Short column.

Investigating the structural behavior of a short-column irregularity under earthquake loads is thus of importance. This importance becomes even greater if there are earthquake-damaged components in a PEF situation. In particular, a detailed evaluation should be made to scrutinize the situation of fireproofed moment-resisting steel and RC structures after earthquake when there is a short column irregularity, since any damage to the fireproofing material in the steel or RC structures can significantly change the PEF resistance. This issue will be discussed further in Chapter 6.

It is also notable that, irrespective of the irregularity type, it is often believed that irregular structures with convex plans are more seismic-resistant than those with concave plans [99]. To consider a plan convex, any arbitrary line drawn between two arbitrary points in the plan shall remain within the plan; otherwise, the plan is considered as concave. Some examples of convex and concave plans are shown in Fig. 2.24.

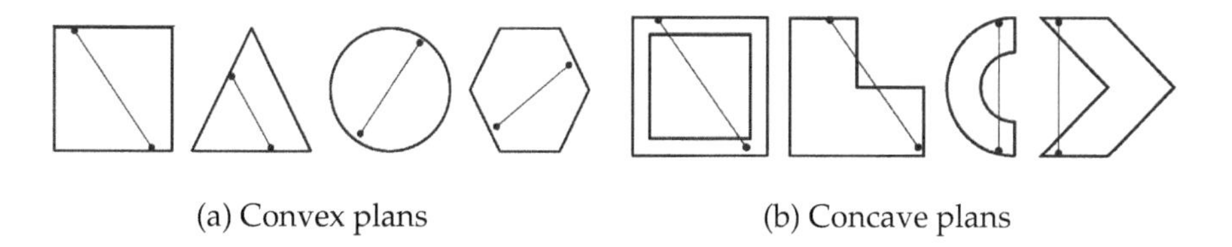

Fig. 2.24: Examples of convex and concave plans [100].

2.4 Performance-based fire safety design

The primary objective of fire safety provisions is to minimize the risk of casualties during a fire event. After that, fire safety regulations also provide additional requirements to reduce financial loss. Hence, it is important to define the objectives of the fire safety regulations in order to satisfy the degree of acceptable risk. The objectives outlined can be met through various measures including: adopting remedies to control the ignition; providing precautionary rules to prepare people for the need to evacuate from a building on fire; controlling combustible materials and fire growth; rapid extinguishment in order to confine the extent of fire damage; and establishing adequate urban facilities, such as fire stations, to respond quickly to fire before it can get out of control. Regarding structural stability under a fire event, the fundamental objective is that the structure will maintain its stability for a reasonable time, which will differ from one case to another.

To meet these fundamental requirements, different methods may be adopted. Fire design provisions are derived from two methodologies: prescriptive-based (also called deemed-to-satisfy) and performance-based. In terms of prescriptive-based fire codes, most requirements prescribe the solutions, but no information is provided concerning the background conditions. Most emphasis in prescriptive-based fire provisions is placed on the structural elements, stating that the elements shall have adequate capability to withstand applied loads in case of a fire event for a predefined time without collapse. In most codes, a 45-minute fire-resistance is considered a minimum need, followed by higher levels of resistance, going even up to five hours. The time required to meet the defined fire resistance in RC members is provided by the concrete cover, such that the thicker the cover is, the higher the level of resistance reached. It is supposed that, based on the concrete cover determined, the temperature at rebar could be limited to below critical temperature, which is the temperature at which rebar loses 50% of its strength. For conventional and pre-stressed rebars, the critical temperatures are assumed to be 593°C and 426°C, respectively. Table 2.7 provides data for the fire resistance of RC members, based on Eurocode 2. As is seen, increasing the concrete cover has a major role in providing more fire resistance.

Table 2.7a: Fire resistance rating of columns [101]

Standard fire resistance (min)	*Column exposed on more than one side*		*Column exposed on one side*	
	Minimum dimensions (mm)	*Minimum cover (mm)*	*Minimum dimensions (mm)*	*Minimum cover (mm)*
30	200	32	155	25
60	250	46	155	25
90	350	53	155	25
120	350*	57	175	35
180	450*	70	230	55
240	450*	75	295	70

* Minimum 8 bars

Table 2.7b: Fire resistance rating of non-load-bearing walls (partitions)

Standard fire resistance (min)	*Minimum wall thickness (mm)*
30	60
60	80
90	100
120	120
180	150
240	175

Table 2.7c: Fire resistance rating of load-bearing walls

Standard fire resistance (min)	*Column exposed on more than one side*		*Column exposed on one side*	
	Minimum dimensions (mm)	*Minimum cover (mm)*	*Minimum dimensions (mm)*	*Minimum cover (mm)*
30	120	10	120	10
60	130	10	140	10
90	140	25	170	25
120	160	35	220	35
180	210	50	270	55
240	270	60	350	60

Table 2.7d: Fire resistance rating of simply supported beams

Standard fire resistance (min)	*Minimum dimensions (mm) Possible combinations of "a" and "b min"*			
30	b min = 80	120	160	200
	a = 25	20	15	15
60	b min = 120	160	200	300
	a = 40	35	30	25
90	b min = 150	200	300	400
	a = 55	45	40	35
120	b min = 200	240	300	500
	a = 65	60	55	50
180	b min = 240	300	400	600
	a = 80	70	65	60
240	b min = 280	350	500	700
	a = 90	80	75	70

a = minimum cover

Table 2.7e: Fire resistance rating of continuous beams

Standard fire resistance (min)	*Minimum dimensions (mm) Possible combinations of "a" and "b min"*			
30	b min = 80	160		
	a = 15	12		
60	b min = 120	200		
	a = 25	12		
90	b min = 150	250		
	a = 35	25		
120	b min = 200	300	450	500
	a = 45	35	35	30
180	b min = 240	400	550	600
	a = 60	50	50	40
240	b min = 280	500	650	700
	a = 75	60	60	50

a = minimum cover

For steel structures, limiting the critical temperature (for example, to 550°C) can be achieved either by covering the profile surface with insulation materials, or by encasing the members with concrete, as shown in Fig. 2.25. The relevant insulation materials vary from some traditional construction materials, such as cementitious-based materials and gypsum-based materials, to more recently used materials, such as sprayed fire-resistant materials (SFRMs). The latter requires less labor to assemble than the traditional construction materials, therefore lowering cost.

(a) Insulation materials (photo courtesy of H. Salehi) (b) Encasing with concrete

Fig. 2.25: Fireproofing of steel components.

The times listed in the above tables are also applicable when fireproofed steel structures are considered. These times are mainly based on experience and past fire losses, although they have also been exaggerated to provide more safety to inhabitants. The times shown are then used as a minimum fire resistance for a particular building. For instance, a four-story school building must have at least a two-hour fire resistance, irrespective of the building's specifications (such as its area and the number of students). Although this emphasis is important, it is also very superficial as it can be interpreted differently. For instance, it is not clear whether the loss of any single element in a fire event will necessarily lead to structural collapse, and, assuming that every single element has adequate fire resistance, does that mean the structure as a whole will also have sufficient fire resistance? As for the first question, since there is often satisfactory redundancy built into structures, losing the load bearing of a single element should not result in a global collapse, as it is expected that there would be alternative load paths. Fig. 2.26, as an example, shows a fire-damaged structure which remained stable after a severe fire event, even though one single element has severely been damaged. The stability of this structure after the fire can be correlated to the load redistribution of the structure after a single element has lost its load-bearing capacity.

The shortcoming highlighted by the second question can be seen from the fact that element-based provisions are usually based on the boundary conditions of the elements at ambient temperature [102]. Under a fire event, however, there will be different boundary conditions, resulting in uncertainty as to whether the structure will remain stable. The tragic collapse of World Trade Center Towers (WTCT) after the 9/11 terrorist attack is an example of such a failure, in which the loss of the load-bearing capacity of a limited number of elements resulted in the progressive collapse of the towers. Investigating the structural response under progressive collapse is a fascinating subject that cannot be addressed here.

Fig. 2.26: Fire-damaged building (photo courtesy of M. Bentnabi).

Another deficiency in the prescriptive-based codes, which can be addressed here, is that they do not consider the building as a whole. For example, there is no consideration given to the existence of non-structural components. It is easily understood that the failure of a structural component may result in structural collapse, but it should also be considered that the failure of a non-structural component, such as a partition, can result in the fire spreading from one cell to adjacent cells in a short period. This in turn can endanger the stability of the structure.

Most prescriptive-based fire codes provide minimum requirements for the protection of a generic occupancy, requiring no high level of engineering knowledge to be understood. For example, the number of fire detectors or fire extinguisher systems, or the defined fire resistance in a particular occupancy can be determined using prescriptive-based fire codes. Although employing these approaches may be applicable as a minimum requirement, the possibility of loss for a particular design is not clear. Therefore, the fire extinguishers required for a given occupancy may be different if determined by various people who assume different extents of damage. As well, many of the prescriptive-based codes are complex and restrictive because they are continuously imposing new regulations on top of the existing ones. Prescriptive-based fire codes do not consider the cost-effectiveness of a design (this might relate to the fact that providing safety for a building's occupants is the priority). While providing adequate safety is undoubtedly important, it can meantime reduce the flexibility available for making further improvements. Furthermore, recent investigations have proven that using prescriptive-based provisions will not necessarily result in higher safety.

Rather, there is evidence showing that the fire resistance of modern open-plan buildings under a realistic situation might be less than the predefined level of fire resistance for those buildings designed based on the prescriptive-based provisions [103]. The significance of this is further underlined in Chapter 3, where different fire curves are discussed. The latter shortcoming of the prescriptive-based provisions can also pertain to the fact that they implicitly say that there is only one means of providing adequate safety. This is, of course, contra to the performance-based approach, where diverse means might be employed for addressing a single concern.

To overcome these deficiencies, performance-based provisions (also called objective-based provisions) are increasingly being employed. There are two aspects to the decision to use performance-based requirements – rationality and accountability [104]. The former refers to providing a higher level of fire safety, based on rational engineering methods (along with the increased complexity of modern buildings, empirical methods seem to be less efficient). The latter relates to quantifying the fire risk, in order to provide a more realistic situation and thus a more economic solution. This economic aspect is not considered in the prescriptive-based provisions, at least explicitly. Using performance-based fire requirements, a fire scenario is described by quantification of the fire's characteristics, such as load, size, severity, and duration. The variation of temperature over time is also calculated. In addition, heat transfer is carefully considered, as it is an important mechanism. As well, to find the most critical situation, several fire scenarios are considered. The scenarios encompass fuel load density and the various compartments (although they would not necessarily be the worst structural scenarios). The structural failure criteria are then determined, according to the defined performance levels, which vary from case to case. For example, assuming that mid-span deflection is a failure criterion, it may differ between a residential building and a health care center, since they might have different performance levels.

While all of these points provide a level of flexibility to a designer, they also create a rather complicated situation, as there are more requirements to recognize the key principles of the performance-based attitude.

This is possibly the reason that the performance-based provisions for fire are much more recent than those for earthquake, as the former require a considerable number of precautionary actions to be considered prior to use. As explained earlier, the performance-based approach clearly has some disadvantages. For example, it is very challenging to define a quantifiable level of safety when it comes to performance criteria. While employing prescriptive-based provisions needs no specific engineering knowledge, the adoption of performance-based codes requires an adequate level of education. As well, most prescriptive-based provisions are developed and imparted by authorities and, as such, are mandatory. It is difficult, however, to evaluate consent for performance-based codes with established requirements. Meantime, the tools used for quantification need to be validated when it comes to the performance-based provisions. Figure 2.27 illustrates the conceptual hierarchies of prescriptive- and performance-based fire safety requirements, which are very similar to the concepts illustrated in Fig. 2.4.

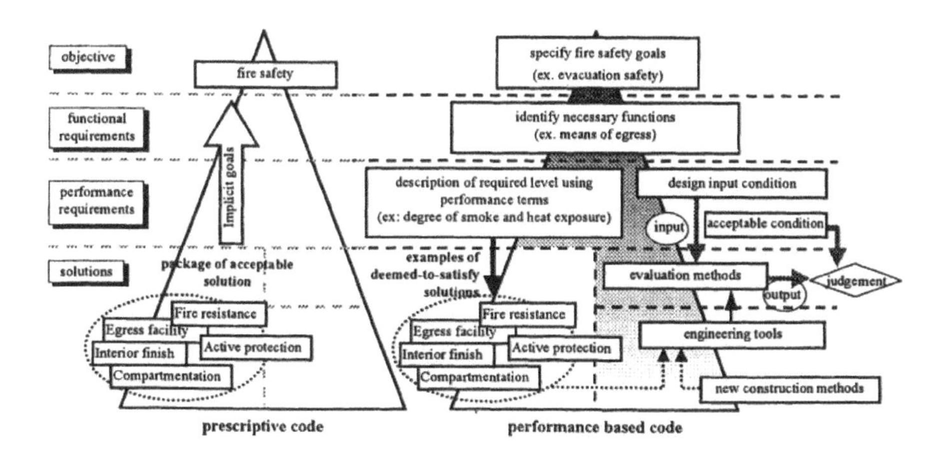

Fig. 2.27: Comparison between prescriptive- and performance-based fire safety provisions [104].

The use of performance-based provisions in building design is undertaken through either definite or probabilistic techniques. The former is used to determine fire development, structural response under fire loading, and evacuation of occupants from the fire, while the latter is used to estimate the level of life or property risk. The results of these techniques are compared with the performance criteria as to whether they are acceptable or not. In the following sections, these criteria are explained.

2.4.1 Deterministic techniques

These methods are used as engineering tools to understand fire behavior, structural response, and the time that it takes to evacuate occupants when there is a fire. Figure 2.28 shows the typical phases in a fire event, including ignition, heating and cooling. Most combustible materials have threshold values for radiant heat flux and surface temperature, after which the materials will ignite. Tables 2.8 and 2.9 [105, 106] show some data regarding the threshold values for ignition and the heat flux range for ignitability, respectively.

The priority regarding fire ignition is thus to reduce ignition prior to exceeding the thresholds. If, however, ignition is triggered, the fire will start to develop and will proceed in three phases. The first phase is pre-flashover (also called fire growth), and its rate is considered a parameter for evaluating whether the performance defined in a fire safety design has been met.

Table 2.8: Threshold values for ignition

Material	*Radiant heat flux for ignition (kW/m^2)*		*Surface temperature for ignition (°C)*	
	Pilot	*Spontaneous*	*Pilot*	*Spontaneous*
Wood	12	28	350	600
Chipboard	18			
Hardboard	27			
Polyoxymethylene	17			
Polymethylene	12			

Table 2.9: Heat flux range for ignitability

Ignitability	*Heat flux range (normal value) (kW/m^2)*
Easy (thin curtains)	≤14.1 (10)
Normal (upholstered furniture)	14.1-28.3 (20)
Hard (thick wood)	>28.3 (40)

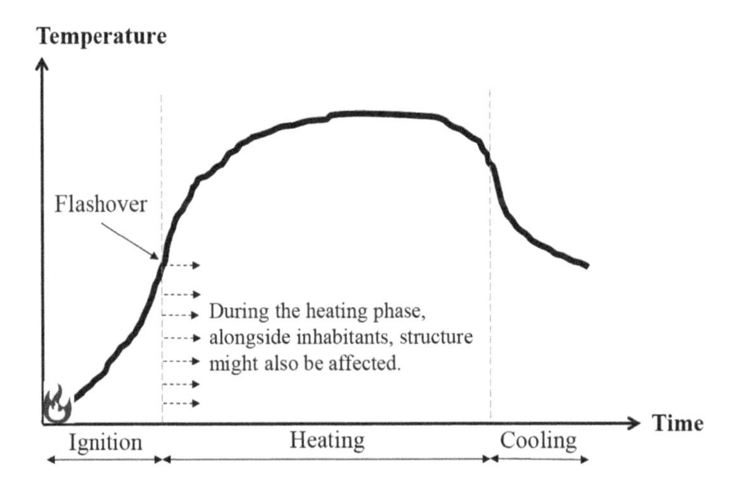

Fig. 2.28: Development phases in a fire.

There are a few methods used to determine the rate of pre-flashover, including test results and computer simulations. One of the most widely used methods is the t-squared (t^2) method. In using this, the heat release rate, $Q = Q_0(t/t_g)^2$, continuously increases until either the combustible material entirely burns out, or Q reaches a peak value, where Q is the heat release rate, and Q_0 is the reference heat release rate, often assumed to be 1 MW. The fire growth rate (t_g) depends on the types of combustible materials, which might be very slow to ultra-fast. Usually, fire growth rates below 600 ($s/MW^{1/2}$) and above 75 ($s/MW^{1/2}$) are considered as slow and ultra-fast, respectively [107]. The values of both combustible materials and ventilation conditions are the governing parameters for the fire to grow until its Q reaches a peak value. The maximum Q is also dependent on the occupancy type. For example, the maximum Q in an office building is assumed to be 250 (kW/m^2) with a medium fire growth rate, while in a store building, the maximum Q is 500 (kW/m^2) with a rapid growth rate of fire.

The second stage of fire development is called post-flashover, and occurs when flame radiation influences the entire compartment. In this case, there is a rapid transition from a localized fire to ignition of all of the combustible materials inside the compartment. It is believed that flashover occurs when the average temperature of a hot gas layer goes beyond 550-600°C, or when the heat flux reaches 20 kW/m^2. Flashover is considered an important stage, because after that, the temperature throughout the compartment rises to even higher than 1000°C, resulting in the fire developing fully. During the heating phase, along with inhabitants, structural systems might also be affected. Once there is a fully developed fire, it is expected that fire protection facilities would respond correctly to the fire before it can spread. Additionally, the protection facilities should satisfy the defined objective levels, through established fire-extinguishing systems, such as sprinkler systems, or by providing adequate fire resistance to the structural and non-structural components. It is understood that structural elements, such as beams, columns, and floor slabs, play a significant role in providing adequate resistance against fire loads. Non-structural components, such as internal and external partitions and firewalls, can also hamper the fire's spread. It bears repeating that a building's performance depends on the performance of both structural and non-structural components, i.e. the whole building as an entity. Thus, close consideration must be given to seeing how all components behave under a fire. The structural and non-structural fire resistance criteria are defined in terms of stability, integrity, and thermal isolation. It is often required that the structural components, such as beams, columns, floors, and load-bearing walls, meet all of the mentioned criteria, while the non-structural components are not required to meet these, as shown in Table 2.10. This table can be related to Table 2.7 and Fig. 2.25, where the fire resistance rating of structural components was mentioned.

Table 2.10: Fire resistance criteria for structural and non-structural components [108]

Building element	*Stability*	*Integrity*	*Insulation*
Partition		required	required
Load-bearing wall	required	required	required
Floor	required	required	required
Beam or column	required		
Fire-resistant glazing		required	

When performance-based provisions are used, the critical temperature criterion is not sufficient, since material nonlinearity is not considered in the process of evaluation. Thus, more criteria need to be considered. This deficiency is highlighted with the consideration that deflection, the rate of deflection, creep strain, and transient strain all have influence on the fire response – particularly in beam components [109]. Furthermore, life safety as an important objective in fire safety design is not directly involved in the critical temperature criterion, while it is obviously paramount that occupants can safely evacuate. To address this point, limiting deflection and the rate of deflection in beam elements under fire condition can improve the safety of residents prior to failure. In this respect, BS 476, as an example, mentions limits to deflection and rate of deflection. Deflection is limited to $L/20$, in which L is the length span in mm. As for the rate of deflection, it is limited to $L^2/9000d$ (mm/min), where d is the effective depth of beam in mm [110].

In some references, the tenability conditions for evacuating occupants are also mentioned as one measure for providing adequate life safety. In the BSI code [105], toxic combustion products, smoke obscuration, and heat are listed as determining factors. For example, it is mentioned in the BSI code that a 5-minute exposure to carbon monoxide at the level of about 6000 ppm can lead to a feeling of exhaustion, but that if the level increases to about 12000 ppm, it can lead to death. In the case of a 30-minute exposure to carbon monoxide, the respective levels decrease to about 1400 ppm and 2500 ppm. The minimum visibility in a room varies from at least 2 m to 10 m, depending on the room's dimensions. Radiation, convection, and conduction as means of heat transfer can also be considered as tenability criteria. For instance, if convection and conduction create temperatures in a compartment of over 190°C, and 60°C, respectively, then tenability is at risk. As for radiation, a heat flux higher than 2.5 kW/m^2 is a limiting condition for tenability.

In addition to the abovementioned failure criteria, there is also a time-dependent criterion, which is determined in order to allow occupants to evacuate from the building prior to untenable conditions being reached [111]. For example, Malhotra [112] has suggested the times shown in Table 2.11 as the critical times for tenability. It is also worth noting that, in practice, the evacuation times in Table 2.11 are doubled in order to provide more safety to occupants. As is seen, the evacuation time increases as the protective facilities increase. This clearly shows how all parts of a building's components influence the defined performance level.

The third phase of a fire's development is called decay phase, or cooling. When more than 70% of the combustible materials are consumed, the fire diminishes, and the temperature declines with time.

2.4.2 Probabilistic criteria

In addition to the definite criteria explained above, it is also important to take probabilistic criteria into consideration. This is because fire, in general, is a

Table 2.11: Critical times for reaching untenable conditions [112]

Type of zone	*Critical time to reach untenable conditions in means of escape (min)*
Unprotected fire zone (zone of occurrence of fire)	
Regular sized room ($\leq$ 100 m^2)	2.2.5
Larger compartment or room (height > 4 m)	4-6
Partially protected zone (zone with heat- and smoke-resisting barriers for a limited time)	
Natural smoke expulsion	5
Pressurization or extraction system	10
Fully protected zone (zone where protection remains acceptable for the whole duration for which protection is required in the building)	
Natural smoke expulsion, no lobby	30
Natural smoke expulsion, lobby	45
Pressurization or extraction system	60

random-based event, which can be influenced by numerous uncertainties, such as human behavior, distribution of combustible materials in a compartment, the effectiveness of established fire extinguisher systems, and occupancy type. Adding consideration of these uncertainties can provide a better estimation of the fire safety of a building. They can be quantified by performing a risk assessment in which both probabilistic and determinative calculations are used. A risk, in general, is accounted for by using the formula $R = P \times L$, in which R, P, and L stand for risk, the probability of exceedance, and loss, respectively. While this formula has a quantitative nature, it can also be expressed qualitatively in nine groups, as shown in Table 2.12.

Table 2.12: Qualitative expression of risk level [113]

	Probability of exceedance		
Probability of loss	*High value*	*Medium value*	*Low value*
High value	High risk		
Medium value		Medium risk	
Low value			Low risk

Determining the fire risk for life safety, the probability of exceedance and loss must be taken into account. For the likelihood of exceedance, there are numerous uncertainties; for the likelihood of loss, it is highly dependent on the interface between the fire event itself and the fire protection facilities, such as egress ways and sprinklers. Therefore, in order to properly assess a fire risk, many combinations between the probability of exceedance and loss should be considered. This, in turn, is a very complicated process. There are two widely used methods for evaluating the likelihood of exceedance – the time-dependent event tree and fault tree techniques. These techniques are each strong in that they consider the reliability of established fire protection systems (such as sprinkler and detector systems), as well as the fuel and compartment characteristics [114].

The application of an event tree technique includes several steps. The first step is to identify a starter event and its probability of occurrence. The second step is to determine the different components influenced by the first step. Then, the binary statuses of either success or failure for each component are assumed. In order to reduce the computational process, any illogical accident sequence is removed. In the next step, the probability of statuses, i.e. success or failure, are assigned. Using a Boolean expression, where the results of each component are expressed in a value of either true or false, the probability of each component sequence is calculated. These processes are shown in Fig. 2.29 for a two-component system. From the figure, the summation of the success and failure probability is determined using Equation 2.7, where $P(S)$ and $P(F)$ stand for success probability and failure probability, respectively.

$$P(S) + P(F) = 1 \text{ or } P(S) = 1 - P(F) \tag{2.7}$$

Assuming the sequence of each component has no effect on the others and using an AND logical Boolean expression, the probability of exceedance can be determined using Equation 2.8.

$$P(S_e.F_i.S_i) = P(S_e.AND.F_i.AND.S_i) = P(S_e).P(F_i).P(S_i) \tag{2.8}$$

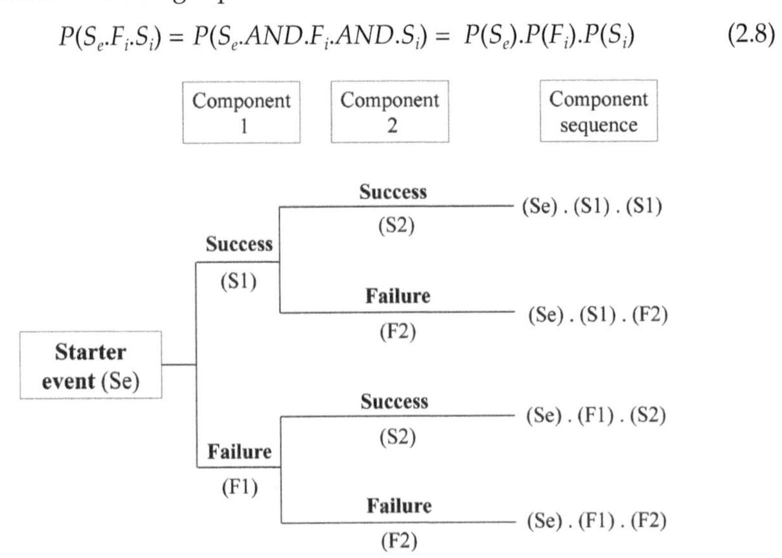

Fig. 2.29: Event tree procedure for a two-component system.

When there is a very low probability of failure in a component (often in the range of 10^{-4}), the statuses of success or failure can be ascribed Yes or No, respectively, representing a binary system.

Figure 2.30 shows an example use of the event tree technique, where the working reliability of established fire-protection facilities, such as sprinklers, automatic detectors, manual detectors, exhaust fans, and emergency doors is considered. There would be 18 scenarios and 17 states, as shown in the figure. As the 18th scenario means there is no fire event, it can be considered as the initial state. The correlation between all the scenarios and states can thus be written as a matrix with 400 arrays. Sprinklers, in general, are designed for two stages: take action, and then, control the fire. For the activation, the possibility of being successful is low at the early phase of the ignition, while as the fire grows the cumulative action probability increases. For the control phase, the fire can be much more easily controlled at the ignition phase than after it has developed.

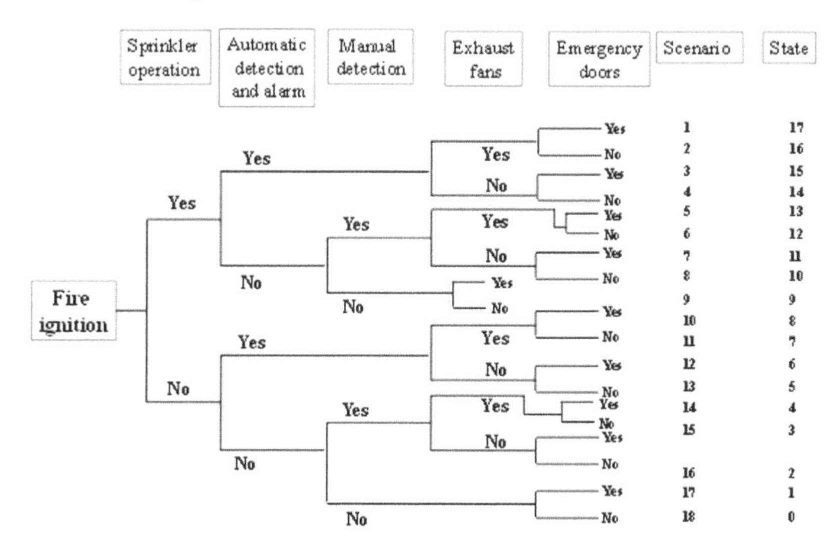

Fig. 2.30: Use of event tree technique to analyze fire scenarios.

The success possibility of the sprinkler can be calculated using the fault tree technique, as shown in Fig. 2.31. This is done by multiplying the likelihood of sprinkler activation and the probability of the sprinkler extinguishing the fire. A similar process can be performed to determine the likelihood of success of the fire detection systems, by multiplying the operational reliability of the detector and the likelihood of detector activation. From a different point of view, if the sprinkler and detection systems are not successful (and provided that the fire is not yet developed), there is a probability that occupants will be warned by fire wardens, i.e. via manual detection, and evacuation will be in train. Nevertheless, at the early stage of a fire, it is hard to inform residents because it can be hard to detect the signs of fire. The probability of successful manual detection, therefore, increases with time. In addition to the previously mentioned facilities to suppress a fire event, another tool is to use smoke exhaust fans. The significance of using exhaust fans in buildings mainly pertains to the threat imposed by smoke to the occupants, as also pointed out earlier [105]. The role of exhaust fans is to remove the layer of smoke and thereby provide further time for residents to evacuate. Nevertheless, most likely the exhaust fans will fail if other automatic systems (i.e. sprinklers and detectors) and manual systems are unsuccessful. Therefore, the probability of exhaust fans operating successfully also increases with time.

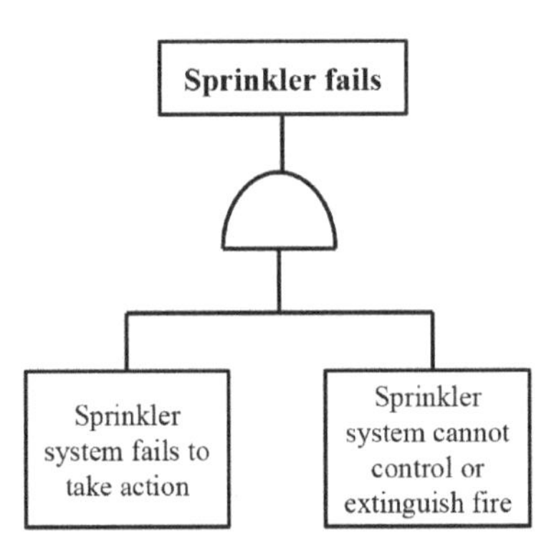

Fig. 2.31: Use of fault tree technique to analyze the performance of sprinklers.

Performing either of the event tree or fault tree analyses will result in finding the times taken until untenable conditions (t_u) and occupant evacuation conditions (t_{oe}) are reached. It is evident that the occupants will be at critical risk if $t_u > t_{oe}$; thus, all of the mentioned facilities are established in order to provide a condition in which $t_u < t_{oe}$. Nevertheless, as both t_u and t_{oe} are influenced by many uncertainties, it is difficult to quantify their relations. As for $t_{u,}$ it is governed by the state of the fire itself, including the ignition, growth, smoke extent, and development, all of which are dependent on the fire and compartment characteristics. These factors are simultaneously based on some variables, such as fire load and ventilation of the compartment, which are also affected by uncertain conditions.

While all stages of a fire are threatening to occupants, in terms of the life safety point of view, the ignition phase is the most important – after that stage, it becomes more difficult to control a fire event. Using the *t-squared* formula, a range of ignition times (t_i) for various fire growth rates, such as very slow, medium, and fast, are plotted. The uncertainty of t_i to t_u can then be determined through a probabilistic distribution, e.g. lognormal simulation. On the other hand, for the t_{oe}, there are many more uncertainties, as it is highly dependent upon the actions of humans under stressed conditions, which can be highly unpredictable. The occupants' actions may vary when they are informed by fire detectors, either automatic or manual. Once occupants have processed this information, they decide to leave the building, which may take some time. Finally, the occupants leave the building. The travel time, particularly, depends on personal characteristics, such as personality, power of judgement and ability to move, as well as on the smoke direction. Other factors such as the building's features, e.g. building's size, volume of combustible materials, exit signs, and egress pathways, are also important to consider. The occupants' responses to fire may also be influenced by groups of people who are simultaneously traveling from the building [115]. It is worth mentioning that the occupants' distribution has a probabilistic nature, which further adds to the level of uncertainties. Figure 2.32 shows the correlation between the mentioned parameters and how they can affect the evacuation time.

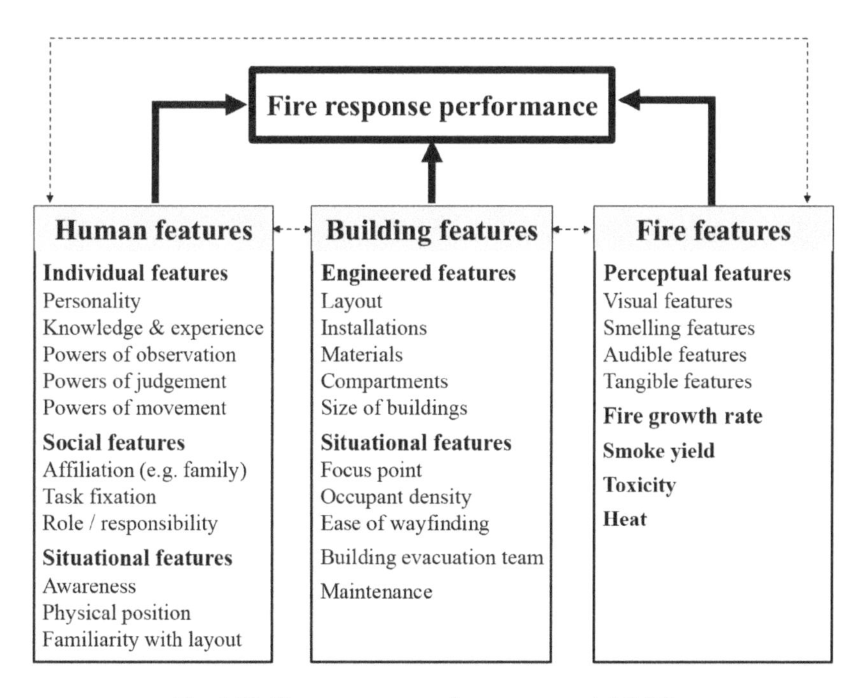

Fig. 2.32: Fire response performance model [116].

There have been some investigations to quantify fire growth rates, detection times, and evacuation times [117-119]. Charters *et al.* [120], for instance, conducted a stochastic-based investigation into fire safety in commercial buildings, based on a database of real fires. They mainly took into account the number of people in the building when the fires began. Other aspects, such as fire rates and fire detection, were also included in their study. The densities of occupants per m^2 were then compared with the design limits of 0.5 people per m^2 (for shops) and 1.5 people per m^2 (for malls). The fire growth rates were determined using the *t-squared* formula for different categories, and the probability distribution was then plotted. For the detection time, the probability distribution was determined and plotted in one-minute intervals, as shown in Fig. 2.33. The figure shows that most fires are detected over the first few minutes, though the distribution is extended over a wider range of time.

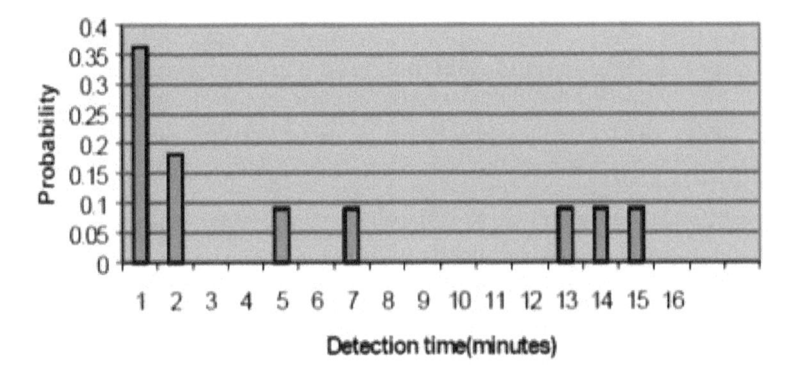

Fig. 2.33: Probability distribution of detection times [120].

The probability of pre-movement times was also determined for some fires, and the results showed that most pre-movement times are lower than 2 minutes, although there were some times between 5 and 7 minutes. The probability distribution for the 2.minute pre-movement fires are shown in Fig. 2.34.

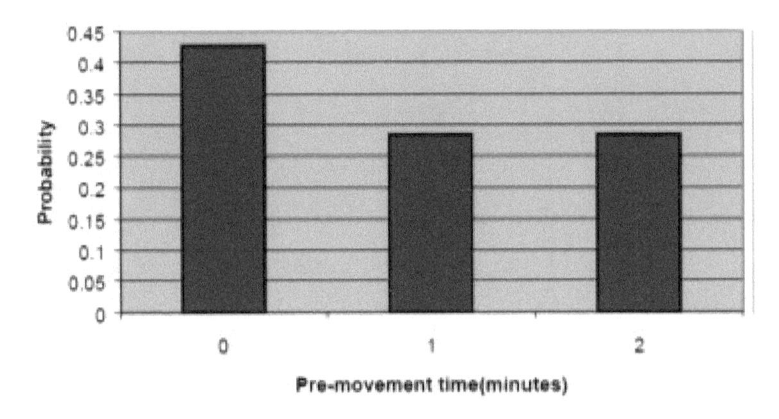

Fig. 2.34: Probability distribution of pre-movement times [120].

Based on the above results, Charters *et al.* suggested a time less than 2 minutes for a live warning message, 3 minutes for a recorded message, and 6 minutes for warning sounders.

To determine the probability of detection time and travel time, there was some work conducted by He *et al.* [121] . They considered that the detectors would be activated when the smoke layer was less than 5% of the ceiling height. For the occupant response, they allocated notional scores to their capabilities, such as alertness, mobility, familiarity with fire events, etc.

All of the calculations above are meaningful if an acceptable level of risk is defined in advance. Acceptable risk depends upon the objectives, which vary from case to case. In other words, the tolerable damage from a fire event will be determined to compare with the results of the fire safety design. One way to address the acceptable level of risk is to use the $R = P \times L$ formula, as pointed out earlier. This means that reducing the probability of exceedance and the probability of loss, either concurrently or separately, can help to limit fire damage to a tolerable level.

It is evident that providing 100% safety to occupants is not possible, due to the existing uncertainties. While there have always been arguments about how to determine acceptable fire damage, the statistics from urban structures under fire events in the past provide some well-accepted guidance. Based on the statistics of residential building fires in Japan, Tanaka found that the number of annual fires was about 19000, and that related casualties were about 7800, i.e. the casualty rate per event was 0.41. A similar study was also conducted on other occupied buildings over the years 2001-2003 by Tanaka [122], as shown in Table 2.13. It is evident that the values mentioned above cannot be directly used worldwide. For example, the average risk of death from fire in residential buildings in China is 4.95×10^{-8} deaths/year m^2, which might differ from that in other countries.

Table 2.13: Fire statistics for several typical occupants [122]

Type of occupancy	*Number of facilities*	*Average area (m^2)*	*Number of fires/ year*	*Number of fires/facility ($\times 10^{-3}$)*	*Number of deaths/ year*	*Deaths/fire ($\times 10^{-2}$)*
Dwelling house	45,258,400 (a)	93	19,093	0.42	1280	6.75
Restaurant, Bar	87,328	243	667	7.63	3.67	0.55
Shop, Market	142,356	616	500	3.51	4.00	0.80
Hospital, Clinic	61,586	1005	154	2.50	0.33	0.22
Hotel, Inn	75,458	942	180	2.39	3.67	2.03
Amusement	18,058	936	145	8.05	0.33	0.23
School	131,448	1131	393	2.99	1.33	0.34
Warehouse	323,701	324	753	2.33	4.00	0.53
Office building	405,729	426	844	2.08	9.33	1.11
Mixed use	581,310	-	3,778	6.49	96.3	2.55

a: Number of household units

Overall, the probabilistic-based approaches are not employed in prescriptive-based fire design as widely as in the definite-based approaches, the reason for which pertains to the absence of a precise risk level tolerable to society. As with the performance-based criteria, however, they are preferred because it is possible to quantify the acceptable risk levels.

2.5 Performance-based approach to post-earthquake fire

A building's structure may experience various extreme loads during its lifetime, including earthquake and fire. Most, if not all, building codes consider the loads of these events separately, i.e. either earthquake or fire. Nevertheless, as shown in Chapter 1, there is always a strong possibility of a PEF occurring, creating a multi-hazard event. Therefore, the interaction between earthquake loads and fire loads should be considered, in order to provide a precise evaluation of the performance of a structure under PEF loads. A PEF performance-based approach includes several different analyses: (1) hazard analysis, (2) structural and non-structural analysis, (3) damage analysis, and (4) loss analysis, as shown in Fig. 2.35. In the following sections, these analyses are introduced.

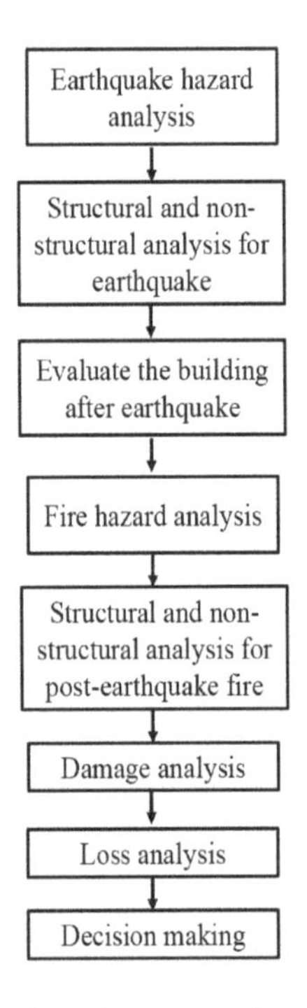

Fig. 2.35: A performance-based post-earthquake fire approach [123].

2.5.1 Hazard analysis

The first step of the performance-based PEF analysis is to perform a hazard analysis. Hazard is the potentiality of an event, which can be quantified by probability-based methods. As for earthquake, it is a natural hazard, and its probability is related to its specifications, such as the distance from faults, site conditions, and the magnitude. A subsequent fire, however, is mostly a technical hazard, and its probability can be analyzed using partial statistics-based approaches. The terminology 'partial statistics-based' refers to the point that, along with the probability of the earthquake itself, the influence of synthetic factors on the intensity of the phenomenon will also be considered. It is evident that the magnitude of a fire relies on different factors, such as construction materials, a building's occupancy type, arrangement of internal partitions, the density of the buildings in an area, the condition of fire extinguisher facilities, the ability of rescue teams to extinguish a fire, the quality of the infrastructure, etc. Also, there are other factors, called influential natural factors (such as wind speed, wind direction, season, time, and climate conditions) that may exaggerate or, by contrast, alleviate the PEF effect. Based on the recorded experiences of PEF (see Chapter 1), there are numerous reasons for fire to occur, some of which are short circuits, overturning of cooking stoves, leaking gas pipes, etc.

2.5.2 Structural and non-structural analysis

This step includes the assessment of both structural and non-structural components when subjected to the design earthquake. According to performance-based seismic codes, buildings are designed to meet a particular level of performance. Thus, the structural and non-structural state, when subjected to the design earthquake, can be assessed. The damage to structural components is not only measured in fiscal units but also in numerical terms, such as relative story drift and hinge rotation. As the seismic damage to the building occurs prior to the subsequent fire, the building's condition requires a re-evaluation before the fire hazard can be ascertained (the earthquake damage may affect the level of fire risk). As a particular case, the condition of the fire protection systems, both active (e.g. sprinklers, smoke detectors, and exhaust fans) and passive (e.g. fireproofing materials) should be taken into consideration. Active systems have been shown to be vulnerable when faced with strong earthquakes, as was mentioned in Chapter 1. LeGrone [124] has provided a summary of damage to fire sprinkler systems following earthquakes that occurred in the US from the 1930s to 2000, showing that, on average, each system sustained around 40% minor and 20% major damage after an earthquake, with the rest of the system remaining almost intact. Similar data is reported by Chen *et al.* [125], regarding the damage sustained by active systems, such as sprinklers, in buildings of Kobe, Japan, after its devastating earthquake in 1995, as shown in Table 2.14. The report, however, did not differentiate minor from major damage.

Table 2.14: Damage to active systems in Kobe City after the 1995 earthquake [125]

Type of fire protection system	*Sprinkler system*	*Indoor fire hydrant*	*Foam extinguishing system*	*Halogenated extinguishing system*	*Automatic fire alarm system*	*Emergency generator unit*	*Fire doors*
Percentage of damage system (%)	40.8	23.7	24.1	10.5	20.1	16.0	30.7

It is worth noting that most fireproofing materials (as passive fire response facilities) are used to isolate the surface of structural components in order to hamper the heat transfer inside the cross-sections, thereby providing further time to withstand applied loads. Evaluating the condition of fireproofing materials is of particular importance in steel profiles, as any damage to these materials after an earthquake can significantly change the heat propagation inside the cross-sections. Some of the most relevant fireproofing techniques for steel structures are spray-applied materials, spray foam insulation, board insulation, and concrete encasement, as shown in Fig. 2.36 (also see Fig. 2.25).

(a) Board insulation

(b) Expandable color insulation

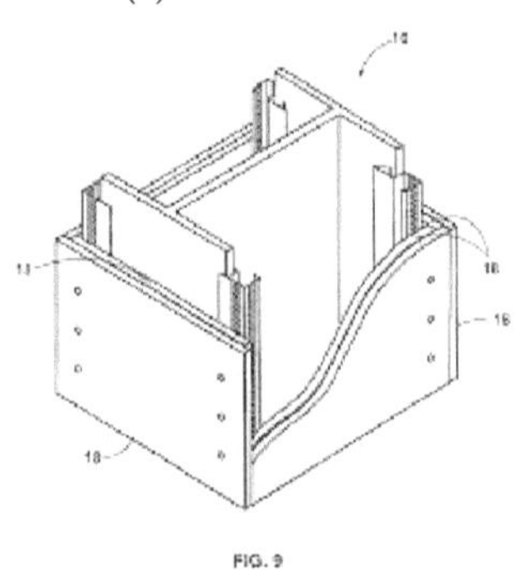

(c) Concrete encasement

(d) Spray-applied materials

Fig. 2.36: Some fireproofing techniques for steel structures (photo courtesy of H. Salehi).

2.5.3 Earthquake damage analysis

Structural and non-structural damage may be expressed in a qualitative fashion, such as 'no damage', 'minor damage', 'moderate damage', 'extensive damage', and 'collapsed', or in a quantitative style, i.e. a number between zero and one. For non-structural damage, sometimes the qualitative expression is mentioned between two boundaries of the functionality of the system, e.g. 'intact' and 'non-functional'. Robertson and Mehaffey [126] used semi-quantitative terms to indicate that, when the results of the assessment are larger than 0.5, the situation is considered to be high risk. Results between 0.25 and 0.5 are usually seen as a risky situation, in which serious attention needs to be paid to control the risk. Results lower than 0.25 usually mean that the risk is under control and no urgent mitigation strategy is required to be adopted. Table 2.15 shows Robertson and Mehaffey's suggestion for non-structural damage assessment.

Table 2.15: Non-structural damage assessment [126]

Seismic fire factor	*Parameters*	*Reliability index (%)*
0.25	Sprinkler system fed from dedicated emergency water tanks gravity fed or from below grade storage via fire pump powered by twin back generator sets designed to resist the maximum considered earthquake	80-100
0.5	Sprinkler system fed from on-site storage facilities powered by backup generator set designed to operate under the design earthquake	60-80
0.7	Sprinkler system fed from municipal supply water likely to sustain significant pressure loss and localized failures at the design earthquake	35-60
1.00	Sprinkler system fed from municipal water supply considered likely to fail rapidly at the design earthquake	0-35

2.5.4 Loss analysis

Loss analysis is essentially the evaluation of direct and indirect economic losses, some of the most important of which are shown in Fig. 2.37. As seen, direct loss includes the costs of building and infrastructure repair and replacement, and deaths and injuries. Indirect loss, on the other hand, is the economic consequences of the damage to buildings and infrastructure, such as the need for temporary and permanent accommodation, interruption to business, etc. [127]. Both direct and indirect economic losses can be grouped into monetary and non-monetary.

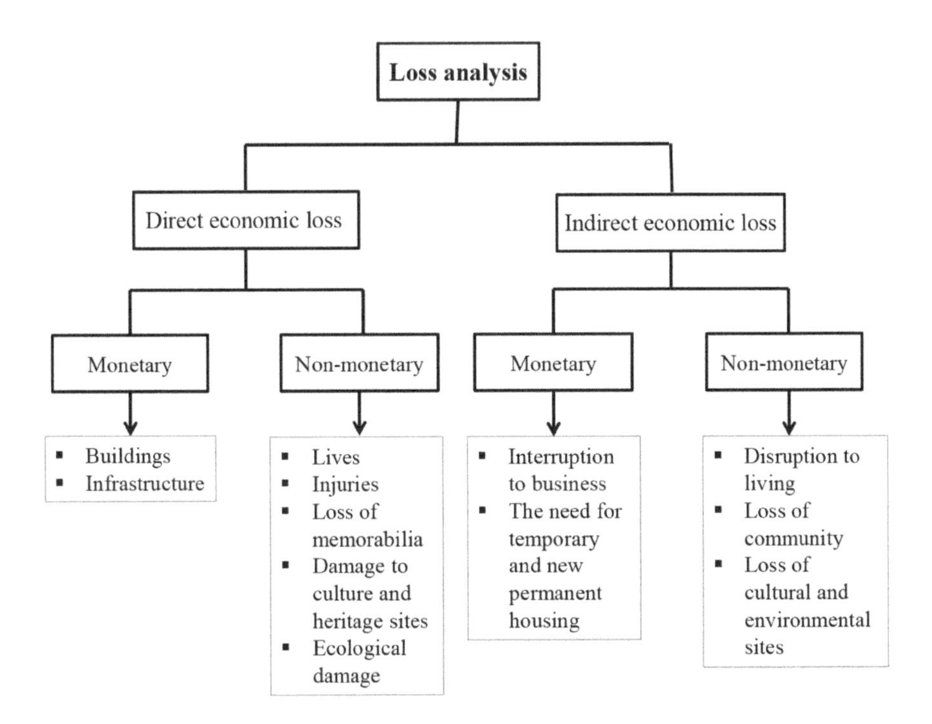

Fig. 2.37: Types of economic losses from earthquakes.

2.5.4.1 Direct economic loss

Direct loss includes damage sustained by structures and infrastructures that can be estimated through damage analysis. Regarding earthquake damage, the formulation of an earthquake loss model can not only act as a tool to predict the economic impact of future events, but also as a means to adopt risk management strategies at macro and micro scales. The macro-scale strategies are those that are implemented in a region, including all of its built facilities. The micro-scale strategies, on the other hand, are implemented at the building scale. At the macro scale, employing a loss model (via which the damage to the built facilities can be predicted) is important to the emergency response after an earthquake. At the micro scale, there might be a need to mitigate earthquake risk through revising the seismic codes for the design of buildings. It is evident that making such a revision may result in additional cost in the construction of a new building, or in the retrofit of an existing building. This extra cost, however, can be justified by a quantitative comparison between the potential losses that might consequently be avoided.

Ideally, earthquake loss models would encompass all possible hazards, i.e. primary, secondary, and tertiary hazards. The primary earthquake hazards are ground shaking, landslides, liquefaction, and surface rupture.

The secondary hazards include tsunami, seiche, flooding, and fire. Tertiary hazards refer mostly to long-term hazards, such as social and environmental problems. The existing earthquake loss models, however, mostly include only the earthquake hazard, because it is believed that while the loss model size increases, the relative effects of the secondary hazards decrease [128]. An earthquake loss model should include the cost of rehabilitation and/or reconstruction of buildings when subjected to the design earthquake. The correlation between the hazard earthquake and the resulting damage is known as 'structural vulnerability', signifying the level of sensitivity of the effect of the hazard [129]. Several methods can be used to assess the structural vulnerability under earthquake loading, hence predicting the earthquake loss. These methods are mainly divided into three groups: 'empirical', 'analytical', and a combination of both, called 'hybrid'. The empirical methods are based either on damage probability matrices or on vulnerability functions [130]. The damage probability matrices are mainly presented based on damage observed by experts, where the damage distributions are given in a range between 0 and 1. This kind of presentation might be easy to understand and possibly the result in a specific area could be generalized to similar areas in order to predict the effects of future earthquake. However, it cannot demonstrate a realistic presentation, as a large number of simplifications might be considered. On the other hand, vulnerability functions are based on a quantity of damage survey data, and information collected in such a way that the correlation between the earthquake and the building's response is given using a normalized vulnerability index. The information collected would include most of a building's characteristics, such as configuration, construction materials, and structural components [131]. Although when comparing the damage probability matrices, more parameters are involved in the vulnerability functions, they are yet dependent on expert judgements that have a degree of uncertainty.

Analytical methods are used to provide further details about the vulnerability of buildings, mainly by directly involving various building characteristics and hazards, and are classified into several groups, some of which are analytical vulnerability curves and damage probability matrices, hybrid methods, and collapse mechanism-based methods [132]. The analytical vulnerability curves and damage probability matrices mainly use computer simulations for dynamic loads, the potential earthquake, and the quantification of the structural response to such an earthquake [133]. These are, however, very time-consuming processes, and the results cannot be directly applied to different regions. If there is a lack of damage data for a particular area, hybrid methods might be used as an alternative [134]. In that case, the post-earthquake damage statistics are combined with analytical approaches. Using statistics can evidently reduce the computational process and thus the time consumed. Collapse mechanism-based methods are mainly used to determine the failure mechanism of masonry buildings via mechanical concepts, such as shear strength and tensile strength [135, 136].

Physically, the modes of collapse are grouped into in-plane and out-of-plane failures. Figure 2.38 shows examples of out-of-plane failure.

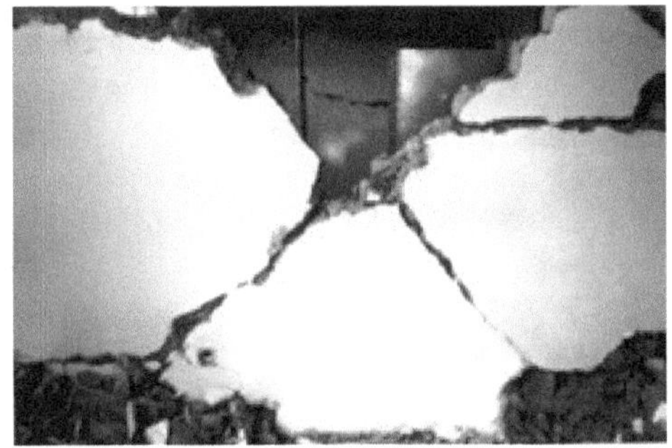

Fig. 2.38: Examples of out-of-plane failure (photo courtesy of H. Salehi).

Using the methods explained above, the rate of death and injury might then be quantitatively ascertained. In FEMA174 and 228 [137, 138], the rate of death and injury is associated with the earthquake intensity, occupancy type, the number of residents over day or night, and structural type. The value of a statistical life is obtained from the previous taxes a person might have paid and, hence, will not be able to pay in the future because of early death. Dunn and Sonnenfeld [139] used the rate of homelessness caused by an earthquake as a means to reach the above aim. Rice and Cooper [140] defined the value of a person in terms of his/her economic worth as a productive member of society, and this amount will vary according to age, sex, color, and degree of educational accomplishments. Sturm and Guinier [141] used different economic content to estimate life loss. Using the human capital approach combined with the consumer expenditure approach, Landefeld and Seskin [142] suggested a value for a life. While the results of such studies might be different in terms of the values proposed, they all point to the fact that such estimation is a very complicated process, with several factors involved in the analysis [143].

2.5.4.2 Indirect economic loss

Indirect losses include all losses that are not caused by a disaster directly, but which still occur as consequences of the disaster. These losses often last longer than the direct losses. While almost all direct losses can be addressed within a few months following a disaster, responses to indirect losses may take several years. In addition, a disaster may have a permanent influence on the economic activities of an overwhelmed area, most likely because the economy-dependent sectors very quickly have to find a safer place to reinvest, and leave the disaster-affected area. Furthermore, there is always a possibility that new economic sectors that arise during the reconstruction activities will grow in such a way that they might replace the original economic units [144]. While this replacement can be seen as providing a new opportunity, it is also a threat to the previous activities. There are also reports showing that small businesses are more threatened by a disaster than large businesses. Tierney [145] conducted a study on the Northridge earthquake in 1994 in Los Angeles, and found that a large number of small businesses had no choice but to close, for reasons that did not relate to the structural earthquake damage.

Taken together, these studies show the long-term effects of natural disasters. Figure 2.39 differentiates schematically the direct and the indirect impacts of natural disasters.

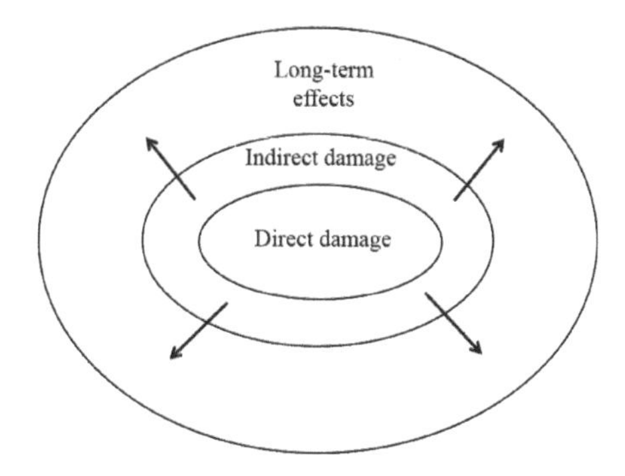

Fig. 2.39: Direct and indirect disaster effects.

Similarly to direct losses, indirect losses are also grouped into two categories: monetary and non-monetary losses [146]. Monetary losses mostly pertain to business interruptions and capital damage, although some references also include the impact of poverty. One way to account for such losses is to use Gross Domestic Product (GDP) [147]. There are reports of economic loss due to major earthquakes, for example, in Nicaragua in 1972, Guatemala in 1976, and El Salvador in 1986, where those economies lost around 40%, 18% and 31%, respectively, of their GDP [148]. The economy of Turkey was affected after the 1999 Kocaeli earthquake, showing a reduction of about 3% of GDP [149]. There are, however, some deficiencies in using such a measure. For example, in large countries, GDP and the scale of the disaster may not be proportionate, because even a very large disaster in one region may have a very small effect on GDP [150]. For example, Hurricane Andrew, considered one of the most destructive disasters in the US, caused only a 0.4% reduction in the GDP [151]. As well, it seems that GDP is more sensitive to change in developing countries than in more economically established countries [152, 153]. In addition to economic loss, there are other types of indirect effects of disasters, some of which are listed here [154]:

- An increase in the operational cost due to possible damage sustained by infrastructure
- A decrease in the output of services due to possible interruption to industry
- A loss of income due to lack of ability to provide urban services, such as electricity
- A reduction in suppliers' activities if they cannot find alternative outlets for their products.

Non-monetary indirect loss has mostly a long-term effect on the macroeconomic variables in a country. For instance, some reports have shown that the national economy may decline due to reduced contributions made by different sectors, such as the education and health sectors. The employment rate, balance of salaries and inflation rates may also change in the long term after a disaster. In some reports, mandatory emigration has been mentioned as one long-term effect of disasters. In addition to the economic impacts, either macro or micro, some findings have called attention to the importance of the social and psychological effects of disasters. Investigations have shown that disasters can cause a broad range of adverse psychological problems, such as fatigue, depression, and chronic anxiety [155]. These studies have also pointed out that children, elderly people, and disabled people are more vulnerable to such problems than other groups [156]. It is evident that these mental health problems cause disruption to the normal functioning of society, and thus have a long-term effect on the country's output. Overall, disasters such as earthquakes and subsequent fires are believed to be unquantifiable in terms of their long-term impacts on the capabilities of communities. While the importance of such effects cannot be ignored and deserve deep scrutiny, further discussion of these is left to others, as it is beyond the scope of this book.

2.5.5 Decision making

In cases of disastrous events such as earthquake and fire, a series of acceptable risks can be defined as being on either an individual or society scale [157]. At the individual level, the acceptable risk can be narrowed down to saving the inhabitants' lives, regardless of the type of disaster. A higher level of safety can also be considered as acceptable individual risk, for example, saving possessions. At society scale, the important question is whether the result of the performed loss analysis is acceptable in terms of its effect on the society as a whole. While it would seem that acceptable social risk can be found by performing a sensitivity analysis on the social benefits versus the social costs, it is widely accepted that this process of evaluation is not easily made explicit [157, 158]. Usually, the concept of risk in a social perspective is reduced to the number of casualties. If the results of the loss analysis go beyond the acceptable risk, then a series of risk mitigation strategies is adopted, either to reduce the risk level or to transfer it. As with the risk mitigations of PEF, the strategies can be implemented on two scales, macro and micro. Macro-scale strategies are adopted to improve the PEF response of urban facilities, such as fire stations, and to facilitate the transportation services. Micro-scale strategies, on the other hand, are mostly focused on individual buildings. An important preventive strategy is to reduce the risk of ignition. Examples of these strategies are listed here.

- A further review of previous PEF events shows that short circuits have played a significant role in causing ignitions. One strategy to respond to PEF risk is thus to utilize systems that disconnect electricity automatically, immediately after an earthquake.
- In line with the previous point, leaking and breakage of gas pipes have also been a common cause of ignition after earthquake. A shut-off valve system which can automatically disconnect the gas supply would thus be advantageous.
- The high density of buildings, particularly in distressed regions, can speed up the spread of fire. The buildings' configurations and the weather conditions can additionally exacerbate the fire's spread and turn it into a conflagration. Making a serious revision to existing urban regions, such as widening thoroughfares, could therefore reduce the fire spread.
- As pointed out in Chapter 1, there are numerous reports showing that disruption to water supply systems after earthquakes has commonly hindered the response to PEFs. Forecasting such an issue and proposing a substitute solution could therefore reduce the risk of a fire's spread.
- The vulnerability of fire detection and extinguisher systems after earthquake has often been an issue. This vulnerability can be compounded if low water pressure is an additional issue after earthquake. Proposing some alternative remedies to address this issue is thus of importance.
- Transportation networks have been shown to be vulnerable to earthquakes, mainly due to debris enclosure. This problem not only causes some people to become trapped, but also hampers their rescue, as well as making evacuation from risky areas more difficult. Addressing this issue is therefore of significance.
- Communication systems, such as mobile and landline systems, have been shown to sustain significant damage from large earthquakes. This can cause problems to rescue teams, when they cannot receive reliable and timely information, and therefore have difficulties arriving at the scene where they are most needed. Substitute systems, such as satellite-based communication systems, could be used as a means of solving this problem.
- Workable plans for evacuating inhabitants from a burning building, or people from an area besieged by fire, must be important parts of both micro- and macro-scale strategies.
- Increasing the PEF resistance of both existing and proposed buildings would ameliorate some of the abovementioned problems. These strategies are explained in detail in Chapter 7.

Chapter 3

Fire Definition

3.1 Introduction

This chapter describes the methods available to simulate a fire in a compartment and the behavior of construction materials under elevated temperatures. Modeling of fire is a complex process, which is done in two parts: the method by which the temperature produced by the fire is accounted for, known as design fire, and the temporal and spatial properties of the considered temperature rise applied to the structural elements, known as boundary conditions. This step is known as the heat transfer mechanism, and is explained in the following sections. Based on these processes, the thermal and mechanical responses of the structural components at elevated temperatures can be studied. By applying the failure criteria, as explained in the previous chapter, it is then possible to see whether a structure has failed.

3.2 Design fire

For a fire to initiate, three basic components have to be in place simultaneously: oxygen, combustible material and a source of heat [159]. In terms of development and as per observations from real events, a fire can be extinguished due to suppression or lack of oxygen, or inadequacy of combustible material. Otherwise, a fully developed fire is expected, as shown schematically in Fig. 3.1.

As is seen in the figure, the phases can be divided into three. The first phase, termed ignition (pre-flashover) represents a localized fire, which can continue to develop in the presence of adequate oxygen and combustible materials. If the fire does continue to develop, and the temperature reaches around 550 °C, there is a transition point called flashover, after which the heating phase starts. During the heating phase, the fire is completely developed and most available combustible materials burn, thus releasing heat. The duration of the heating phase depends on various factors, the most important of which are the thermal inertia of the combustible materials (fuel-controlled fire) and

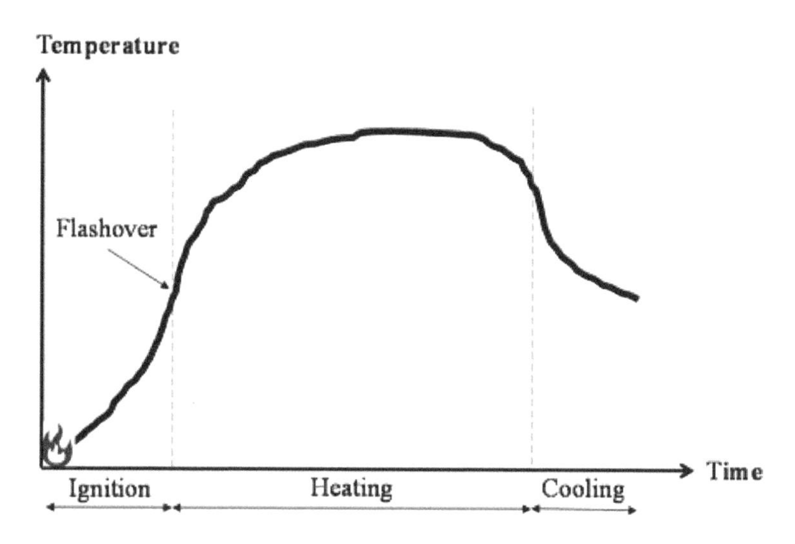

Fig. 3.1: Development phases in a fire.

the availability of oxygen (ventilation-controlled fire) [160]. On the other hand, the heat release rate (HRR) is a function of the burning rate, which correlates with the dimension and location of the openings. Indeed, for specific amounts of combustible materials inside a compartment, it is the size and position of the openings that determines whether the heating phase is fuel-controlled or ventilation-controlled, consequently resulting in different HRRs. If a fire cannot be suppressed before flashover, the heating phase becomes the most serious part of a fire. In that case, and in terms of structural failure, the structural elements must have adequate fire resistance to avoid collapse. The last phase is the decay (cooling) phase. Once more than 70% of the combustible materials have been consumed, or when there is no longer adequate oxygen, the fire diminishes and the temperature declines with time. Structural behavior in the cooling phase is also of importance, particularly in the joints of steel structures. During the decay phase, extensive tensile forces (particularly in steel structures) are created in axially restrained beams, and a rapid collapse might occur. This results from the fact that bolts, welds and structural members do not use the same materials as the structure, leading to different rates of temperature loss [161].

The failure of structural components under the cooling phase has enjoyed some attention from researchers in the past, focusing mostly on the strain reversal occurrence of steel structures [162, 163]. Iu *et al.* [164] performed a study on the response of steel structures under the cooling phase, based on the plastic hinge concept. The plastic deformation of the structural components over the cooling phase was monitored using the incremental-iterative Newton-Raphson method, while the mechanical properties and the thermal creep deformation of the components were taken into account. Wang *et al.* [165] proposed a model for describing the residual stresses and deformations of steel structures under the cooling phase, by making some improvements to the previous work conducted in [162, 163]. Yang *et al.* [166] investigated the behavior of concrete-filled steel tubular columns under natural fires, with more emphasis particularly on the cooling phase and post-fire loading. They found that there are some influential factors on the residual strength of the columns, such as load ratio and temperature evolution versus time. Dimia *et al.* [167] investigated the possibility of failure of reinforced concrete (RC) columns during and after the cooling phase of natural fires. They found that the main reason for failure occurring particularly during the cooling phase was the different rates of temperature loss for the external and internal layers of the columns' cross-sections. They also performed a comparative analysis to see how the columns would behave under various fire durations, and found that the columns were more vulnerable under short durations and low slenderness ratios. In another study, Du *et al.* [168] analytically investigated the behavior of concrete-encased steel columns under eccentric loading, when those columns were subjected to heating and cooling phases. To do that, they used the ISO834 fire curve for heating phases of 60- and 90-minute durations, while a linear cooling phase was subsequently included to the curves. While they recorded considerable lateral deformation to the columns during the heating exposure, the deformations increased further over the cooling phase, with the possibility of failure thus increasing. Du *et al.* also found that there are a number of factors that can influence the columns' responses under fire exposure, such as load ratio, slenderness ratio, and duration. Gernay and Franssen [169] recently investigated the performance of steel, RC, and timber columns under cooling phase. Defining a performance-based indicator, the vulnerability of the columns to delayed failure under cooling phase was carefully controlled. They showed that, by using the defined indicator, they were able to provide further information for organizing structural systems and their possible tendencies to cooling-based failure.

The design fire scenario should cover various aspects of a compartment, such as the type of occupancy, the amount of combustible materials and the size and location of openings. There is however a very long history of different proposals for the design fire.

3.2.1 Time-temperature curves

The first documented attempt to record a fire test occurred in the 1790s. The test was conducted by a number of architects in London, using two different materials, iron plates and gypsum, as a means of fireproofing building floors. The temperature evolution was recorded while the test was being conducted. Many decades later, in the 1870s, when modern concrete was being gradually introduced as a structural material, some tests were conducted by Hyatt [170], who determined the fire resistance of RC floors. It was in the 1890s that consideration was given to conducting fire tests instead of merely measuring the fire resistance ratings of materials. Although there were numerous fire tests conducted over the 1890s (such as on the Equitable Building in Denver, the Wainwright Building in St. Louis, and the Vereinigung von Feurversicherungs-Gesellschaften in Berlin [171]) the first systematic tests were introduced in 1896 in New York [171, 172]. To achieve that, numerous tests were implemented in small compartments, with dimensions of 4.3 m length, 3.4 m width, and 3.0 m height. The compartments were filled with wood fuel, as per the belief that this could provide a close model for real fire. The tests were continued for about five hours, to the point where temperatures arrived at around 1093 °C. Based on those test results, the first fire curve was developed in 1899 in New York. The first national attempt to standardize a fire curve was made in 1904 by the American Society for Testing and Materials (ASTM), and resulted in the development of the first standard fire curve in 1906, named the ASTM E119 fire curve. The ASTM E119 curve was based on worst-case fires in enclosures, and ignored the heating and cooling phases of a real fire. Equation 3.1 represents the ASTM E119 curve, where T (°C) is the temperature at time t (hr) and T_0 is the ambient temperature (often assumed to be 20 °C).

$$T = 750[1 - e^{-3.79533(t_h)0.5}] + 170.41(t_h)^{0.5} + T_0 \tag{3.1}$$

By the 1920s, it had become clear that real temperatures in full-scale fire tests differed from those measured over the ASTM E119 curve [173]. While the ASTM E119 curve was being introduced (mostly in the US), other countries were also working on similar tests, e.g. Canada, Britain, Germany, and Sweden. In 1914, Ingberg [174] received extensive support from the National Bureau of Standards (NBS) to thoroughly investigate the fire problems in the US. Over some years, he supervised more than 1,000 tests, resulting in the collection of substantial and detailed information regarding the fire performance of various construction materials [175]. Ingberg monitored fires in office-based rooms and provided a condition in which all of the fire loads were allowed to burn out. To achieve that, ventilation was adjusted to the rooms in such a way as to provide the most severe result. A comparison was also made between Ingberg's tests and the standard time-temperature curve above a threshold temperature. The time-temperature relation is determined using Ingberg's equation, as shown in Equation 3.2. Table 3.1 shows fire load to an equivalent time in the standard furnace.

$$T = kL'' \tag{3.2}$$

where T is in minutes, L'' is the fire load (wood) in kg/m^2, and k is almost 1.

In addition, Ingberg performed a number of fire tests on full-scale multi-story buildings in order to monitor the structural response under real conditions. The buildings were furnished with both combustible and non-combustible materials, as is common in everyday situations. The severity of the fires was then scrutinized carefully, providing a great deal of helpful information towards fire codes and design approaches. In 1928, Ingberg made a comparison between the condition and duration of two full-scale

Table 3.1: Fire load to an equivalent time in the standard furnace [174]

Combustible content (wood equivalent) (kg/m²)	*Equivalent (kJ/m² × 10⁻⁶)*	*Standard fire duration (h)*
49	0.90	1
73	1.34	1.5
98	1.80	2
146	2.69	3
195	3.59	4.5
244	4.49	6
293	5.39	7.5

buildings (a two- and a five-story building) under fire, with the standard furnace exposure. The results were very different. Based on Ingberg's work, the Fire Resistance Classification of Building Constructions was then written in 1942 [176]. This document covers four types of construction: fireproof, incombustible, exterior-protected and wood. In addition, the fire severity of various types of buildings, and the fire-resistance rating of construction materials were also estimated.

In 1946, delegates of 25 countries, mainly from Europe and North America, held a meeting in London aimed at establishing an international organization that could produce uniform industrial standards and set a common benchmark for use around the world. The International Organization for Standardization (ISO) was officially established in 1947. It now has over 160 countries worldwide as its members. In 1975, ISO developed a fire curve, the ISO 834 fire curve, which is slightly different to ASTM E119. The time-temperature relation is accounted for using Equation 3.3, based on the ISO 834 curve. In the equation, t is in minutes.

$$T = 345 \log_{10}(8t + 1) + T_0 \tag{3.3}$$

Nevertheless, comparative investigations have shown that specimens exposed to either the ASTM E119 or ISO 834 curves would show fire resistances with not more than about 5 minutes difference. This similarity mainly pertains to the fact that the ISO 834 curve was developed using the same data as that used for the ASTM E119 curve [177].

During recent decades, more focus has been paid to compartment fires, highlighting the paramount importance of various ventilation parameters, fuel load, and the thermal inertia of boundaries [178]. One of the first studies on this basis was conducted by Kawagoe and Sekine in 1963 [179]. Considering different ventilation parameters, they plotted different fire curves versus time. Equation 3.4 represents the time-temperature relation as per Kawagoe and Sekine's work.

$$T = kL''\left(\frac{A_t}{A_v\sqrt{h}}\right)^{0.23} \tag{3.4}$$

where T is temperature (°C), k is 1.06 and L'' is fire load (kg/m^2). The term $A_t / A_v\sqrt{h}$ is the ventilation parameter, in which A_t is the area of the internal envelope, A_v is the ventilation area, and h is the height of the ventilation opening, e.g. window or door. The ventilation parameter varies from 5 to 30 (m$^{-1/2}$).

A study was conducted in the UK [180] to underline the role of fire load per unit of ventilation area. This research was the basis of Law's work, where the ultimate possible temperature of very well fireproofed steel elements and insulated floors were compared to that of the standard fire [181]. To do that, Law prepared a number of small compartments (with 0.5 m, 1.0 m and 1.5 m heights), and allowed the entirety of internal combustible materials (wood) to burn out. Law found that temperature variations inside the compartments are independent of the dimensions of the ventilations. The time-temperature relation can be found through Equation 3.5.

$$T = \frac{kL''A_F}{\sqrt{A_v(A_t - A_F - A_v)}} \quad (3.5)$$

where k is 1.0 and A_F is floor area of the compartment. The other parameters are the same as introduced above. For a very well insulated floor, A_F is not considered.

Magnusson and Thelandersson [182] proposed a method for determining the time-temperature curve for standard brick or concrete compartments, considering different ventilation parameters and fuel loads per unit. Defining an opening factor, $A_v/A_t\sqrt{h}$ ($m^{1/2}$), the fuel load per unit, q_t (MJ/m^2), and a calorific value of 18 MJ/kg for wood (as fuel), they developed Equation 3.6.

$$q_t = 18L''A_F/A_t \quad (3.6)$$

Soon after, Pettersson [183], inspired by Law's work, proposed a numerical time-temperature curve. The main difference from Law's equation was to involve the height of the ventilation (h) and the thermal properties of the compartment by applying the factor k_f, as represented in Equation 3.7.

$$T = 1.21{k_f}^{0.5}L''A_F/(A_v\sqrt{h}A_t)^{0.5} \quad (3.7)$$

An important study was conducted by Harmathy and Mehaffey [184], in which they focused on post-flashover compartment fires and the fire spread between different floors in multi-story buildings. They defined a normalized heat load, H_N (s$^{0.5}$K), which is the heat absorbed over the surface area of the concrete element during fire exposure. Equation 3.8 shows the time-temperature relation based on Harmathy and Mehaffey's studies.

$$T = 6.6 + 9.6\times10^{-4}H_N + 7.8\times10^{-9}{H_N}^2 \; 0 < H_N < 9\times10^4 \quad (3.8)$$

In the equation, H_N is accounted for by Equation 3.9, where □ is the proportion of heat evolution in the compartment and is determined by Equation 3.10.

$$H_N = 10^6(11.0\delta + 1.6)L''A_F / [A_t(K\rho c)^{1/2} + 1810(A_v\sqrt{h}L''A_F)^{1/2}] \quad (3.9)$$

$$\delta = 0.41(H^3 / A_v\sqrt{h})^{-1/2} \text{ or 1, whichever is the less,} \quad (3.10)$$

where H is the height of compartment, K is thermal conductivity, ρ is density, and c is specific heat. The term $(K\rho c)^{1/2}$ refers to the thermal inertia of the compartment boundaries, which for concrete is 2190 ($Jm^{-2}s^{-1/2}K^{-1}$).

In 1981, based on Magnusson and Thelandersson's work, Wickström [185] developed the concept of parametric fires, which is well-outlined in Eurocode 1. In Wickström's work, the fire curve is divided into two phases, heating and cooling, in which the heating phase is adopted from the ISO 834 curve with the difference that a sum of exponential terms is taken into account. The cooling phase, on the other hand, is determined linearly, considering three different duration times of fire.

In addition to the abovementioned time-temperature fire curves, there are also different curves that have been developed for a particular occupancy type. For example, the hydrocarbon fire curve is used for a compartment when there are very fast-burning flammable materials, and the external fire curve is used when a compartment is exposed to fire from its external sides. These equations are represented by Equations 3.11 and 3.12, respectively [186].

$$T = 1080(1 - 0.325e^{-2.167t} - 0.675e^{-2.5t}) + T_0 \quad (3.11)$$

$$T = 660(1 - 0.687e^{-0.32t} - 0.313e^{-3.8t}) + T_0 \quad (3.12)$$

While more time-temperature-based curves were developed during the 1990s and later, more attention was paid to developing fire curves that more closely represented real fires. The reason for seeking new curves mostly related to the limited knowledge regarding the temporal development of temperature in actual room fires when the ASTM E119 curve had been developed. At that time, there was also a lack of understanding of fire dynamics. The standard fire curve was initially developed in order to idealize the time-temperature relations in furnaces (as it was supposed to symbolize a severe fire). It is noteworthy that "furnace temperature" has no certain meaning, since it depends largely on the placement of the thermocouples [187]. Figure 3.2 shows some of the fire curves for a two-hour period, as explained above.

3.2.2 Parametric curves

To overcome the deficiencies of the ASTM E119-based fire curves, parametric fire curves seemed to be a suitable alternative. Parametric fire curves involve the energy and mass balance equations for the fire compartment. To do that, fire load, thermal properties, ventilation conditions, and compartment geometry are considered. The result of such an approach can thus be very

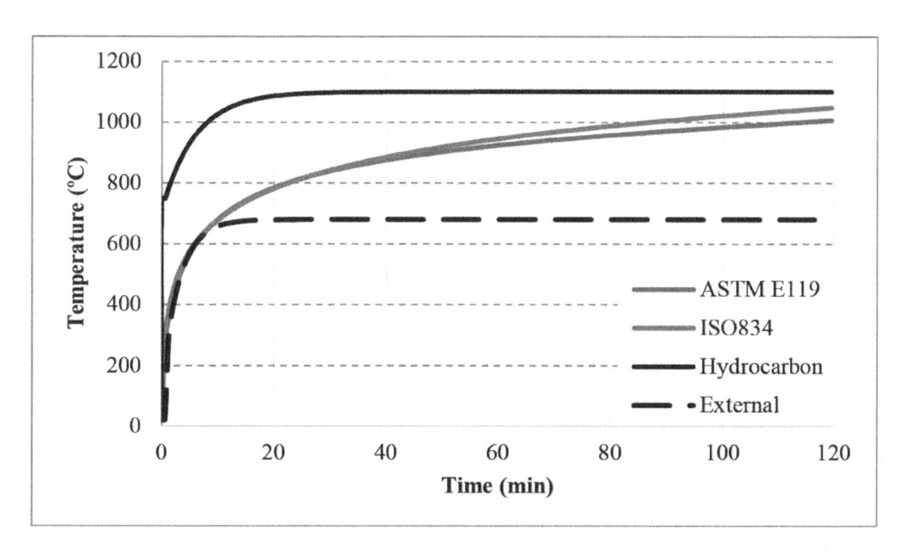

Fig. 3.2: Some fire curves.

rigorous. In 1991, Eurocode 1: Actions on Structures Exposed to Fire [188] released the first codified parametric fire curve. The variation of temperature over time is defined using Equation 3.13.

$$T = q_{fd}\, c' w_f y_{n1} y_{n2} \tag{3.13}$$

where q_{fd} is the design fire load density per unit floor area (MJ/m²), c' is the conversion factor, which involves the thermal properties of the enclosure (MJ/m²), w_f is the ventilation factor, and y_{n1} and y_{n2} are the safety factors. The above equation was proposed as a draft, with some limits to floor areas and the thermal properties. It was not until 1993, however, that amendments were made to the draft and, as a result, parametric fire curves were developed in Annex A of Eurocode 1. The parametric fire curves (also called natural fire curves) are valid for covered areas smaller than 500 m² when there is no opening in the roof. In addition, the maximum compartment height is limited to 4 m [189]. To draw a natural fire curve, several input data are needed:

- Thermal characteristics of the compartment, such as walls and roof
- Geometric characteristics, such as the area of the compartment and the opening
- Fire load density ($q_{f,d}$) in MJ/m² which is derived from the available sources
- The rate of fire growth, i.e. slow, medium or high, which is also provided by the codes.

The application of the method starts with:

- Wall factor b, which is the thermal inertia and which accounts for the thermal properties of the enclosure using Equation 3.14.

$$b = \sqrt{c\rho\lambda} \tag{3.14}$$

where c, ρ and λ stand for specific heat (J/kgK), density (kg/m^3) and thermal conductivity (W/mk), respectively.
When different materials are used in a compartment, an average b factor is used, as represented in Equation 3.15, where A_i and b_i relate to each layer.

$$b = \frac{\sum b_i A_i}{\sum A_i} \tag{3.15}$$

- Parameter O, which accounts for the opening in the vertical walls using Equation 3.16

$$O = A_v \sqrt{h_{eq}} / A_t \tag{3.16}$$

where A_v is the area of the openings, A_t is the total area of the enclosure (walls, ceiling and floor including the opening) and h_{eq} is the average height of the openings and is accounted for using Equation 3.17.

$$h_{eq} = \Sigma A_{vi} \,.\, h_i \,/\, \Sigma A_{vi} \tag{3.17}$$

- Expansion factor Γ, which is a dimensionless parameter and is calculated using Equation 3.18.

$$\Gamma = [(O/0.04)/(b/1160)]^2 \tag{3.18}$$

When Γ is more than 1, the temperature in the heating phase increases sharply and is always more than that in the ISO834 fire curve. However, the duration of the heating phase is often short. By contrast, when Γ is smaller than 1, the temperature in the heating phase is lower than that in ISO834, with a longer duration.
- Design fire load density, which is calculated using Equation 3.19.

$$q_{t,d} = q_{f,d} \times (A_{roof}/A_t) \tag{3.19}$$

where $q_{t,d}$ is the fire load density relating to the area of enclosure, $q_{f,d}$ is the occupancy of the fire compartment, A_{roof} is the floor area, and A_t is the total area of the enclosure.
- $q_{f,d}$ is accounted for using Equation 3.20, involving fire fighting measures, both those that are available inside the compartment and those that are provided by the urban facilities.

$$q_{f,d} = \delta_{q1}\delta_{q2}\delta_n m q_{f,k} \tag{3.20}$$

where δ_{q1} is the risk of fire activation, δ_{q2} is the risk of fire activation owing to the type of occupancy, δ_n is the fire fighting measure factor, m is the combustion factor (which is assumed to be 0.8 for cellulosic material), and $q_{f,k}$ is the characteristic fire load density. Tables 3.2 to 3.4 provide information about these factors.

Table 3.2: Factor δq_1 for various floor areas [190]

Compartment floor area, A_f (m^2)	δ_{q1}
25	1.10
250	1.50
2500	1.90
5000	2.00
10000	2.13

Table 3.3: Factor δq_2 for different occupancies [190]

Occupancy	δ_{q2}
Museums, swimming pools	0.78
Residential buildings and offices	1.00
Industrial buildings	1.22
Chemical laboratories	1.44
Manufacturing of fireworks or paints	1.66

Table 3.4: Characteristics of fire load densities for different occupancies [190]

Occupancy	*Average (MJ/m^2)*	*80% fractile (MJ/m^2)*
Dwellings	780	948
Hospital rooms	230	280
Hotel bedrooms	310	377
Offices	420	511
Shops	600	730
Manufacturing and storage	1180	1800
Public space	100	122
Libraries	1500	1824
School classrooms	285	347

In order to take into account the effect of active fire fighting, the factor δ_n is applied to Equation 3.20. This factor is based either on available fire-extinguishing systems established in the building, such as detection systems and sprinklers, or on those brought and used by professional fire brigades and rescue teams. The first row of Table 3.5 shows the proposed values of δ_n in a normal situation when the facilities are available, while the second row represents data when these facilities are unavailable.

Table 3.5: Firefighting measures [188]

Firefighting	*Automatic water extinguishing systems*	*Water supply*	*Automatic fire detection*	*Automatic fire alarm*	*Automatic transmission*	*An onsite fire brigade*	*An off-site fire brigade*	*Safe access routes*	*Normal fire fighting devices*	*Smoke exhaust systems*
	δ_1	δ_2	δ_3 or δ_4		δ_5	δ_6 or δ_7		δ_8	δ_9	δ_{10}
If available	0.61	0.7	0.87	0.73	0.87	0.61	0.78	1.0	1.0	1.5
If not available	1.0	1.0	1.0	1.0	1.0	1.0	1.0	1.0	1.5	1.5

- Depending on the rate of fire growth, the shortest possible duration of the heating phase (t_{lim}) is determined using Table 3.6.

Table 3.6: The shortest possible duration of the heating phase [188]

Growth rate	*HRR*	t_{lim} *(min)*	t_{lim} *(hour)*
Slow (public areas)	250	25	0.417
Medium (residential and light commercial areas)	250	20	0.333
Fast (libraries and shopping centers)	500	15	0.250

- The longest duration of the heating phase (t_{max}) is determined using Equation 3.21

$$t_{max} = 0.2 \times 10^{-3}\, q_{t,d}/O \tag{3.21}$$

- The accounted for t_{max} is then compared with t_{lim}, which can result in two different situations. If $t_{max} > t_{lim}$, then the heating phase is ventilation-controlled, meaning that oxygen has the major role in fire development. By contrast, if $t_{max} \leq t_{lim}$, then the heating phase is fuel-controlled. The procedure for accounting for the temperature in the heating and cooling phases is shown in the following equations. When the fire is ventilation-controlled for the heating phase, the temperature evolution is determined using Equation 3.22.

$$T_g = 20 + 1325(1 - 0.324e^{-0.2t^*} - 0.204e^{-1.7t^*} - 0.472e^{-19t^*}),\ t^* = \Gamma t \quad (3.22)$$

T_{max} is given where $t^* = t^*_{max}$

During the cooling phase, depending on the value of t^*_{max}, the following equations are used.

$$t^*_{max} \leq 0.5: T_g = T_{max} - 625(t^* - t^*_{max}) \quad (3.23)$$

$$0.5 < t^*_{max} < 2.0: T_g = T_{max} - 250\,(3 - t^*_{max})\,(t^* - t^*_{max}) \quad (3.24)$$

$$t^*_{max} > 2.0: T_g = T_{max} - 250(t^* - t^*_{max}) \quad (3.25)$$

where $t^*_{max} = \Gamma t_{max}$

When the fire is fuel-controlled for the heating phase, prior to using Equations 3.22 to 3.25, the opening factor and the expansion factor have to be modified, using Equation 3.26 and 3.27.

$$O_{lim} = 0.1 \times q_{t,d}/t_{lim} \quad (3.26)$$

$$\Gamma_{lim} = \left(\frac{O_{lim}/0.04}{b/1160}\right)^2 \quad (3.27)$$

If $O > 0.04$, $q_{t,d} < 75$ and $b < 1160$, Γ_{lim} is multiplied by factor k, as shown in Equation 3.28.

$$k = 1 + \left(\frac{O - 0.04}{0.04}\right)\left(\frac{q_{t,d} - 75}{75}\right)\left(\frac{1160 - b}{1160}\right) \quad (3.28)$$

3.2.2.1 Example of Eurocode 1 parametric fire curve

A rectangular compartment with dimensions of 11.6 m by 11.6 m with a floor-to-ceiling height of 4.0 m is assumed (Fig. 3.3). The dotted lines in the figure represent the openings. The fire curves are plotted for two different occupancies, one residential and the other educational. Thus, two different fire load densities are considered. The availability (or non-availability) of fire fighting measures is also involved in the calculation process. The floor and the ceiling are made of normal-weight concrete, and the compartment walls are built with standard bricks. Table 3.7 shows the thermal characteristics of the materials in the compartment.

Table 3.7: Thermal characteristics of the considered materials in ordinary environmental conditions

	Specific heat (c) (J/kg K)	*Material's density (ρ) (kg/m³)*	*Thermal conductivity (λ)(W/m K)*
Floor and Roof	1000	2400	1.6
Walls	840	1600	0.7

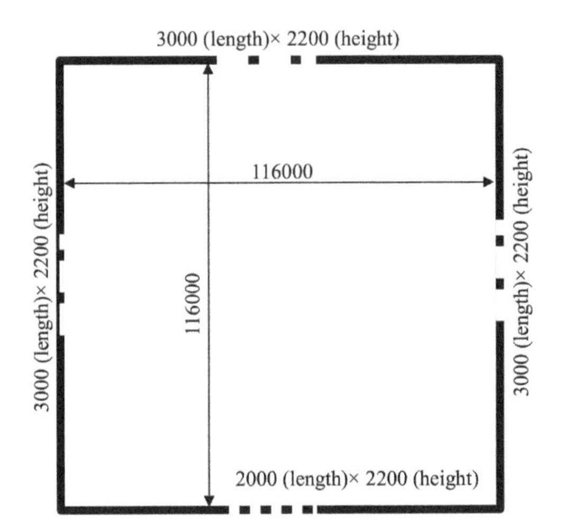

Fig. 3.3: Compartment plan (dimensions are in mm).

The fire load density is calculated using Equations 3.19 and 3.20; this allows for the type of occupancy and the influence of active measures. The area of the roof and floor is 269.12 m^2, the opening area is 24.2 m^2, and the total area including floor, roof, walls, and openings is 454.7 m^2.

$$q_{t,d} = q_{f,d} \times (A_{roof+floor}/A_t) \tag{3.19}$$

$$q_{f,d} = \delta_{q1}\delta_{q2}\delta_n m q_{f,k} = \Pi\delta m q_{f,k} \tag{3.20}$$

For this example, δ_{q1} is 1.54, and δ_{q2} is 1.0. The combustion factor is assumed to be 0.80 for most cellulosic materials. $q_{f,k}$, or the so-called "characteristic fire load density" can be calculated using the values given in Annex E of Eurocode 1; for residential buildings and schools, these densities are 948 MJ/m^2 and 347 MJ/m^2, respectively. In order to include the effect of active fire fighting measures, the factor δ_n is applied in Equation 3.20. This factor is based either on available fire-extinguishing systems established in the building, such as detection systems and sprinklers, or on those brought and used by professional fire brigades and rescue teams. Here, they are included for both possibilities – when they are available and when they are not, as shown in Table 3.8.

The application of the method starts with defining several factors, as follows:

- Wall factor b is defined using Equation 3.14, which determines the thermal properties of the enclosure

$$b = \sqrt{c\rho\lambda} \tag{3.14}$$

Using the information given in Table 3.7, the b factor of the compartment shown in Fig. 3.3 is therefore calculated, which is 1653.2 J/m^{2}S$^{0.5}$K.
Parameter O accounts for the opening in the vertical walls using Equation 3.16

Table 3.8: Firefighting measures

Firefighting measures	*Automatic water extinguishing systems*	*Water supply*	*Automatic fire detection*	*Automatic fire alarm*	*Automatic transmission*	*An onsite fire brigade*	*An off- site fire brigade*	*Safe access routes*	*Normal fire fighting devices*	*Smoke exhaust systems*	*Sum*
	δ_1	δ_2	δ_3 or δ_4		δ_5	δ_6 or δ_7		δ_8	δ_9	δ_{10}	$\Pi\ \delta$
Available	0.61	0.7	0.87	~~0.73~~	0.87	0.61	~~0.78~~	1.0	1.0	1.5	0.30
Not-available	1.0	1.0	1.0	~~1.0~~	1.0	1.0	~~1.0~~	1.0	1.5	1.5	**3.47**

$$O = A_v \sqrt{h_{eq}} / A_t \tag{3.16}$$

For the compartment here, the calculated opening factor is 0.079.

- Expansion factor Γ is accounted for using Equation 3.18

$$\Gamma = [(O/0.04)/(b/1160)]^2 \tag{3.18}$$

Using the above equation, the expansion factor of 1.92 is accounted for, which shows a sharp increase in the heating phase.

Defining the two parameters of the shortest and the longest possible heating phase, which are shown with t_{lim} and t_{max}, respectively. The shortest possible duration of heating phase is dependent on the fire growth rate, from slow to fast. Here, assuming a medium fire growth, a value of 0.333 hours is adopted. For the longest time, t_{max}, Equation 3.21 is used.

$$t_{max} = 0.2 \times 10^{-3}\, q_{t,d}/O \tag{3.21}$$

Accounting for the values of $q_{f,d}$ for both the educational and residential occupancies, using Equations 3.19 and 3.20. The summary is shown in Table 3.9.

Based on data provided in Table 3.9, the t_{max} parameters are now accounted for. If $t_{max} > t_{lim}$, then the fire is ventilation-controlled; otherwise the fire is fuel-controlled.

Table 3.9: Summary of thermal load calculations

	Educational occupancy		*Residential occupancy*	
	$q_{f,d}$ *(MJ/m²)*	$q_{t,d}$ *(MJ/m²)*	$q_{f,d}$ *(MJ/m²)*	$q_{t,d}$ *(MJ/m²)*
Firefighting measures are available	104	61	284	168
Firefighting measures are not available	1204	710	3290	1941

Table 3.10: The values of t_{max}

	Educational occupancy t_{max} (hr.)	*Residential occupancy* t_{max} (hr.)
Firefighting measures are available	0.15	0.43
Firefighting measures are not available	1.80	4.91

It is evident from Table 3.10 that, for residential occupancy, the fire is ventilation-controlled whether fire fighting measures are available or not, because for both situations t_{max} is larger than the assumed t_{lim}. For educational occupancy, however, the fire is fuel-controlled when the fire fighting measures are available, but it is ventilation-controlled when the fire fighting measures are not available. When the $t_{max} < t_{lim}$, some modifications have to be made to the opening factor (O) and the expansion factor (Γ), as shown in Equations 3.23 to 3.25. Finally, the temperature evolution during the heating and then cooling phases are plotted, as shown in Fig. 3.4. As seen, the presence or absence of the fire fighting measures can significantly influence the fire's temperature and duration. The dotted lines in Figs. 3.4a and 3.4c are imagined lines when it is assumed that the fire is suppressed prior to reaching flashover (assuming that flashover occurs around 550°C). Therefore, if the sprinklers, for example, work properly, the fire will never fully develop. When there are no extinguishing systems available, however, the fire will be terminated when almost all of the fuel load is consumed.

Although the natural fire curve proposed by Eurocode 1 is relatively simple to use, in some cases it cannot demonstrate a realistic time-temperature curve, neither in the heating phase nor in the cooling phase. This may be because, for example, the maximum temperature of the fuel-controlled compartments is set to an exact time, e.g. 20 minutes. In addition, while tests performed have shown that it takes a few minutes for a fire to fully develop, the ignition phase is not considered in the Eurocode 1 method. On the other hand, HRR is a function of the burning rate, which is correlated with the dimensions and locations of the openings. In addition, given a specific amount of combustible materials inside a compartment, it is the openings that determine whether the heating phase is fuel-controlled or ventilation-controlled, thus resulting in different HRRs [159]. Although HRR is explained in Eurocode 1 Annex E, no sequential connection is made between HRR and the code [189]. This, therefore, results in a non-realistic design fire that contradicts the concept of performance-based design.

To overcome this deficiency, Zehfuss and Hosser [189] proposed a more realistic natural fire curve, called iBMB, that considers both HRR and the ignition phase. According to their method, using the consumed fire load density (q'', MJ/m^2), i.e. the total available combustible material in a compartment, HRR is determined, and is then written as a function

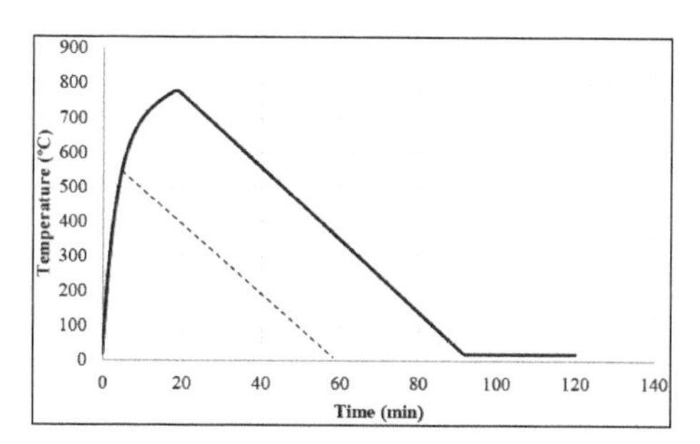

(a) Educational occupancy, fire fighting measures are available

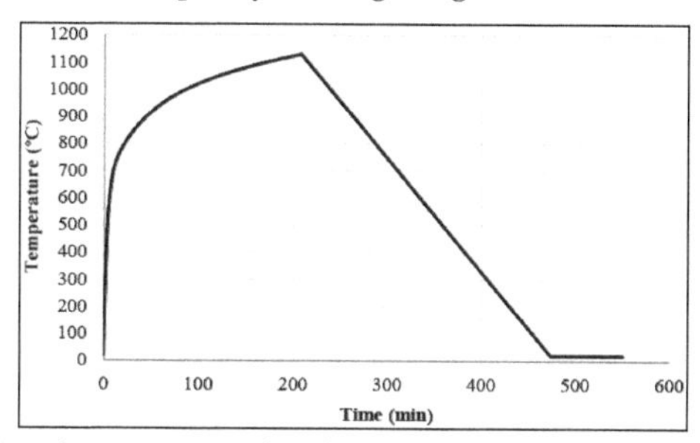

(b) Educational occupancy, fire fighting measures are not available

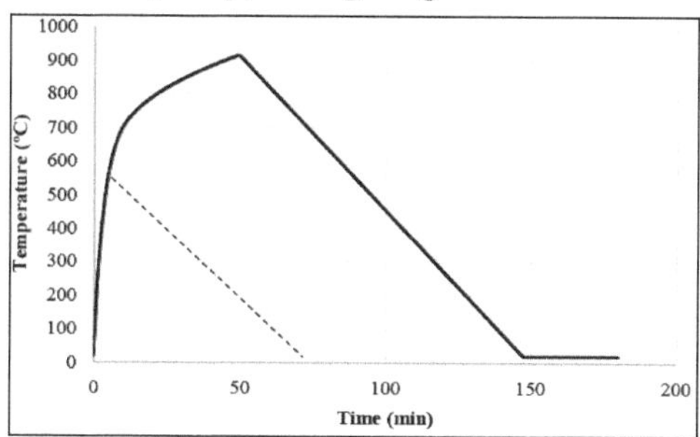

(c) Residential occupancy, fire fighting measures are available

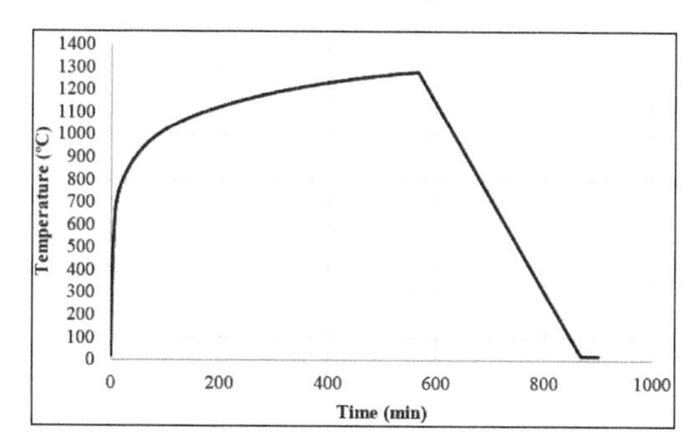

(d) Residential occupancy, fire fighting measures are not available

Fig. 3.4: Temperature evolution and decline for the example of Eurocode 1 parametric fire curve.

of time versus temperature [191]. One of the advantages of this method is its applicability in the modeling of successive fires. Most residential and commercial buildings are subdivided into separate rooms, with internal walls and partitions that hamper the rapid spread of fire. Therefore, in this situation, a growing fire spreads sequentially from one room to the next, until the entire floor is engulfed in fire. In this respect, iBMB provides a suitable base to model a more realistic fire in different conditions. In general, an iBMB fire curve is divided into three distinctive parts: ignition, heating and decay, which are concurrently linked with the HRR of a compartment, as is schematically shown in Fig. 3.5. As is seen, the curve has a quadratic function of time from beginning to t_1. The HRR then remains constant until more than 70% of the fire load is consumed, which corresponds to t_2. Finally, t_3 corresponds to the end of the cooling phase, when the total fire load has been consumed and the HRR therefore declines to 0. The times of t_i are then accounted for, based on the fire load consumed. In addition, in terms of being fuel-controlled or ventilation-controlled, the value of temperature at each step can be separately determined. The application of this method is explained below.

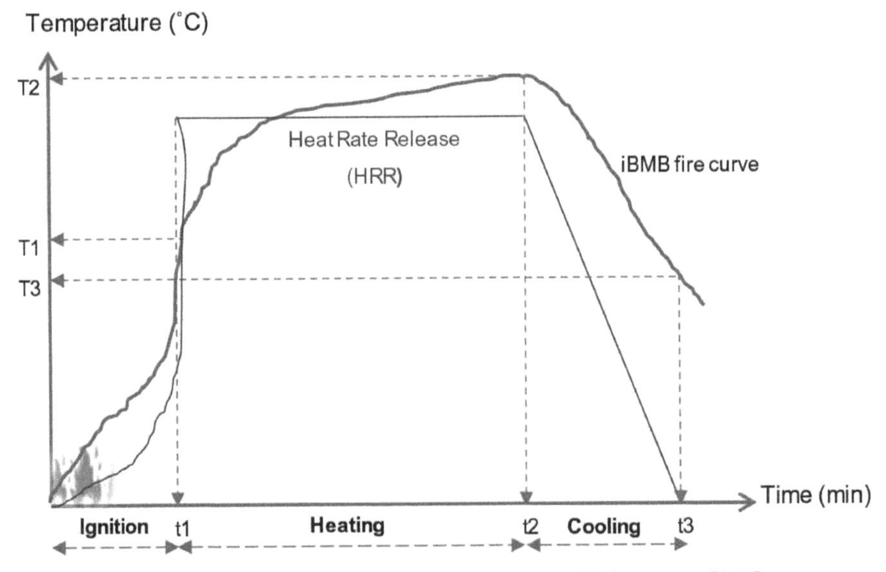

Fig. 3.5: The HRR and the corresponding iBMB fire curve [189].

As mentioned earlier, fire often grows as a successive process, initiating in one cell and then spreading to other cells. Therefore, the HRR is first calculated for the cell from which the fire initiates. A new HRR is then determined for the extended cell, i.e. as the partitions break down. This approach ensures that new geometry and larger openings are taken into account. The partitions' failure times can be extracted from codes, such as ASTM E119 [192]. Figure 3.6 schematically shows the rate of heat release in a partitioned compartment.

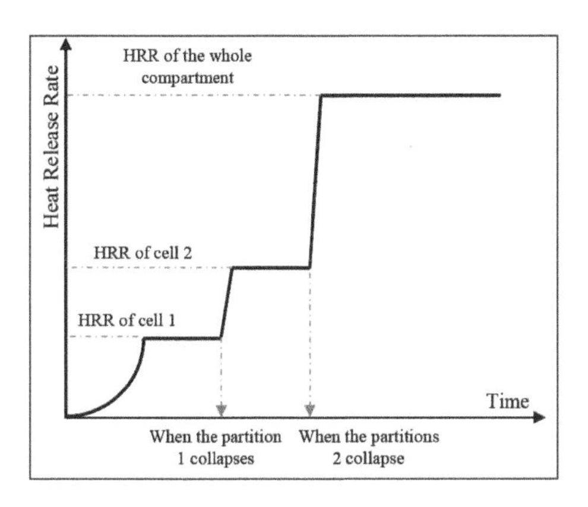

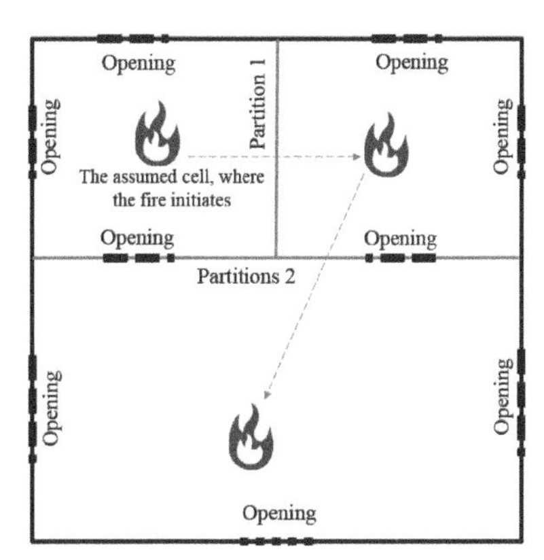

Fig. 3.6: HRR development in the case of successive fire in a partitioned compartment.

The heat release rate is determined using Equation 3.29, where $\dot{Q}$, $\dot{m}$, γ *and* H_{net} stand for the HRR, the mass flow rate, the combustion coefficient, and the net calorific value, respectively. The combustion coefficient and the net calorific value are often obtained from wooden fire loads, which are assumed to be 0.7 and 17.3 MJ/kg, respectively.

$$\dot{Q}(t) = \dot{m}(t)\gamma H_{net} \tag{3.29}$$

As mentioned earlier, the ventilation conditions can change the HRR, resulting in either fuel-controlled or ventilation-controlled modes. If the fire is ventilation-controlled, then the burning rate ($\dot{m}$) is determined through Equation 3.30, in which A_w and h_w are the area of the openings in the compartment and the average height of openings, respectively. The equation is based on wooden fire loads, which are commonly referenced in accounting for the design fire.

$$\dot{m} = 0.1A_w h_w^{1/2} \tag{3.30}$$

Substituting Equation 3.29 in Equation 3.30 with the mentioned values for γ and H_{net}, the maximum HRR ($\dot{Q}_{max,v}$) can therefore be rewritten by Equation 3.31.

$$\dot{Q}_{max,v} = 1.21A_w h_w^{1/2} \tag{3.31}$$

On the other hand, if the fire is fuel-controlled, then the maximum HRR ($\dot{Q}_{max,f}$) is determined by Equation 3.32, where A_f is the floor area of the compartment.

$$\dot{Q}_{max,f} = 0.25A_f \tag{3.32}$$

As shown in Fig. 3.5, the HRR $\dot{Q}(t)$ from 0 to t_1 changes quadratically, which is generally determined using Equation 3.33, where t_g is the time of fire growth (for ordinary buildings, often assumed to be 300 sec).

$$\dot{Q}(t) = (t/t_g)^2 \tag{3.33}$$

Therefore, based on Equations 3.31 and 3.32, the maximum in a compartment with a defined fire load density is determined using Equation 3.34.

$$\dot{Q}_{max} = Minimum\ (\dot{Q}_{max,v}\ \&\ \dot{Q}_{max,f}) \tag{3.34}$$

For the adoption of the iBMB fire curve, a maximum fire load density needs to be selected as a reference, e.g. $q'' = 1300\ \mathrm{MJ/m^2}$ for ordinary buildings. The HRR diagram for various areas is then plotted based on the selected reference. Depending on the fire being fuel-controlled or ventilation-controlled, the temperature ($T_{i\,=\,1,\,2,\,3}$) corresponding to time ($t_{i\,=\,1,\,2,\,3}$) is determined (Fig. 3.5). If the fire is ventilation-controlled, then Equations 3.35 to 3.39 are employed; otherwise, i.e. in the case of a fuel-controlled fire, Equations 3.40 to 3.43 are employed, as shown in Table 3.11. These equations are based on a reference fire load density of $q'' = 1300$ MJ/m^2.

Table 3.11: Temperature-time equations for fuel- or ventilation-controlled compartments

Ventilation-controlled fire		*Fuel-controlled fire*	
$T_1 = -8.75 \times 1/O - 0.1\ b + 1175°\mathrm{C}$	(3.35)	$T_1 = 24000k + 20°\mathrm{C}$ for $k \leq 0.04$ and 980°C for $k > 0.04$	(3.40)
$T_2 = (0.004\ b - 17) \times 1/O - 0.4\ \mathrm{b} + 2175°\mathrm{C} \leq 1340°\mathrm{C}$	(3.36)	$T_2 = 33000k + 20°\mathrm{C}$ for $k \leq 0.04$ and 1340°C for $k > 0.04$	(3.41)
$T_3 = -5.0 \times 1/O - 0.16\ \mathrm{b} + 1060\ °\mathrm{C}$	(3.37)	$T_3 = 16000k + 20°\mathrm{C}$ for $k \leq 0.04$ and 660°C for $k > 0.04$	(3.42)
$O = A_w h_w^{1/2}/A_t$; A_t = the total area with openings	(3.38)	$k = [(\dot{Q}^2/A_w h_w^{1/2}/A_T b)]^{1/3}$; A_T = the net area	(3.43)
$b = \sqrt{c\rho\lambda}$	(3.39)		

Using the equations above, the HRR for the values lower than the reference value at each time is determined for the three stages of Fig. 3.5. For parts of $0 \leq t \leq t_1$, $t_1 \leq t \leq t_2$ and $t > t_2$, the temperature is accounted for using Equations 3.44 to 3.46, respectively.

$$T = \frac{T_1 - T_0}{t_1^2} t^2 + T_0 \tag{3.44}$$

$$T = (T_2 - T_1)\sqrt{\frac{t - t_1}{t_2 - t_1}} + T_1 \tag{3.45}$$

$$T = (T_3 - T_2)\sqrt{\frac{t - t_2}{t_3 - t_2}} + T_2 \tag{3.46}$$

Accordingly, the HRR for each section is accounted for. For the first section, Q_1 is determined using Equation 3.47.

$$Q_1 = \int_0^{t_1} (\frac{t}{t_g})^2 dt \tag{3.47}$$

It is well-accepted that the maximum HRR occurs when more than 70% of the combustible materials has been consumed, thus the HRR in the second section, $Q_{2,x}$, is accounted for by using Equation 3.48, in which Q_x is the fire load and is ascertained by Equation 3.49.

$$Q_{2,x} = 0.7Q_x - Q_1 \tag{3.48}$$

$$Q_x = q_x A_f \tag{3.49}$$

where q_x is the fire load density, which is lower than the reference fire load density (q'') of 1300 MJ/m^2.

Finally, the temperature at the third section changes logarithmically, which is accounted for using Equation 3.50.

$$T_{3,x} = (T_3/\log_{10}(t_3 + 1)) \log_{10}(t_{3,x} + 1) \tag{3.50}$$

By making insertions from Equation 3.50 to Equation 3.46, the temperature at the third branch of Fig. 3.5 is ascertained using Equation 3.51.

$$T = (T_{3x} - T_{2,x})\sqrt{(t - t_{2,x})/(t_{3,x} - t_{2,x})} + T_{2,x} \tag{3.51}$$

Meanwhile, as mentioned earlier, in the iBMB fire curve, the flashover process is also considered. The HRR at the ignition phase is accounted for using Equation 3.52, where $\dot{Q}_{fo}$ stands for the rate of heat release in the flashover process.

$$\dot{Q}_{fo} = 0.0078A_T + 0.378A_W\sqrt{h_w} \tag{3.52}$$

Accordingly, the time taken for the fire to develop is determined using Equation 3.53.

$$t_{1,f0} = \sqrt{t^2{}_g Q_{fo}} \tag{3.53}$$

3.2.2.2 Example of iBMB fire curve

The compartment shown in Fig. 3.7 (with dimensions of 15.0 m by 15.0 m, and floor-to-ceiling height of 3.0 m) is considered. To account for the thermal actions, it is assumed that the occupancy type is residential and, as such, the fire load density is assumed to be 948 MJ/m^2. It is also assumed that some fire fighting facilities are available, as shown in Table 3.5. The floor and the ceiling are made of normal-weight concrete, the external walls are built with standard bricks, and the internal partitions are made of gypsum board. For the partitions, a fire resistance of 35 minutes is assumed [193]. Table 3.12 shows the thermal characteristics of the materials in the building. In addition, it is assumed that the fire initiates from the partitioned room and then spreads to the whole floor. Using the above-explained equations, the fire curve can be plotted. The calculations are shown in summary in Table 3.13. The fire curve is also plotted for the situation in which there are no internal partitions; thus, the compartment is an open flat. This situation might be more applicable when the partitions have already collapsed prior to the fire initiating, e.g. after an earthquake when there is a high possibility of internal partitions collapsing.

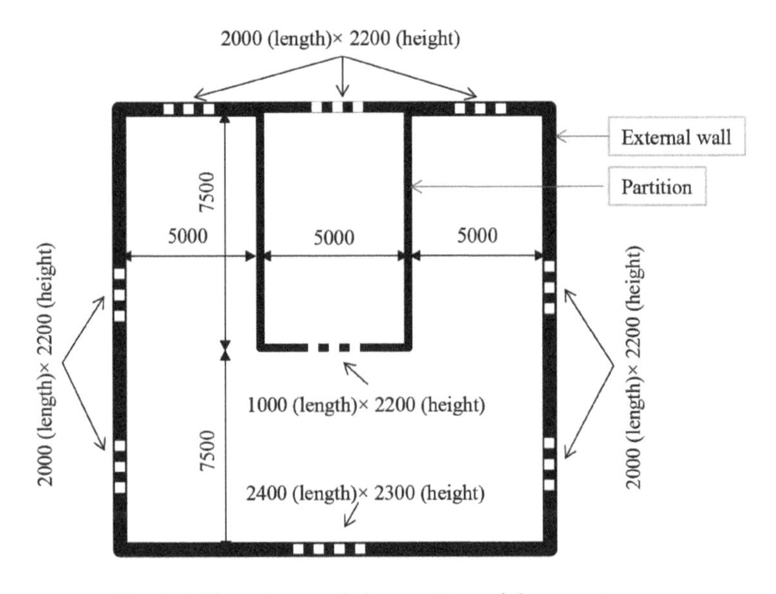

Fig. 3.7: Plan view and the position of the openings.

Table 3.12: Thermal characteristics of the considered materials

Description	*Specific heat (J/kg K) (c)*	*Material's density (kg/m³) (ρ)*	*Thermal conductivity (W/mK) (λ)*
Floor and roof	1000	2400	1.6
External walls	840	1600	0.7
Partitions	852	743	0.276

Figure 3.8 shows the time-temperature curves according to the iBMB method, compared with the ISO834 standards fire method. In this figure, the sharp increase in the partitioned floor temperature occurs when the partitions fail (assumed to be 35 minutes), which results in a swift increase in the released energy, owing to the fast ignition of the fire loads on the rest of the floor.

Table 3.13: Summarized calculations in the partitioned floor

$A_{f(room)} = 37.50\ m^2$; $A_{f(floor)} = 225.0\ m^2$	
$A_{w(room)} = 6.60\ m^2$; $A_{w(floor)} = 33.88\ m^2$	
$A_{t(room)} = 150.0\ m^2$; $A_{t(floor)} = 630.0\ m^2$	
$A_{T(room)} = 143.4\ m^2$; $A_{T(floor)} = 596.1\ m^2$	
$h_w = 2.2\ m$	
Ventilation factor$_{(room)} = 9.79\ m^{3/2}$; Ventilation factor$_{(floor)} = 50.25\ m^{3/2}$	
Opening factor$_{(room)} = 0.0653\ m^{1/2}$; Opening factor$_{(floor)} = 0.0798\ m^{1/2}$	
q'' (decreased by active measures) = 663 MJ/m^2	
Total fire load room $Q_{663} = 24885$ MJ; Total fire load floor $Q_{663} = 149175$ MJ	
Averaged thermal properties $b_{(floor)} = 1717$; Averaged thermal properties $b_{(room)} = 1224.5$	
Floor: $\dot{Q}_{max} = \text{Min}(\dot{Q}_{max,v}\ \&\ \dot{Q}_{max,f}) = \text{Min}(60.80\ \&\ 56.25) = 62.39$: Thus Fuel-controlled	
Room: $\dot{Q}_{max} = \text{Min}(\dot{Q}_{max,v}\ \&\ \dot{Q}_{max,f}) = \text{Min}(11.85\ \&\ 9.38) = 7.59$: Thus Fuel-controlled	
$Q_{1300\ (reference)} = 48750$ MJ	$Q_{1300\ (reference\text{-}remained)} = 257575$ MJ
$Q_{663\ (reference)} = 24863$ MJ	$Q_{663\ (remained)} = 130971$ MJ
$t_1 = 10$ min; $Q_1 = 800$ MJ	$k_{new} = 0.0395$
$Q_2 = 33325$ MJ; $t_2 = 69$ min	$Q_{2(remained)} = 181103$ MJ
$Q_3 = 14625$ MJ; $t_3 = 122$ min	$Q_{3(remained)} = 75672$ MJ
Room: $Q_{2,663} = 16604$ MJ; $t_{2,663} = 2370$ sec ≈ 40 min	Floor: $Q_{2,663} = 73475$ MJ; $t_{2,663} = 3332$ sec ≈ 56 min
Room: $Q_{3,663} = 7459$ MJ; $t_{3,663} = 3958$ sec ≈ 66 min	Room: $Q_{3,663} = 40012$ MJ; $t_{3,663} = 5540$ sec ≈ 92 min
$T_{1,f} = 908°C$	$T_{1,f} = 967°C$
$T_{2,f} = 1192°C$	$T_{2,f} = 1323°C$
$T_{3,f} = 612°C$	$T_{3,f} = 652°C$
$T_{2,663} = 1145°C$	$T_{2,663} = 1195°C$
$T_{3,663} = 609°C$	$T_{3,663} = 724°C$

3.2.3 Zone models and localized fires

Zone models are also used for compartment fires. The theory of zone models is based on the assumption that the temperature inside a compartment can be divided into two layers: the upper layer contains hotter gas, which is close to the ceiling; and the lower layer contains cooler gas, which is close to the floor [194]. The assumption is thus more applicable for a pre-flashover fire, when the fire is still local and is in an early stage. In small compartments, it is acceptable that while the fire is developing, the thickness of the upper layer increases and, by contrast, the thickness of the lower layer decreases in such

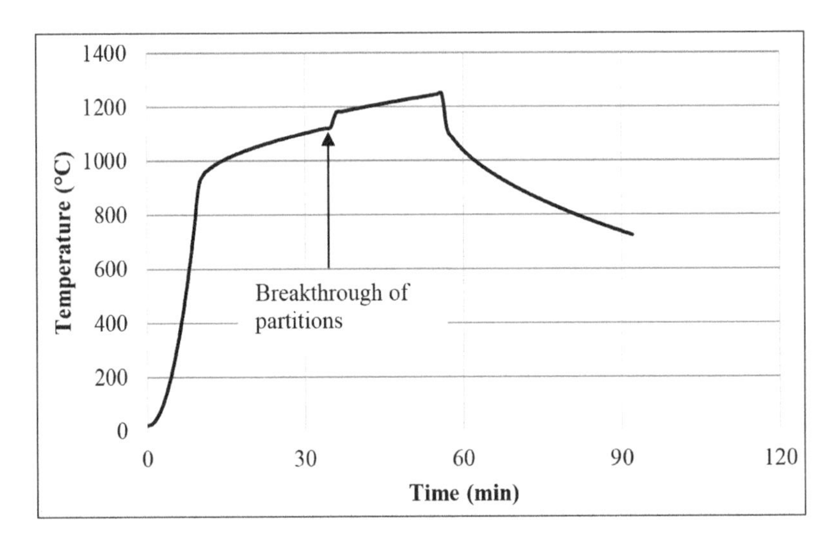

Fig. 3.8: The time-temperature curve based on the iBMB method.

a way that, when the fire is completely developed, the temperature inside the compartment is almost uniform, i.e. there is only a hot gas layer. This process is shown in Fig. 3.9.

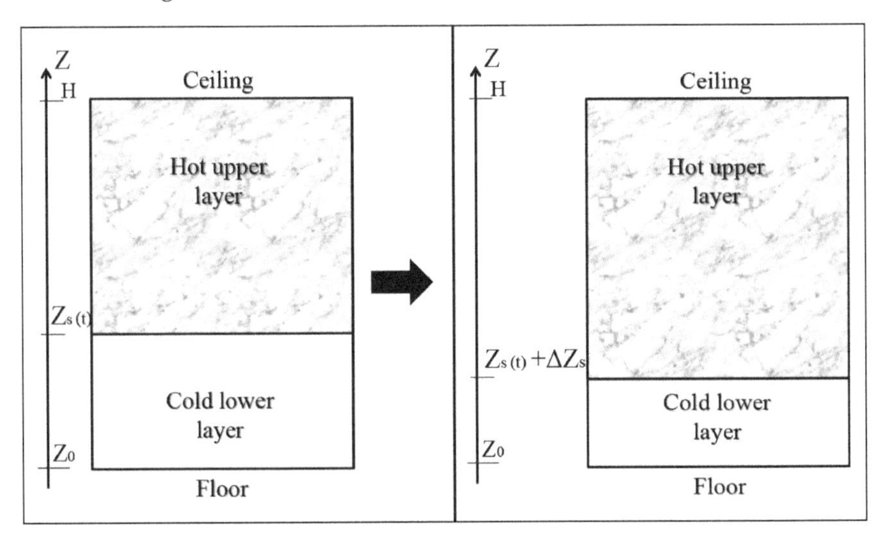

Fig. 3.9: Two-zone models.

In the figure, the vertical axis (Z) corresponds to the wall height. Over the time interval (Δt), a part of the wall height (ΔZ) is added to the upper layer. In addition, it is assumed that there is a horizontal interface between each layer. Applying the equilibrium of mass and energy correlated with time, the temperature evolution is then determined. The application of zone models in compartment fires has been well developed in a computer tool called OZone developed by Cadorin, *et al.* [195].

In addition to determining the evolution of gas temperatures by two-zone models, there are also some empirical models to estimate smoke propagation in compartments. Referring to section 2.3.1 of Chapter 2, the life safety level is a function of the smoke layer height, the toxic gas concentration, and the radiative heat flux. In two-zone compartment fires, the upper layer can particularly influence beam and slab elements. While the thermal actions of such an effect can be calculated using advanced computer tools as pointed out, there are also some simplified models to estimate the flame height and thus the temperature evolution in the elements. There are a number of equations to account for the flame height, such as those developed by McCaffrey [196], Zukoski *et al.* [197], Heskestad [198] and Hasemi *et al.* [199]. The equations proposed are highly dependent on the HRR and the diameter of the base of the fire, although the diameters of fires are generally smaller than the flame height [200]. The Heskestad and Hasemi methods are allowed in Annex C of EN1991-1-1-2, provided that the diameter of the fire (D) is less than 10 m and the HRR (Q) is less than 50 MW. The application of these methods is explained below. Heskestad's method is used when the flame length (L_f) does not affect the ceiling of the compartment ($L_f < H$), as shown in Fig. 3.10. The flame length and the temperature in the plume along the symmetrical vertical axis are given by Equations 3.54 and 3.55.

$$L_f = -1.02D + 0.0148Q^{2/5} \tag{3.54}$$

$$T_{(z)} = T_0 + 0.25Q_c^{2/3}(z - z_0)^{-5/3} \leq 900°\text{C} \tag{3.55}$$

where Q_c is the convective part of the HRR (W), assumed to be $0.8Q$, and T_0 is the ambient temperature, assumed to be 20°C.

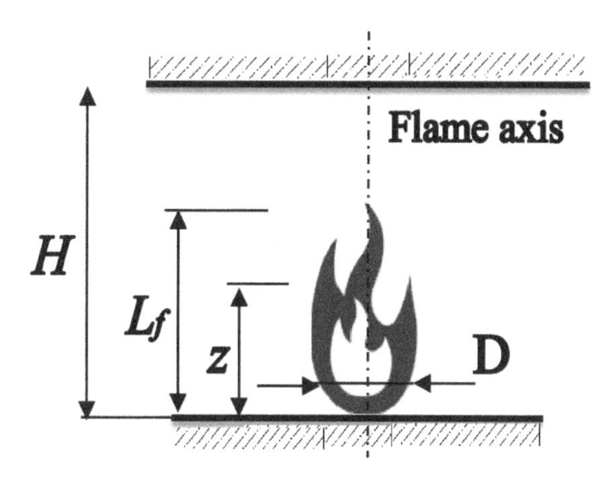

Fig. 3.10: Flame not affecting the ceiling.

On the other hand, when the flame reaches to the ceiling ($L_f > H$), Hasemi's method can provide an estimate for the evaluation of a local fire effect on the beams and slabs. Referring to Fig. 3.11, the heat flux $\dot{h}$ (W/m²) received by the surface area at the ceiling level is accounted for using Equation 3.56.

$$\begin{aligned} \dot{h} &= 100000 \quad \text{if } y \leq 0.3 \\ \dot{h} &= 13600 - 121000\, y \quad \text{if } 0.3 < y < 1.0 \\ \dot{h} &= 15000 y^{-3.7} \quad \text{if } y \geq 1.0 \end{aligned} \tag{3.56}$$

where y is a parameter which is determined using Equation 3.57. When there are several local fires, the total heat flux is determined using Equation 3.58.

$$y = \frac{r + H + z'}{L_h + H + z'} \tag{3.57}$$

$$\dot{h}_{tot} = \dot{h}_1 + \dot{h}_2 + \ldots \leq 100000 \tag{3.58}$$

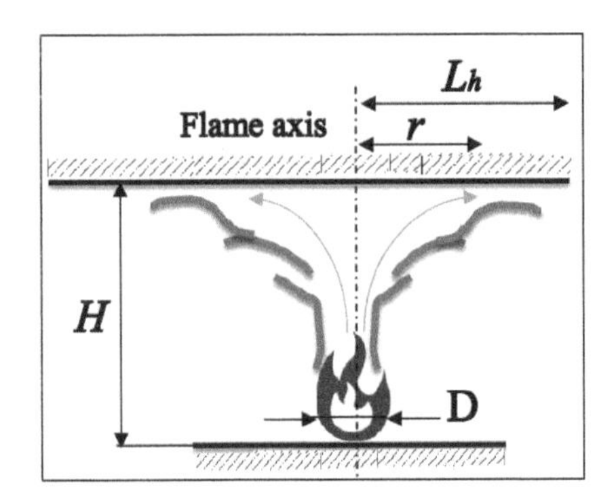

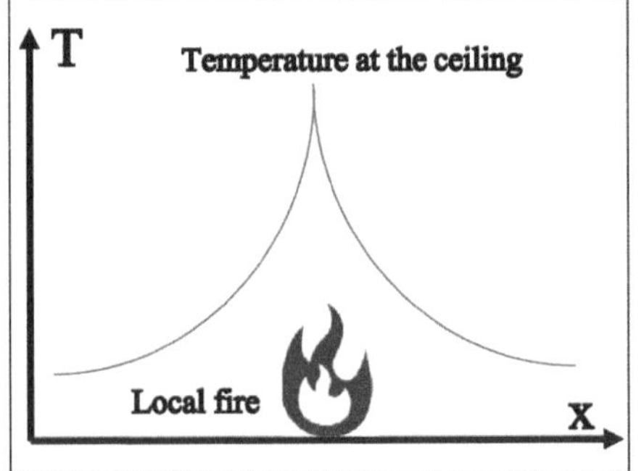

Fig. 3.11: Flame affecting the ceiling.

In Equation 3.57, r is the horizontal distance between the vertical axis of the fire and the point along the ceiling where the thermal flux is calculated, L_h is the horizontal flame length (which is accounted for using Equation 3.59), and z' is the vertical position of the virtual heat source and is accounted for using Equation 3.54.

$$L_h = (2.9H(Q^*_H)^{0.33}) - H \tag{3.59}$$

where Q^*_H is a dimensionless HRR and is accounted for using Equation 3.60.

$$Q^*_H = \frac{Q}{1.11 \times 10^6 H^{2.5}} \tag{3.60}$$

$$z' = 2.4D(Q^{*\,2/5}_D - Q^{*\,2/3}_D) \text{ when } Q^*_D < 1.0 \tag{3.61}$$

$$z' = 2.4D(1.0 - Q^{*\,2/5}_D) \text{ when } Q^*_D \geq 1.0$$

where Q^*_D is also a dimensionless number, called the Froude number, and is determined using Equation 3.62.

$$Q^*_H = \frac{Q}{1.11 \times 10^6 D^{2.5}} \tag{3.62}$$

It should also be noted here that the net heat flux, $\dot{h}_{net}$, is the difference between the heat flux received by the surface area at the ceiling level using Equation 3.56, and the heat energy released by the element to the surroundings by convection and radiation. The net heat flux can be determined using Equation 3.63.

$$\dot{h}_{net} = \dot{h} - \alpha_c(T_s - 293) + \varepsilon\sigma(T_s^{\,4} - 293^4) \tag{3.63}$$

where α_c is the convection heat transfer coefficient, T_g and T_s are the gas temperature and surface temperature of the element in K, respectively, ε is the surface emissivity of the element, and σ is the Stephan Boltzmann constant, which is 5.67 10^{-8} W/m²K⁴. The surface emissivity of carbon steel is often taken as 0.7, and for concrete between 0.85-0.94. The convection coefficient varies and depends on the fire curve used. It is 25 W/m²K for an external curve (see Equation 3.11), and 50 W/m²K for a hydrocarbon curve (see Equation 3.12). For the parametric curves and zone models, it is assumed to be 35 W/m²K.

3.2.3.1 Example of localized fire

It is assumed that a vehicle is burning in a car park, as schematically shown in Fig. 3.12. The HRR and the height of the car park are considered to be 6 MW, and 3.2 m, respectively. Here, the heat flux received by two beam elements under the ceiling, located at 4.5 m (point A) and 9.0 m (point B), respectively, from the center of the car, is determined. It is also assumed that the car has a typical rectangular form, and has dimensions of 4.5 m long and 1.8 m wide; thus, the surface area is 8.1 m². That means the equivalent diameter D is 3.2 m.

The calculation starts with determining the length of the flame, in order to check whether the flames are impinging on the ceiling. To do that, Equation 3.54 is used.

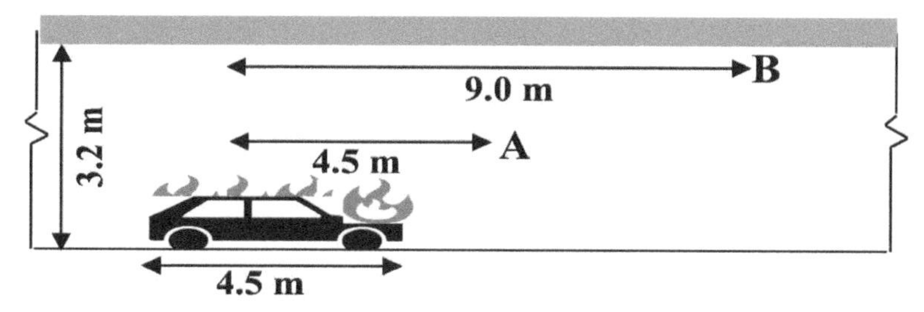

Fig. 3.12: A car on fire in a car park.

$L_f = -1.02\times3.2 + 0.0148(6\times10^6)^{2/5} = 4.35$ m: thus, the ceiling is in touch with the flames.

The Froude number is determined, using Equation 3.62:

$$Q^*_D = 6\times10^6/(1.11\times10^6\times3.2^{2.5}) = 0.295$$

As $Q^*_D < 1.0$, Equation 3.61 is used to determine the location of the virtual source.

$$z' = 2.4\times3.2(0.295^{2/5} - 0.295^{2/3}) = 1.32 \text{ m}$$

Using Equation 3.60, Q^*_H is determined. For this, it should be noted that although the net height was assumed to be 3.2 m, the fire source is not indeed located on the ground but on the chassis of the car. Assuming it is 0.4 m above the ground, the effective height is 2.8 m.

$$Q^*_H = 6\times10^6/\ (1.11\times10^6\times2.8^{2.5}) = 0.412$$

Thus, the horizontal flame length is determined, using Equation 3.59.

$$L_h = (2.9\times2.8\times0.412^{0.33}) - 2.8 = 3.26 \text{ m}$$

Thus, the dimensionless ratio of y for points A and B (see Fig. 3.13) is determined, using Equation 3.57.

$$y = \frac{r + H + z'}{L_h + H + z'} \tag{3.57}$$

$$y_A = (4.5{+}2.8{+}1.32)/(3.26{+}2.8{+}1.32) = 1.17$$

$$y_B = (9.0{+}2.8{+}1.32)/(3.26{+}2.8{+}1.32) = 1.78$$

Because the value of y for both points A and B is more than 1.0 m, the heat flux $\dot{h}$ is determined using the third part of Equation 3.56.

$$\dot{h}_A = 15000\times1.17^{-3.7} = 8390.0 \text{ W/m}^2$$

$$\dot{h}_B = 15000\times1.78^{-3.7} = 1776.4 \text{ W/m}^2$$

Using Equation 3.63, and assuming a surface emissivity of 0.7 and a convection coefficient of 35 W/m^{2}K, the temperatures of the surface at points A and B are, respectively, 219°C and 93°C.

3.2.4 CFD models

Three-dimensional computational fluid dynamic (CFD) is a computerized code, which is increasingly used to simulate a compartment fire [201]. CFD models solve the fundamental equations between fluid dynamics, turbulence, heat and mass transfer, combustion, chemistry, and mechanical systems. The thermal radiation and structural properties of the compartment are also involved in the computation processes, as both are of importance. CFD models are capable of visualizing all stages of a fire's evolution from the beginning to the end, including smoke and heat travel. To do that, a CFD model follows a numerical approach by dividing a fluid area into small parts. These models are, however, very sensitive toward even small changes, which can greatly influence the outputs. This is possibly the reason that most codes allow the use of CFD models only with appropriate caution. There are several CFD software packages, such as FDS, Smartfire, CFX, FLUENT, and PHOENICS. Figure 3.13 shows an example of a fire and smoke simulation in a compartment.

Fig. 3.13: An example of fire and smoke development in a compartment using computer simulation FDS [202].

3.2.5 Vertically traveling fire

The complexity of fire dictates the consideration of various scenarios in order to provide maximum safety to residents. One of the most complex behaviors of fire is seen in multi-story buildings, when several stories are involved in fire. Indeed, when a fire starts from somewhere in a story, it might travel vertically to other stories for several different reasons. The spread can travel externally, e.g. through windows, façades, and cavities, or internally, e.g. through cables. Numerous vertically traveling fires have been documented, such as Knowsley Heights (Manchester) in 1991, Basingstoke in 1992, Irvine (Ayrshire) in 1999, Paddington (London) in 2003, Windsor Tower (Madrid) in 2005, Berlin in 2005, Hungary in 2009, Tamweel Tower (Dubai) in 2012, Istanbul in 2012, Baku (Azerbaijan) in 2015, Asaluyeh (Iran) in 2015, etc. In the following sections, a brief description is given of the reasons for the vertical spread of fire.

3.2.5.1 Fire spread via windows

One way for fire to spread to higher stories is through windows. This subject has been of interest to researchers for decades. When a compartment fire generates a heat flux on a window, that heat flux is absorbed by the glass and causes some cracks to occur along the window's edges. This normally occurs at around 180°C. When the temperature increases to around 300°C, the glass breaks down, allowing the flames to access the outside of the building. If the flames can reach the upper story, then the fire can rapidly extend, as schematically shown in Fig. 3.14. It is generally accepted that flames can extend more than 2 m above a window.

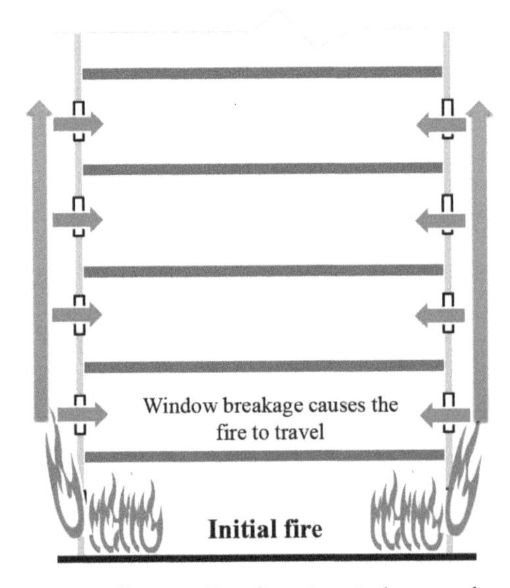

Fig. 3.14: Vertically traveling fire via windows or façades.

3.2.5.2 Fire spread via façades and cavities

Materials used in façades (if sufficiently ignitable) can also act as fire transporters from one floor to the next. The ignition of a façade may occur because of direct exposure to flames after the windows have broken, or be due to receiving radiant heat from a nearby building. Even if combustible materials are not used in the façade, there is still a possibility that the fire can travel to the story above. Investigations and observations have revealed that a fire can simply move up through the cavities between the structural frame and the façade [203]. These cavities can exist either as a part of the construction process, or as a result of delamination. In that case, the cavity acts as a chimney for hot gases and thus the flames can rapidly move higher. Some reports have shown that flames traveling through cavities can extend to 5 to 10 times the original flame length, independent of the façade's material. Figure 3.15 schematically shows fire traveling vertically through cavities.

3.2.5.3 Fire spread through cables

The upward spread of flame through cables is of some concern, particularly for high-rise buildings. There have been numerous reports of vertically

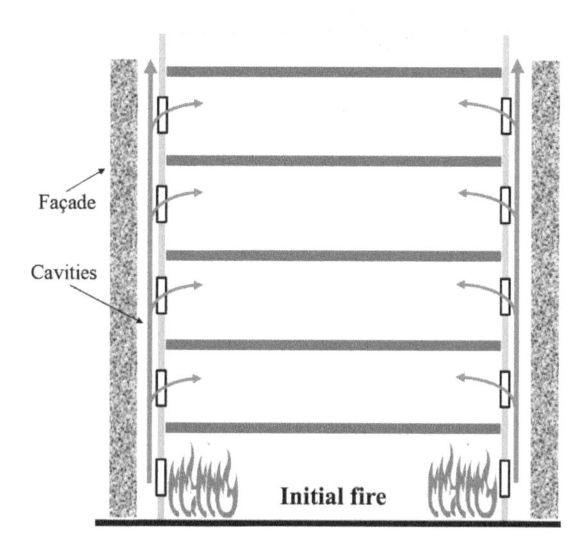

Fig. 3.15: The role of cavities in spreading fire.

traveling fires occurring via cables and their ducts. For example, it has been said that the fire in the WTC spread via telephone lines. There is not yet much theoretical information available about the cable-based spread of fire [204-206]. Amongst those limited studies, Rahkonen and Mangs [207] prepared an informative report about fire events which spread upwards via cables and ducts in several different countries, showing that there is a remarkable degree of uncertainty in the studies of electrical faults. However, one point is obvious – most wires and ducts are covered by flammable materials, such as plastic (Fig. 3.16). If a fire occurs in a story, there is a high possibility that the flammable materials coating the wires located inside the walls will burn, which can lead to a cavity being created that will stretch from one story to the next. As these cavities become very good ventilations, the fire can thus travel rapidly from one story to another.

3.2.6 Structural behavior during vertically traveling fire

It might be assumed that fire travels from one story to the next with almost no delay. In that case, the assumption that all stories will heat up simultaneously, or, if structural failure does not occur during the heating phase, that all stories will cool down simultaneously would be justified. While this assumption may hold for low-rise buildings, in tall structures it would not be completely correct [208]. In tall buildings, although there have been only a limited number of cases with almost concurrent fires on several stories (such as the fire in the WTC), most fires have traveled from one story to another with a significant delay. There might be no structural difference between buildings experiencing simultaneous and delayed fires when only

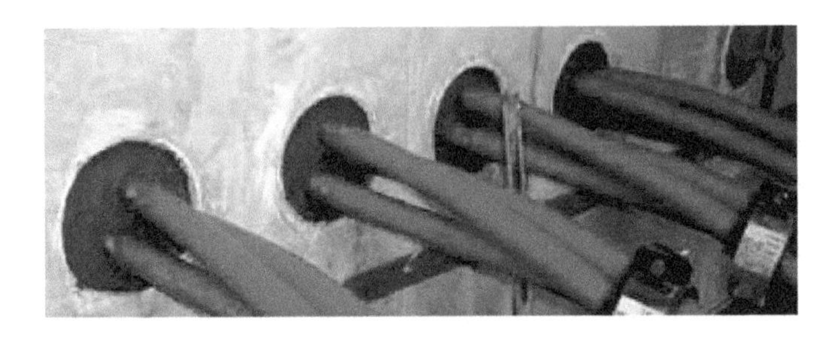

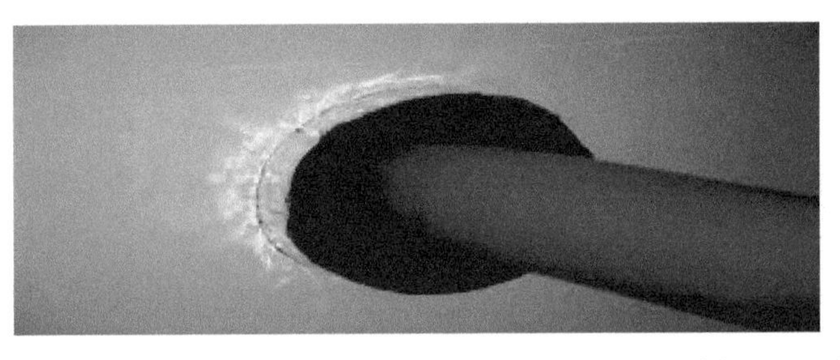

Fig. 3.16: Cables and ducts covered by plastic materials. (http://www.cablejoints.co.uk/sub-product-details/duct-seals-duct-sealing-csd-rise-duct-seal)

the heating phase is taken into consideration. When a story of a building is heating up, it starts to expand, as schematically shown in Fig. 3.17a and b. The outer columns, particularly, begin to deflect downward. While the temperature increases, the strength of the materials decreases, and the story continues to deflect, and the lateral forces of the columns decline. This can lead to global structural collapse.

When the fire travels from one story to the next with a time lag, those stories might experience different phases of heating or cooling and thus, different forces of tension or compression, as shown in Fig. 3.17c. In this case, there is no guarantee that the building will remain stable while the fire in a story is cooling down. There are numerous reports showing that structural behavior under the cooling phase may even be more important than during the heating phase. It is thus evident that the structural behavior under a vertically traveling fire with a time lag is even more challenging than concurrent fires on several stories. To imagine this challenge better, Fig. 3.18 provides an example showing the temperature distribution inside an HE 200 profile and a concrete slab with 80 mm depth, during both heating and cooling regimes. The figure shows that exposing the beam and the slab to fire during the heating phase results in a rapid increase in temperature in all parts of the steel section and in the bottom side of the concrete, and a gradual increase in the center of the concrete slab. During the cooling phase, however, the steel and the surface of the concrete lose their temperatures at a fast pace, while the remainder of the slab continues to become hotter. This pertains to the lower conductivity coefficient of concrete compared to steel, leading to a slower rate of gaining or losing temperature during the analysis.

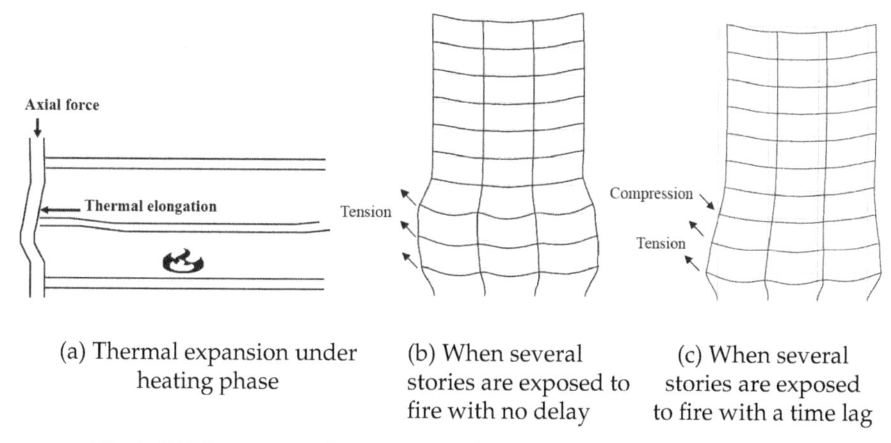

(a) Thermal expansion under heating phase

(b) When several stories are exposed to fire with no delay

(c) When several stories are exposed to fire with a time lag

Fig. 3.17: The structural response under concurrent and delayed fires.

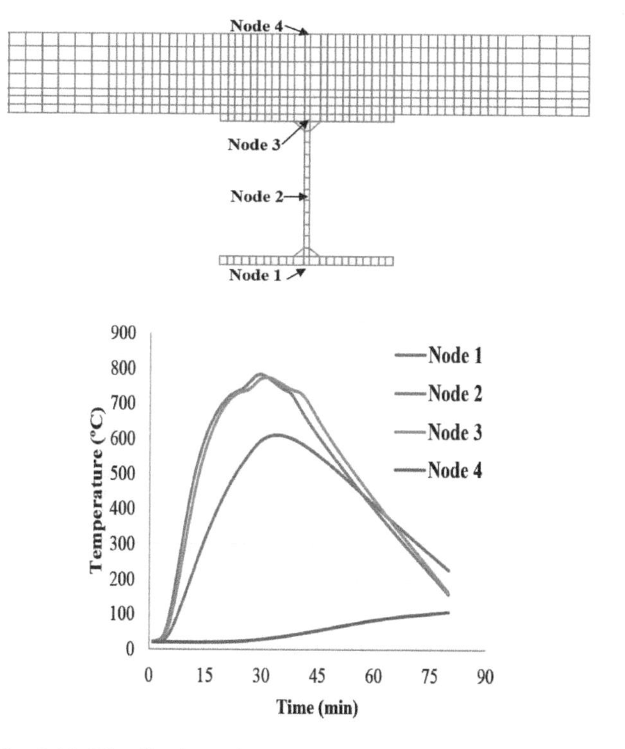

Fig. 3.18: Distribution of temperature inside an IPE profile and the slab above, during heating and cooling phases.

3.2.7 Horizontally traveling fire

What is in common to almost all of the fire scenarios discussed above is that they consider a homogeneous temperature throughout the compartment.

Although this assumption might be justified in small and medium compartments, it is not appropriate for large compartments. Several tests and observations have revealed that combustible materials in large compartments do not burn simultaneously throughout the enclosure, and thus a uniform temperature assumption is faulty. This is possibly the reason why the application of the natural fire curves is confined to compartments with areas up to 500 m^2. Although the majority of buildings designed in the past might have met the scopes of these natural fire methods, most modern buildings go beyond these scopes. This limitation becomes more important with the current tendency to design large open-plan compartments with no partitions. In addition, observations from real fires (such as those that occurred at the Interstate Bank in Los Angeles, the WTC in New York ('9/11'), the Windsor Tower in Madrid, Spain, and the Faculty of Architecture building at TU Delft, Netherlands) have revealed that in large open-plan compartments, the fire may move across the story plates. This means that the combustible materials in the compartment do not burn simultaneously, and as a result, the temperature would not be uniform throughout the enclosure. In large compartments, the combustible materials will be consumed at a rate governed by the existing ventilation and the fire will last much longer than in small compartments.

There has been some interest in investigating horizontally traveling fire for several decades, even before the 1960s, but recently this has received more attention. Gales [209] reported that there have been numerous studies on horizontally traveling fire under a project called *The St. Lawrence Burns*, a joint venture by the National Research Council (NRC) in Canada and the British Joint Fire Research Organization (BJFRO). The results of these studies were kept as inter-organizational reports and were not published externally until 2013. The St. Lawrence Burns project used full-scale fire tests on real urban buildings, in order to monitor the variations of temperature versus time in the compartment fires and then compare the test results with standard fire behavior. The results of those tests showed that the combustible materials are consumed over time and that the fire spreads from one spot to the next, i.e. the fire travels.

One of the first analytical investigations was performed by Bailey *et al.* [162] in 1996, which monitored the behavior of steel-framed structures under the cooling phase. To do that, they considered a large open compartment with nine bays and assumed different fire scenarios. The first scenario was to assume that all of the bays were burning concurrently. That scenario obviously represented a uniform temperature assumption. In the other scenarios, the same fire was applied to the bays but with an offset in time. This new scenario aimed to mimic the traveling fire scenario. The fire was assumed to spread from one bay to the next when the temperature in the first bay reached its peak. In that case, while the first bay, for instance, was experiencing the cooling phase, the adjacent bay was under its heating phase. Although Bailey *et al.*'s work was a pioneering method for considering a non-uniform fire inside a compartment, it was not able to provide a full understanding of the spatial and temporal conditions of the temperature evolution over the other bays. Their work nevertheless showed that more displacements are observed in the middle of bays when their traveling fire scenario is considered than in the case of a uniform fire. This conclusion was restricted to their case, and cannot be generalized to other cases.

In 1992, a broad program of modeling fires in large compartments was commenced at the Heavy Engineering Research Association (HERA) in New Zealand, resulting in a method called "Fire Models for Large Fire Cells" [210]. In the HERA models, large compartments were divided into smaller hypothetical units, and fully developed fires were assumed. The selection of the unit area was first correlated to the fuel load, varying from 50-150 m^2, but they were subsequently set to 50 m^2 for all fuel loads. Every hypothetical unit had two conditions: fire, preheat, smoke logged; or burnt out. This is shown in Fig. 3.19. A strong point in HERA's work was to assume preheating and cooling temperatures in each hypothetical unit. To do that, those temperatures were first assumed to be between 200 and 675°C, later changed to 400-800°C. For a fire to fully develop, it needs access to the outside air. This access can be provided after the windows fail. As pointed out earlier, the breakage of windows would occur at around 300°C. However, HERA's assumption was slightly different; they assumed this would occur at around 350°C. For the fire to spread, they assumed a rate of 1 m/s for fully ventilated conditions, and 0.5 m/s for partially ventilated conditions. While HERA's work was novel in its modeling of non-uniform fire in large compartments, it was meanwhile dependent on many assumptions, including ventilation conditions, fuel distribution and fire size. All of these factors can bring about notable change in the variation in temperature inside the compartment.

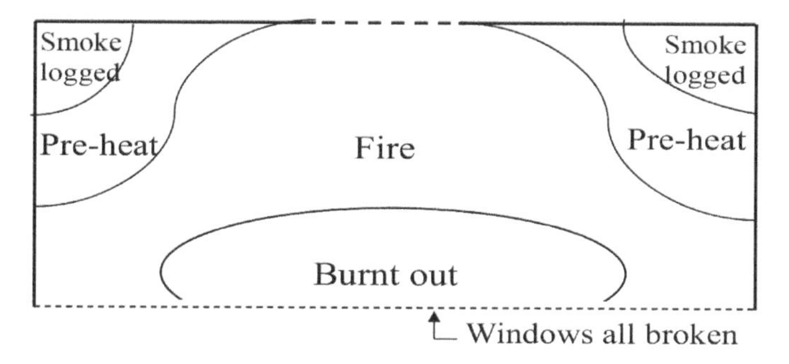

Fig. 3.19: HERA's test [210].

In 1998, Cooke [211] conducted a number of fire tests to monitor the temperature variations over the length of a narrow compartment. The fire loads were provided by uniformly distributed wooden cribs located throughout the compartment. Only one point of ventilation was provided, at one end of the compartment. The tests results showed that the temperature is not uniform throughout the compartment; instead, along with the fire evolution over time, the peak values of temperature were observed to occur near the ventilation source.

In 1999, Kirby *et al.* [212] conducted nine fire tests in a long compartment, with about 23 m length, 6.1 m width and 3.0 m height. The compartment was constructed of different materials – the roof was made of reinforced concrete, the walls were made of lightweight concrete blocks, and the floor was made of dense concrete with about 7.5 cm thickness. Ventilation was provided at only one end of the compartment. To provide the fire load, wooden cribs were used. The tests were conducted in a way that the fuel burnt with a rate of time, thus representing a traveling fire. A test was separately conducted where all fuel loads burnt concurrently. This later test was performed in order to compare the results of non-uniform and uniform burning. Results of the non-uniform burning tests showed that the temperature distribution was not uniform inside the compartment.

One of the most informative fire tests was conducted in a high-rise building in Dalmarnock, UK [213]. The test was mainly aimed at monitoring the variation of temperature inside the highly sensored compartments. To do that, the compartments were first furnished as ordinary offices, and then ignited. All fire stages were observed during the test, including the ignition, the heating, and the cooling phases. The data documented from the test showed that the combustible materials provided inside the compartments did not burn concurrently. The tests clearly showed a non-uniform temperature distribution along the height and across the story plates, some of which are shown in Fig. 3.20.

In the same vein, a full-scale test was conducted by Welch *et al.* [214], as BRE large compartment fire tests. A thermally-insulated compartment with dimensions of 12.0×12.0×3.0 m was established. Wooden cribs were then used as a fire load equivalent to 40 kg/m^2, distributed uniformly throughout the compartment. In addition to two openings being established, numerous thermocouples were installed in a grid pattern within the compartment. As well, some thermocouples were placed over the height of the openings to monitor the temperature variation. The test results showed that a local area of peak temperature travels over time through the compartment, from adjacent to the opening area to far away from it.

From a structural point of view, Ellobody and Bailey [215] investigated the structural performance of a post-tensioned concrete floor subjected to non-uniform temperatures (mostly based on Bailey's previous work as mentioned above). They considered different time-temperature distribution zones throughout the floor, and showed that the time-deflection behavior of the concrete slab is highly dependent on the assumed fire scenarios. Ellobody and Bailey also showed that any change in the heating and cooling phases between zones could result in a cyclic deflection pattern.

The influence of non-uniform temperatures on structural members was investigated by Gillie and Stratford [216]. Based on the Dalmarnock test

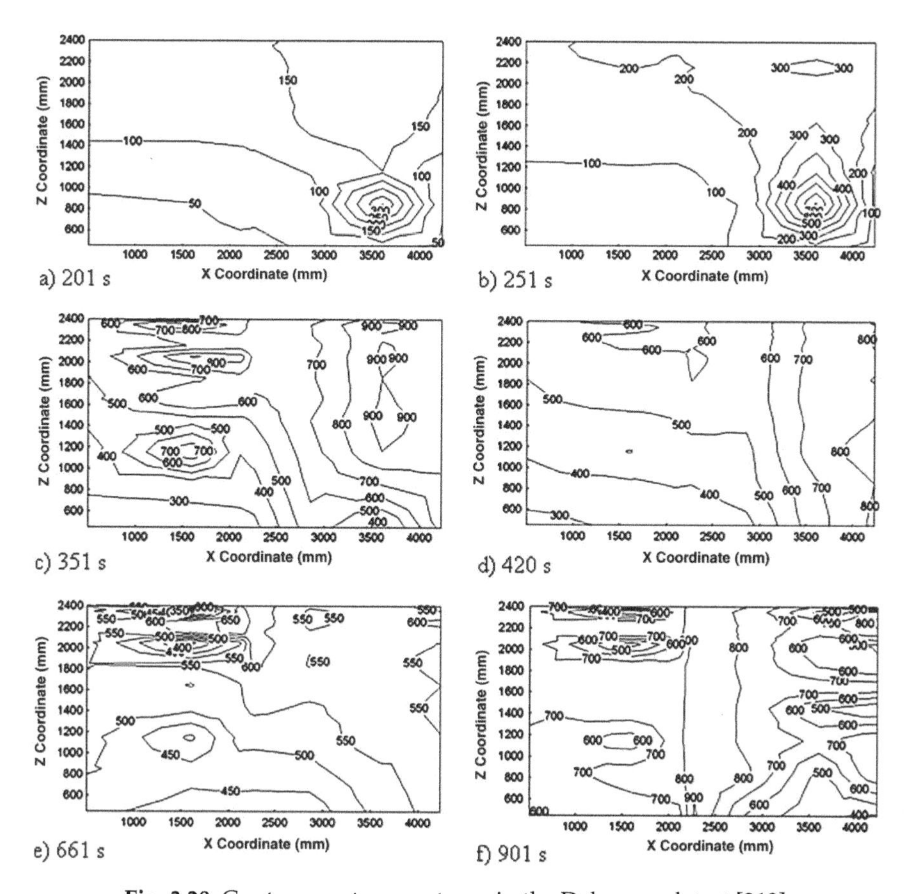

Fig. 3.20: Gas temperature contours in the Dalmarnock test [213].

results (see above), Gillie and Stratford showed that the structural effect of non-uniform temperature on large open areas could not be ignored.

There is also another method for considering traveling fire, proposed first by Stern-Gottfried [217] and then further developed by Stern-Gottfried and Rein [218, 219]. The Stern-Gottfried and Rein traveling fire method has also been recently addressed by Rackauskaite *et al.* [220], where some improvements were made for better addressing the fire dynamic. Based on the Stern-Gottfried and Rein model, the temperatures arising from the fire can be divided into two relative regions, near field and far field, as illustrated in Fig. 3.21. The near field temperature (T_{nf}) refers to the region where the combustible materials are burning, and hence only a portion of the compartment is influenced at any point in time. On the other hand, the far field temperature (T_{ff}) refers to those regions that are yet to burn.

It was shown in Stern-Gottfried and Rein's work that the far field temperature is higher than the ambient temperature (T_{∞}), because of layers of hot gases inside the compartment. The near field size pertains to the

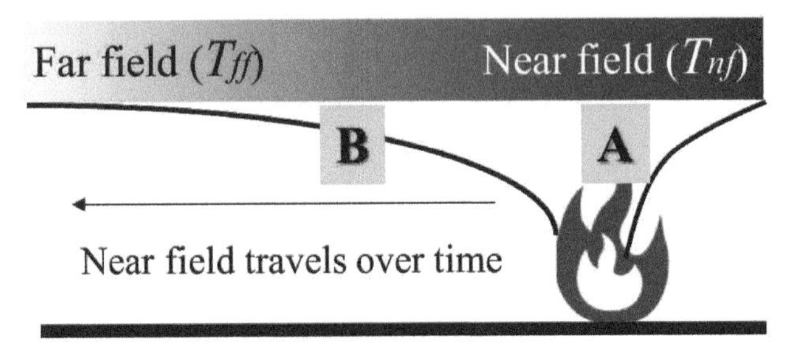

(a) Concept of traveling fire

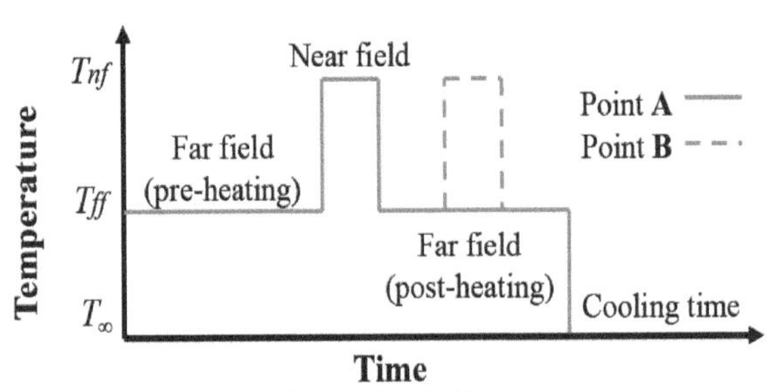

(b) Near field and far field temperatures

Fig. 3.21: Illustration of a horizontally traveling fire [217].

available ventilation, and is defined as an input to the model. In addition, the near field temperature depends on the flame temperature, and thus relies on the type of fuel being consumed. In ordinary structures, such as residential and office structures, the near field temperature is typically assumed to be 800-1000°C for small fires [221], and 1200-1300°C for large fires [49]. The time taken for the fire to travel from one point to the next relies on the fuel load density and the HRR per unit area. From this, and considering a family of fires, the total HRR can be accounted for using Equation 3.64, where $\dot{Q}$ is the total HRR (kW), A_f is the floor area of the fire (m^2) and $\dot{Q}''$ is the HRR per unit area (kW/m^2).

$$\dot{Q} = A_f \dot{Q}'' \tag{3.64}$$

The local burning time of the fire in the area of A_f is calculated using Equation 3.65, where t_b is the characteristic burning time (sec) and q_f is the fuel load density (MJ/m^2).

$$t_b = \frac{q_f}{\dot{Q}''} \tag{3.65}$$

For example, for a fuel load density of 570 MJ/m^2, corresponding to most conventional office building fire loads and the typical HRR value of 500 kW/m^2, the characteristic burning time is about 19 minutes. This time is comparable with results from Walton and Thomas's study and Harmathy's report, in which a well-ventilated and fully developed fire normally takes about 30 minutes [222, 223]. It is evident that the local burning time is not dependent on the burning area. Hence, there is no difference between the various portions of burning areas (since all combustible materials are consumed over the same time, t_b). The fire, then, travels from one point to the next while the total fire load is consumed over time. Therefore, a longer time is needed to consume the entire fire load in a large compartment than in a small area. The total burning duration for an exclusive fire size is introduced by a theoretical maximum time (t_{total}) and is calculated by Equation 3.66, where *RFA* is the ratio of floor area. For instance, if 25% and 100% of the total area are under fire exposure, the total burning durations are 90.25 minutes and 36.81 minutes, respectively.

$$t^*_{total} = t_b(RFA + 1) \tag{3.66}$$

On the other hand, the far field temperature pertains to the compartment's dimensions. This means that the temperature of the gas decreases while the distance from the near field increases. The far field temperature can be determined using closed-form solutions or computer tools such as CFD. One of the most widely used equations is Alpert's equation [224], which is based on a series of tests conducted on large-scale fires. Based on Alpert's equation, the maximum temperature is accounted for using Equation 3.67, where T_{max} is the maximum temperature (K), r is the distance from the center of the fire (m), and H is the height of the floor (m).

$$T_{max} - T_\infty = \frac{5.38(\dot{Q}/r)^{2/3}}{H} \tag{3.67}$$

Alpert presents an equation for the maximum ceiling jet temperatures to define the near field ($r/H \leq 0.18$) and far field ($r/H \geq 0.18$) temperatures. The traveling fire methodology assumes the near field to be at the flame temperature (gas temperature) and does not use Alpert's equation. Should the results of the equation exceed the near field temperature at any point, they are limited at the gas temperature. It is assumed that the fire travels linearly from one of the ends of the plan toward the other end.

Law *et al.* [225] showed that assuming various fire developments would not cause much change to the structural response. This is, of course, based on simplified assumptions and needs to be validated by conducting adequate tests. In the traveling fire method proposed by Stern-Gottfried and Rein, it is also assumed that the far field temperature is uniform along the width of the plan, whereas it varies along the length. This assumption leads to the fire's development being reduced to a one-dimensional problem, which makes it a simple method to use. However, this is also needs to be validated by adequate tests. Based on their method, the far field temperature at any position is determined using a linear distance from the fire, and by assuming a constant spread rate. In Equation 3.68, s and L_f stand for the spread rate (m/s) and the length of the fire (m), respectively.

$$s = \frac{L_f}{t_b} \tag{3.68}$$

Considering fixed local burning (here 19 minutes), Equation 3.67 shows a correlation between the length of the fire and the spread rate, so that the fire travels faster while its scale increases. The position of the fire over time can be traced by dividing the plan into numerous grids of fixed length, with each grid having a unique temperature at a specific time. It should be mentioned that the grid size (Δx) affects the resolution of the far field temperature, the total burning duration, and the resolution of the bay ($L_f/\Delta x$). Each node has a single far field temperature at any given time, at which point it can be unburnt, on fire, or burnt out, as schematically shown in Fig. 3.22. In the figure, Δx, L and L_f stand for the grid size, total length, and fire length, respectively. The far field distance ($r = x_{ff}$) is adapted from the fire's center to the center of the zone being examined (node i). Note that this traveling fire model presumes that columns have the same temperature along their length, calculated based on their distance from the fire's center.

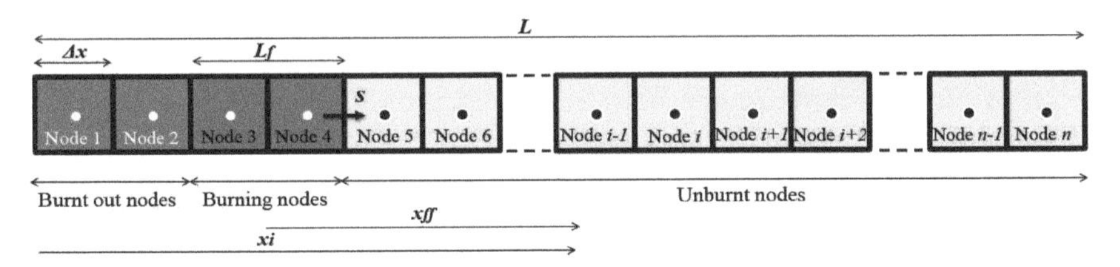

Fig. 3.22: Spatial discretization and associated parameters.

The application of this methodology was used in a study conducted by Law *et al.* [225] on the behavior of concrete frames under traveling fire. Defining a number of arbitrary failure criteria, such as critical rebar temperature (593°C) and the ultimate strain in the tension rebars, they found in which size of fire the most critical result would occur. The results showed that the structural vulnerability was considerably more in a 25% fire size than in other fire sizes.

A study was conducted on traveling fire in large compartments by Horová *et al.* [226], which was a simulation description of a test conducted in September 2011 in the Czech Republic using the Fire Dynamic Simulator (FDS). The test monitored the temperature evolution in a two-story steel-concrete composite structure, with 10.4 m width, 13.4 m length, and 9 m height. The fire load was provided by wooden cribs distributed in a 24 m² area. An opening with no glass was also provided in one of the compartment's sides. In addition, 20 thermocouples were placed in different positions to record the temperature variations over time. The fire was then initiated from one side and gradually spread to the other sides. The results showed that the temperature was highly non-uniform during the fire. A similar study has also recently been conducted by Cheng *et al.* [227] as a numerical approach to monitoring the temperature distribution of structural components under traveling fire. FDS software was used to simulate the fire inside a compartment with dimensions of 12 m length, 9 m width and 4 m height. As usual, wooden cribs were used as the fire load, distributed in the center of the compartment. To monitor the temperature variations, thermocouples were installed at different places, in such a way that all information regarding gas temperature on the surface of beams and ceiling as well as columns could be recorded. The fire started from one side of the compartment and gradually stretched to other sides, until all of the wooden cribs were ignited after about 20 minutes. The test records showed that the temperature was not uniform throughout the compartment; instead, it varied from one point to the next over time.

Behnam and Hashemi Rezvani [228], and Behnam [229] have also investigated the response of seismic-damaged, tall, unprotected steel structures, and multi-story reinforced concrete structures, respectively, under horizontally traveling fire and based on the methodology proposed by Stern-Gottfried and Rein. These studies are explained in Chapter 6.

3.2.7.1 Example of horizontally traveling fire

The traveling fire methodology is applied here to a compartment, as shown in Fig. 3.23. Assuming a fuel load density of 570 MJ/m^2, and the typical HRR value of 500 kW/m^2, the characteristic burning time is about 19 minutes. In order to monitor the temperature variations in the compartment, different fire sizes are assumed. It is assumed that the fire commences from axis A, and moves towards the other axes. Four fire sizes (of 12.5%, 25%, 50%, and 100% of the floor area) are selected. Information for the fire sizes is shown in Table 3.14, including the HRR, the maximum total burning duration, the spread rate, and the near field temperature (which is assumed to be 1200°C). Using Equation 3.59, the far field temperature is calculated. The grid size (Δx) of 1500 mm is selected; hence each span is divided into four quarters, which are supposed to provide adequate resolution for the far field temperature and the total burning time. The gas phase temperatures for beams and columns are then plotted separately, some of which are shown in Fig. 3.24. These curves clearly show that the temperature variation in the longitudinal direction of the case study plan is not constant, and thus, while some nodes are being heated up, others are being cooled down. This variation in the temperature is important, since it may intensify the collapse risk – different temperatures can result in different tensile forces in axially restrained beams.

Table 3.14: The size range of the fire

Fire size (%)	$A_f\,(m^2)$	$\dot{Q}\,(MW)$	$t^*_{total}\,(min)$	*s (m/min)*	$T_{nf}\,(°C)$
12.50	72	36	161.50	0.16	1200
25.00	144	72	90.25	0.32	1200
50.00	288	144	54.62	0.63	1200
100.0	576	288	36.81	1.26	1200

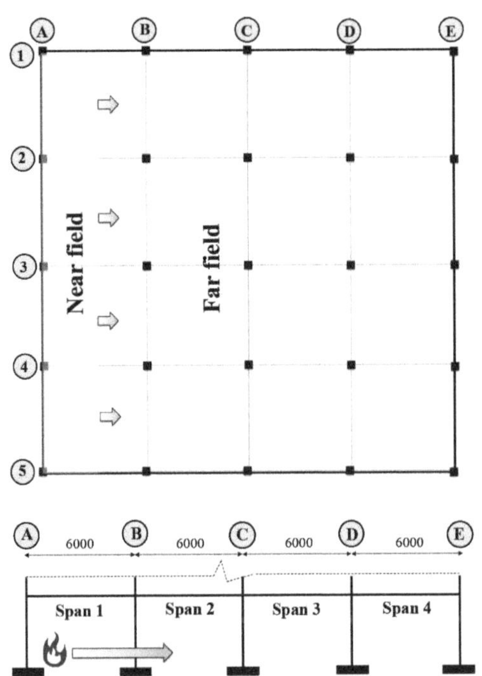

Fig. 3.23: The case study (dimensions are in mm).

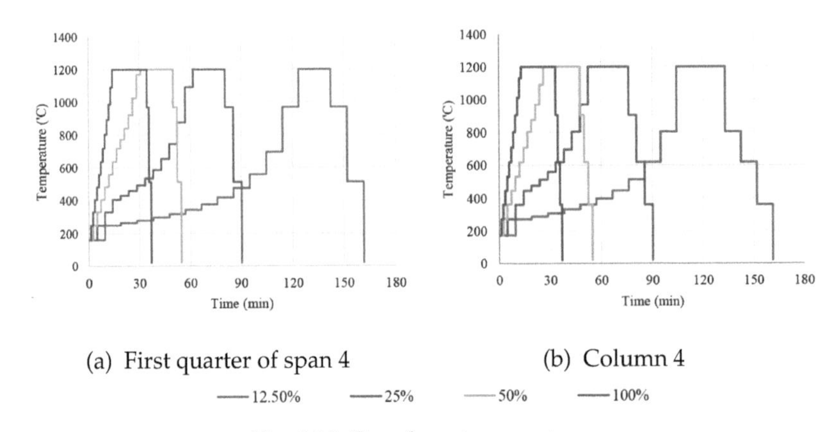

Fig. 3.24: Gas phase temperatures.

Chapter 4

Heat Transfer and Materials Behavior under Elevated Temperature

4.1 Introduction

A fundamental step in predicting the response of structural components under fire exposure is to perform a heat transfer analysis. The previous chapter mentioned how to characterize a fire in terms of its time-temperature manner under different conditions. In some conditions, it is possible to consider a very simple equation, such as standard fire curves. In a more precise condition, natural fire curves are used. While the complexity of conditions increases, a more complex fire characterization would be used, such as CFD models, zone models and traveling fire models. To calculate temperature variations at different points inside the structural elements, a heat transfer analysis is then required. This analysis should be performed throughout the duration of the fire. As the temperature changes with time, the temperature within the structural elements will also change. A thorough understanding of heat transfer to structural elements is of importance in order to provide an accurate prediction of the structural resistance. This chapter describes how temperature transfers from one point to the next, and how materials behave under elevated temperature.

4.2 Heat transfer basics

After deciding on the method used to determine the temperature produced by the source of fire (T_{gas}), the problem of heat transfer reduces to differential equations with certain boundary conditions that allow the calculation of T_s at the surface of the structural elements, and the transfer of heat through the element to the other side. In order to illustrate this in a simple manner, it is assumed that we are dealing with a cubic compartment with insulated walls and floor, and a non-insulated ceiling, named here as the solid. The transfer of fire is therefore only in the vertical direction from T_{gas} at the fire source (floor level) to T_s (bottom of the ceiling) to T (the top surface of the ceiling), as shown in Fig. 4.1, although there is some radiation from wall-to-wall radiative exchange of heat (which is ignored for simplicity).

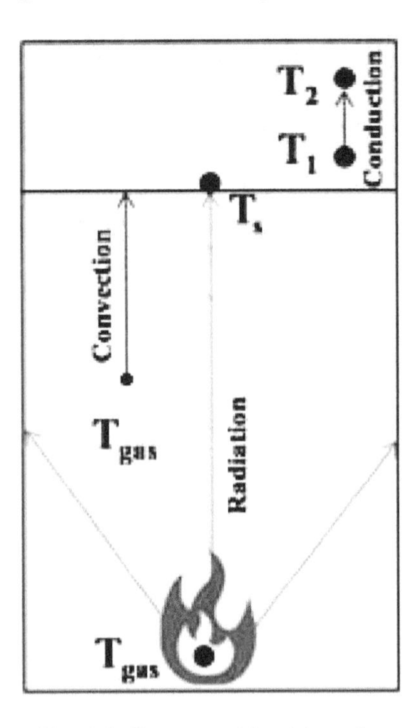

Fig. 4.1: Conceptual heat transfer.

The heat is transferred through one or more of the basic laws, i.e. convection, radiation and/or conduction [159]. The total energy balance, $\dot{q}''_{tot}$ (W/m²), is therefore applied by Equation 4.1:

$$\dot{q}''_{tot} = \dot{q}''_{rad} + \dot{q}''_{conv} + \dot{q}''_{cond} \tag{4.1}$$

where $\dot{q}''_{rad}$, $\dot{q}''_{conv}$ and $\dot{q}''_{cond}$ stand for the energy (heat flux) transferred from the source of fire to its surrounding area by radiation, convection and conduction, respectively.

The total heat flux ($\dot{q}''_{tot}$) provided by the fire is then conducted through the solid to the other side at a gradient (because of the temperature differences between the top and bottom surfaces) according to Equation 4.2.

$$\dot{q}''_{tot} = \dot{q}''_{cond} = -K_i(\partial T/\partial x) \tag{4.2}$$

In Equation 4.2, $\dot{q}''_{cond}$ is the energy transferred by conduction, and K_i is the conductivity coefficient, which depends on the material of the solid.

Convection is ascertained by buoyancy forces between a gas and a solid, due to density variances rising from temperature variations in the air. In other words, the heat transferred by the movement of the gas is measured by convection, as shown in Equation 4.3.

$$\dot{q}''_{conv} = h_c (T_{gas} - T_s) \tag{4.3}$$

In Equation 4.3, T_{gas} is the gas temperature (produced by a source of fire) and T_s is the surface temperature of the solid. The term h_c, which is the convection heat transfer coefficient, is not constant; rather, it varies based on the geometry and the direction of the solid, the properties of the fluid or the gas, and the difference between T_{gas} and T_s [49]. The convection heat transfer coefficient shall be divided into two: one coefficient on the surfaces exposed to fire, and another on the surfaces not exposed to fire [230]. For the surfaces exposed to fire, the coefficient can be introduced using Equation 4.4.

$$h_c = B\left[\frac{g \times P_r}{T \times v}\right]^{1/3} \kappa\,(\Delta T)^{1/3} \tag{4.4}$$

where B is 0.13 (a vertical surface configuration), g is 9.81 m/s^2, Pr is the Prandtl number, T is the absolute air temperature, v is the air viscosity, k is the air conductivity, and ΔT is the temperature difference between the air and the surface. Using the data provided in Table 4.1 and substituting this in Equation 4.4, the values of h_c at various temperatures can be found. This shows that h_c rises in a rapid trend to around 6.7 W/m^2 °C at 200°C and then remains almost constant. Given that more meaningful changes in the material characteristics occur after around 200°C, and that the heat transfer via radiation is much greater than the heat transfer via convection, the coefficient heat transfer value of $\approx$6.5 W/m^2 can be used on the surfaces exposed to fire for simplicity.

The convection heat transfer coefficient for surfaces unexposed to fire is determined using Equation 4.5, where α is a constant assumed to be 2.2. In Eurocode 1, the value of h_c for unexposed surfaces is constant and assumed to be 4 W/m^2. The value is nevertheless assumed to be 9 W/m^2 when the radiation effect on the heat transfer is also included.

$$h_c = \alpha(\Delta T)^{1/4} \tag{4.5}$$

In order to transfer heat via convection, the existence of a medium is required; however, radiative heat transfer needs no material between the furnace and the solid. Energy transfer via radiation occurs through electromagnetic waves, which can be absorbed, conveyed, or reflected at a surface. In other words, radiative heat transfer is a two-way process, by which a receiver (a wall, for instance) can also be a radiator. The summarized radiation is then defined, considering the radiation for the gas ($\dot{q}''_{rad,g}$) and the radiation between a structural element and a wall ($\dot{q}''_{rad,w}$), as represented in Equation 4.6 [232].

Table 4.1: Values of air properties at atmospheric pressure [231]

T (°C)	*Viscosity (10⁶ m²/s), v*	*Conductivity (W/m°C), k*	*Prandtl number, Pr*
-73.0	7.49	0.018	0.739
-23.0	9.49	0.022	0.722
27	15.68	0.026	0.708
77	20.76	0.030	0.697
127	25.90	0.034	0.689
177	28.86	0.037	0.683
227	37.9	0.040	0.680
277	44.34	0.044	0.680
327	51.34	0.047	0.680
377	58.51	0.050	0.682
427	66.25	0.052	0.684
477	73.91	0.055	0.686
527	82.29	0.058	0.689
577	90.75	0.060	0.692
630	99.30	0.063	0.696
680	108.20	0.065	0.699
730	117.8	0.068	0.702
830	138.6	0.073	0.704
930	159.1	0.078	0.707
1030	182.1	0.084	0.705
1130	205.5	0.089	0.705
1230	229.1	0.095	0.705

$$\sum \dot{q}''_{rad} = \dot{q}''_{rad,g} + \dot{q}''_{rad,w} \tag{4.6}$$

$$\dot{q}''_{rad,g} = \Phi \varepsilon_g \sigma (T_{gas}^4 - T_s^4) \tag{4.7}$$

In Equation 4.7, Φ is the configuration factor of the element, which (as a conservative choice) is assumed to be 1.0, and ε_g is the emissivity. Given that the fire itself can be sub-divided into two parts, the furnace and the flames, the emissivity is defined for the furnace and the flame separately. The emissivity of the furnace in the absence of adequate soot particles can be approximated as zero. For the flame, nevertheless, if there is adequate black smoke, it is commonly assumed to be 1.0. The term σ, called the Stefan-Boltzmann constant, is equal to 5.68×10^{-8} W/m²K⁴. Therefore, Equation 4.6 can be rewritten as:

Furnace: $$\dot{q}''_{rad,g} = \Phi\varepsilon_g\sigma\left(T_{gas}^{\ 4} - T_s^{\ 4}\right) \approx 0 \quad (4.8)$$

Flames: $$\dot{q}''_{rad,g} = \sigma\left(T_{gas}^{\ 4} - T_s^{\ 4}\right) \text{ W/m}^2 \quad (4.9)$$

For the second term of Equation 4.6, the radiation is given by Equation 4.10, where T_w, A_1, A_2, $F_{1\text{-}2}$ and L are the wall's temperature, the areas of the solid and furnace, the view factor, and the distance between the walls and the solid, respectively. Moreover, κ, called the extinction coefficient of the gas between the wall and the solid, varies from around zero for the furnace to ∞ for the flames, which in turn results in completely different radiative heat fluxes for the furnace and the flames, as shown in Equations 4.10 to 4.12 [232, 234].

$$\dot{q}''_{rad,w} = \frac{\sigma(T_s^{\ 4} - T_w^{\ 4})}{\dfrac{1-\varepsilon_1}{A_1\varepsilon_1} + \dfrac{1-\varepsilon_2}{A_2\varepsilon_2} + \dfrac{1}{A_1F_{1-2}}} e^{-\kappa L} \quad (4.10)$$

Furnace: $$\dot{q}''_{rad,w} = \frac{\sigma(T_s^{\ 4} - T_w^{\ 4})}{\dfrac{1-\varepsilon_1}{A_1\varepsilon_1} + \dfrac{1-\varepsilon_2}{A_2\varepsilon_2} + \dfrac{1}{A_1F_{1-2}}} \quad (4.11)$$

Flames: $$\dot{q}''_{rad,g} \approx 0 \quad (4.12)$$

Assuming that the view factor ($F_{1\text{-}2}$) is equal to 1.0, and the furnace area (A_2) is considerably larger than the solid (A_1), the total heat flux to the surface of the solid is summarized in Equations 4.13 to 4.16:

Furnace: $$\dot{q}''_{Tot} = h_c(T_{gas} - T_s) + \varepsilon_1\sigma\left(T_s^{\ 4} - T_w^{\ 4}\right) \quad (4.13)$$

Flames: $$\dot{q}''_{Tot} = h_c\left(T_{gas} - T_s\right) + h_r\left(T_{gas} - T_s\right) = h_{Tot}\left(T_{gas} - T_s\right) \quad (4.14)$$

where:

$$h_r = \sigma\left(T_{gas} + T_s\right)\left(T_{gas}^{\ 2} + T_s^{\ 2}\right) \quad (4.15)$$

$$h_{Tot} = h_c + h_r \quad (4.16)$$

On the other hand, fully developed fires are generally assumed to have a very high emissivity, and thus the medium can be considered optically thick. In that case, the radiative heat fluxes will only occur between the gas and the structural element and, therefore, no exchange will exist between the flames. The summation term, consequently, drops to a single term, as shown in Equation 4.17.

$$\sum \dot{q}''_{rad} = h_r (T_{gas} - T_s) \quad (4.17)$$

Therefore, using the equations above, Equation 4.18 can be used as a summary of the heat transfer mechanism.

$$\dot{q}''_{Tot} = h_c \left(T_{gas} - T_s\right) + h_r \left(T_{gas} - T_s\right) = h_{Tot}(T_{gas} - T_s) \quad (4.18)$$

Welch *et al.* [214] showed that the heat flux transferred by convection and radiation changes with time, from around 40 kW/m^2 to 250 kW/m^2, which can significantly change the results of total heat, as schematically shown in Fig. 4.2.

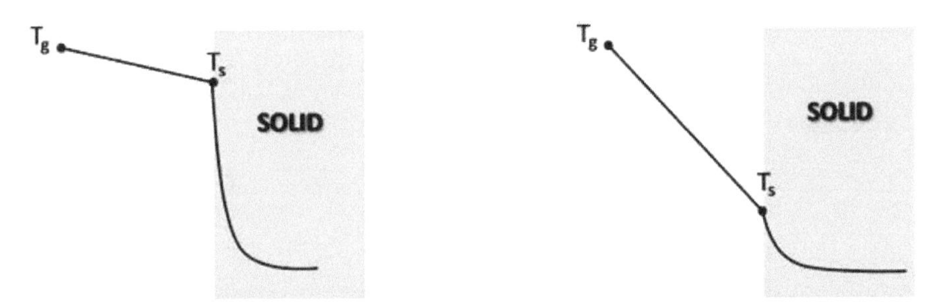

Fig. 4.2: The changes of T_s based on heat flux transferred by convection and radiation.

4.3 The effect of elevated temperature on concrete and steel

Materials' thermal and mechanical characteristics change considerably when they are exposed to fire, as a result of high thermal stresses [235]. These changes consequently affect the structural behavior and the load capacity [236]. In the following sections, the thermal and mechanical behaviors of both concrete and steel (as two most commonly used materials) are discussed.

4.3.1 Thermal and mechanical properties of steel

4.3.1.1 Thermal expansion

The ability of a unit length of a material to expand when its temperature is increased by 1°C is measured by a coefficient called thermal expansion (α). For hot-rolled steel profiles and rebars, a linear increase of the thermal expansion is observed up to 700°C. Then, after a temporary decrease between 700°C and 800°C, which relates to changes in the crystal structure, it again increases. For pre-tensioned steel, a linear increase is observed during the rise in temperature [237], as shown in Fig. 4.3.

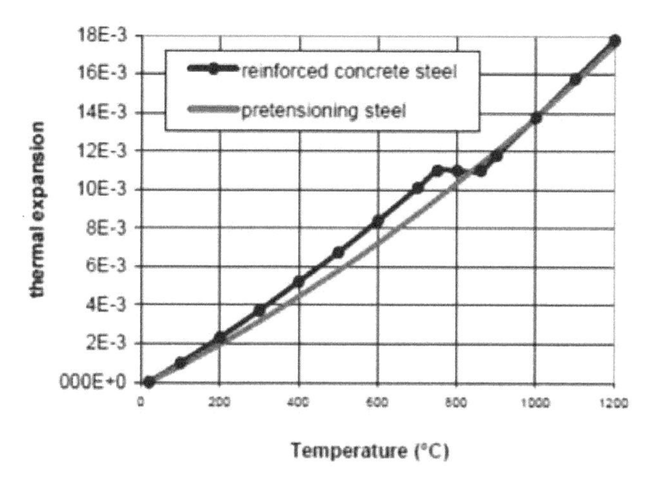

Fig. 4.3: Thermal expansion of steel, mentioned in EN 1992-1-2:2005 Eurocode 2.

4.3.1.2 Poisson's ratio

The ratio of lateral strain to longitudinal strain is called Poisson's ratio (ν), which changes between 0.27 and 0.30 with increasing temperature [238].

4.3.1.3 Creep

Creep is a time-dependent and nonlinear deformation under sustained loading. While creep might be ignored at ambient temperature, it can become significant when a metallic material sustains a high-temperature condition over a long time. In that case, creep can have an adverse effect on the load capacity and can even cause the member to fail. Creep is a microscopic process, and normally occurs in steps: dislocation creep, followed by diffusion creep. Dislocation creep relates to the crystal lattice of the materials and can bring about plastic deformation. It is also very sensitive to the loads applied. Diffusion creep relates to deformation of the crystalline solids due to the distribution of vacancies over their crystal lattice. It can occur at high temperature, and leads to long-term plastic deformation. Compared to dislocation creep, diffusion creep is more sensitive to temperature variations [239]. As fires generally occur over a short period of time, the creep effect can largely be ignored.

4.3.1.4 Stress-strain relationship

Commonly, along with an increase in the temperature, the yield strength decreases. In steel, the mechanical properties start decreasing significantly between 300°C (elastic modulus) and 500°C (compressive strength) [240, 241]. Figure 4.4 shows the stress-strain relationship in hot-rolled bars at high temperatures, as mentioned in Eurocode 2. The strains induced in an element (independent of the material) are measured by Equation 4.19, in which $\Delta\varepsilon$ is total strain, ε_σ is mechanical strain, ε_{Th} is thermal strain, ε_{cr} is creep strain and ε_{tr} is transient strain.

$$\Delta\varepsilon = \varepsilon_\sigma(\sigma, T) + \varepsilon_{Th}(T) + \varepsilon_{cr}(\sigma, T, t) + \varepsilon_{tr}(\sigma, T) \quad (4.19)$$

In the above equation, the creep strain can be ignored and the transient strain relates only to the expansion of the cement paste. Therefore, the total strain in steel structures is created by mechanical strain (due to applied loads) and thermal strain [242].

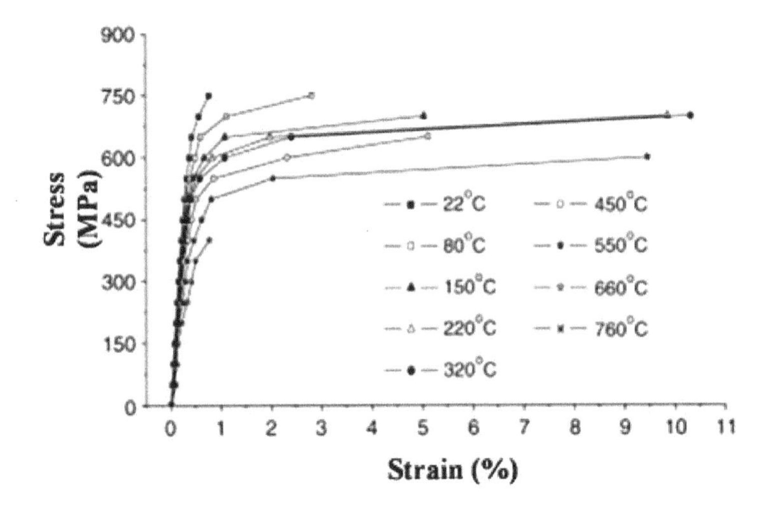

Fig. 4.4: Stress-strain curves for hot-rolled steel at different temperatures [243].

4.3.1.5 Specific heat

Specific heat is defined as the required heat per unit of mass to increase the temperature by 1°C. The specific heat of steel (C_a) varies with temperature, in the way described in Equation 4.20 and Fig. 4.5.

$$C_a = 425 + 0.773\,T - 1.69 \times 10^{-3} + 2.22 \times 10^{-6}T^3 \quad 20\ °C \leq T \leq 600\ °C$$

$$C_a = 666 + 13002/(738 - T) \quad 600\ °C \leq T \leq 735\ °C$$

$$C_a = 545 + 17820/(T - 731) \quad 735\ °C \leq T \leq 900\ °C$$

$$C_a = 650 \quad 900\ °C \leq T \leq 1200\ °C \quad (4.20)$$

4.3.1.6 Thermal conductivity

Thermal conductivity (λ) is defined as the quantity of heat transmitted by a 1°C gradient. Steel is a very conductive material and has a thermal conductivity of 54 W/mK at ambient temperature. However, this decreases linearly when its temperature increases to around 800°C. Beyond 800°C, it remains stable as shown in Fig. 4.6.

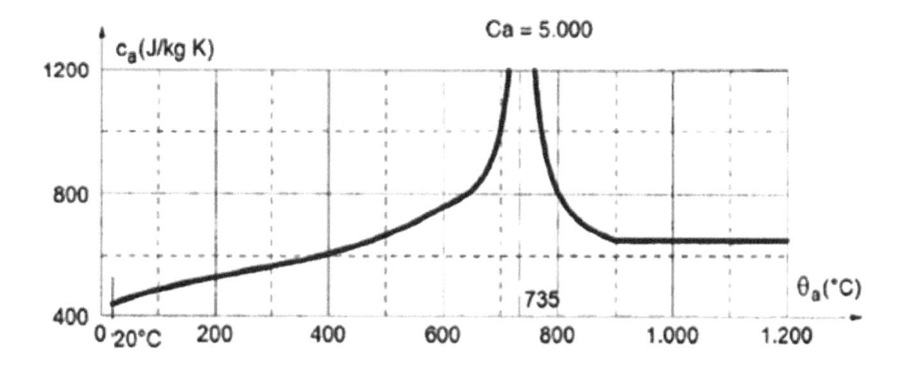

Fig. 4.5: Variations in the specific heat of steel [190].

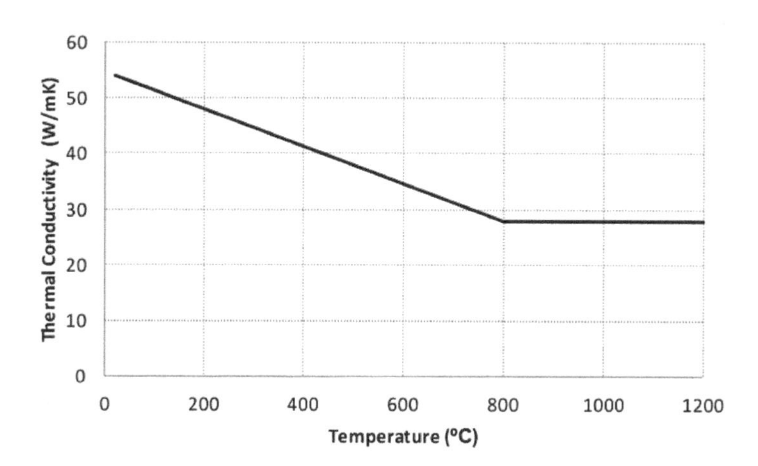

Fig. 4.6: Thermal conductivity of steel [190].

4.3.2 Thermal and mechanical properties of concrete

4.3.2.1 Thermal expansion

The thermal expansion of concrete changes with the aggregate type, the cement paste percentage, the stress level, and the heating rate. Figure 4.7 shows the thermal expansions of different concretes. As can be seen, the thermal expansion of concrete increases in a nonlinear fashion up to around 800°C and then becomes constant. Eurocode 2 suggests a value of 8×10^{-6} between 20°C and 1000°C for concrete.

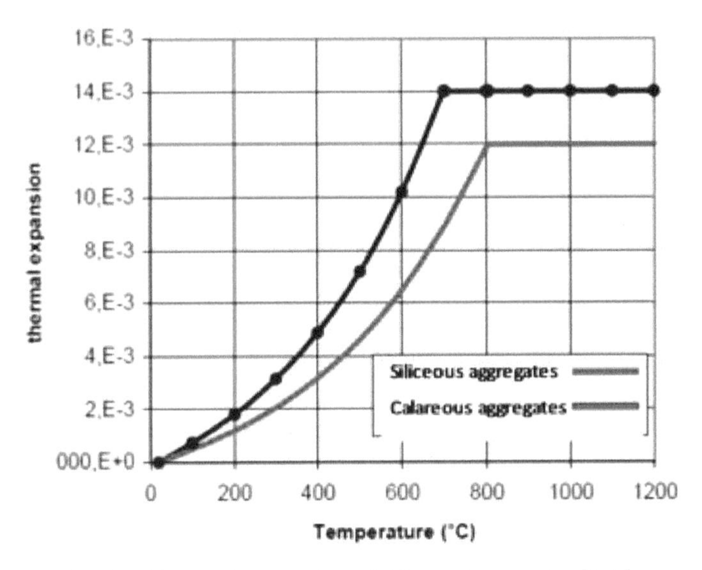

Fig. 4.7: Thermal expansion of different concretes [159].

4.3.2.2 Poisson's ratio

Figure 4.8 shows the changes in Poisson's ratio (v) for concrete at elevated temperature. As seen, Poisson's ratio changes notably with changes of temperature. While at ambient temperature, Poisson's ratio is roughly constant until about 70% of the ultimate stress; at 450°C it is constant to around 20% of the ultimate stress.

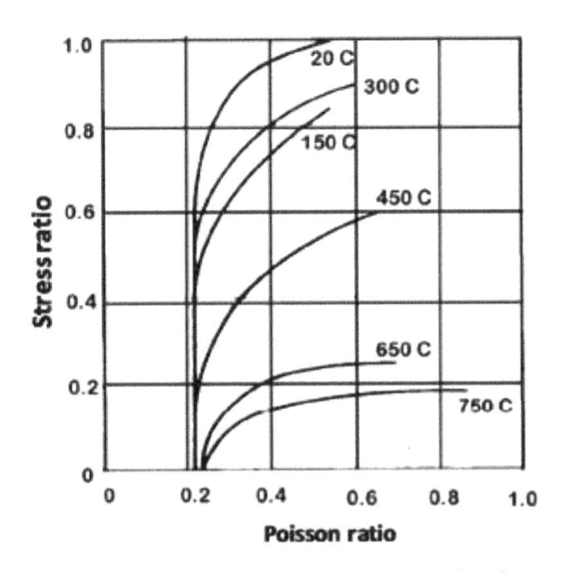

Fig. 4.8: The changes in Poisson's ratio [244].

4.3.2.3 Creep

Creep in concrete comprises two parts, the creep of the cement paste, and the creep of the aggregate [245, 246]. The total strain (ε_{tot}) is then considered as the sum of thermal strain, instantaneous stress-related strain (ε_{σ}) and transient creep strain (ε_{tr}) (and possibly also basic creep strain, ε_{cr}), as represented in Equation 4.21 [247].

$$\varepsilon_{tot} = \varepsilon_{th} + \varepsilon_{\sigma} + \varepsilon_{tr}\ (+\varepsilon_{cr}) \tag{4.21}$$

Figure 4.9 shows the results of a creep test for a preloaded gravel aggregate concrete. As seen, the value of creep below 200°C is almost negligible, and between 200°C and 400°C is not notable. Nevertheless, the value increases considerably at higher temperatures. It is also worth noting that, in a fire, the creep rates are much less important than is long duration loading, since fires normally last a maximum of a few hours only [248].

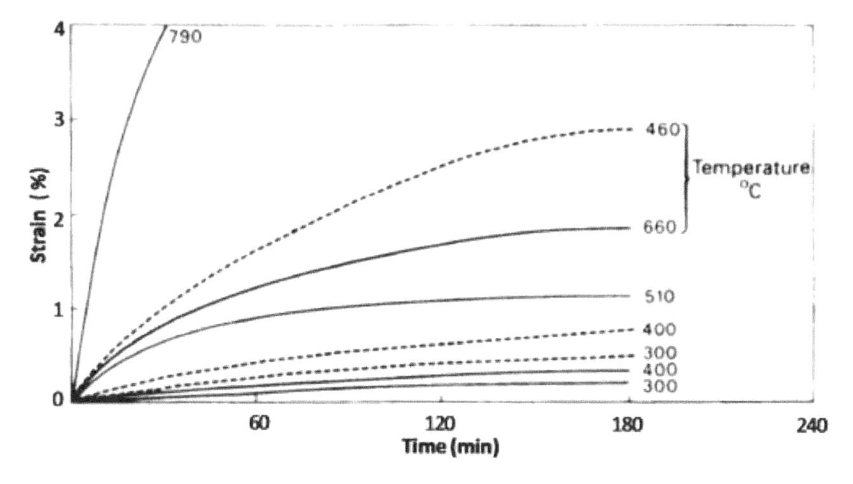

Fig. 4.9: Concrete creep [249].

4.3.2.4 Stress-strain relationship

The compressive strength of concrete and, consequently, the modulus of elasticity, decrease significantly under elevated temperature [250]. While most of the earlier studies were performed on specimens with no applied load, it has been shown that heated and preloaded specimens experience a smaller strength reduction than specimens with no applied load [251]. Other studies have also shown that the type and the size of aggregates, as well as the aggregate-cement ratio, can affect the stress-strain relationship of concrete. Nonetheless, almost all of the past studies point to the fact that concrete is no longer considered a structurally relevant material beyond around 500°C [244, 251-253]. Figure 4.10 shows the stress-strain relationship at different temperatures, developed by Eurocode 2.

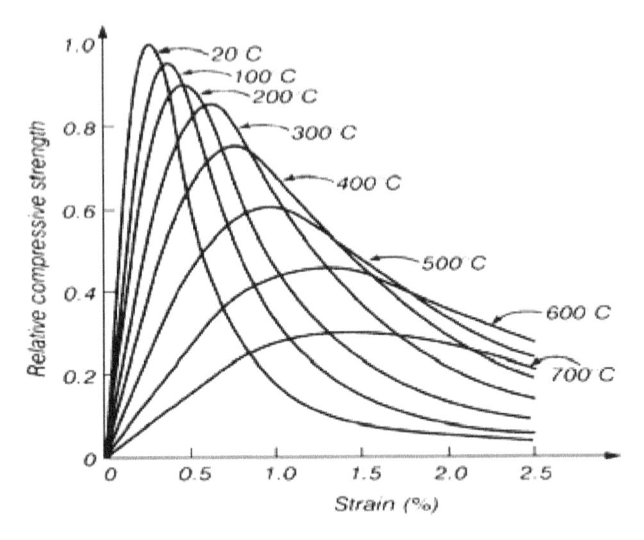

Fig. 4.10: Stress-strain curves for concrete at different temperatures [254].

4.3.2.5 Specific heat

The specific heat of concrete (c_p, J/kgK) depends mostly on its water content and on the type of aggregate. Experiments have shown that between 100°C and 200°C, which corresponds to the temperature needed for water evaporation, the peak of specific heat is observed (Fig. 4.11a). Figure 4.11b shows the variation in specific heat for different types of aggregate. Equations 4.22, developed by Eurocode 2, are used to determine the specific heat of conventional dry concrete.

For $20°C \leq T \leq 100°C$: $c_p = 900$

For $100°C \leq T \leq 200°C$: $c_p = 900 + (T - 100)$

For $200°C \leq T \leq 400°C$: $c_p = 1000 + (T - 200)/2$

For $400°C \leq T \leq 1200°C$: $c_p = 1100$ (4.22)

4.3.2.6 Thermal spalling

Thermal spalling of concrete cover under fire exposure is an important phenomenon, which may occur suddenly and violently, which is brittle and which may lead to a significant decrease in the load-bearing capacity of the structure [255]. This is because, after spalling, the boundaries of the concrete members change in such a way that the reinforcement is directly exposed to the fire and thus is rapidly heated up; hence the strength of the member decreases more quickly [256]. Even today, the mechanism of spalling is not completely understood and thus it has not been quantified [257], although some attempts have been made to somehow quantify thermal concrete

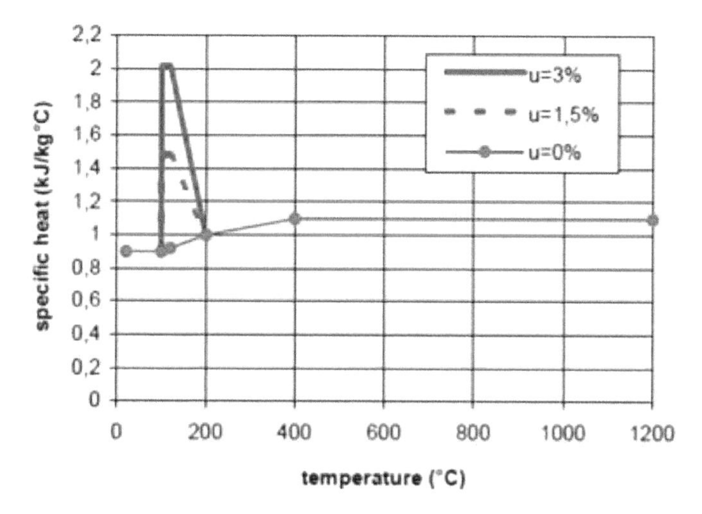

(a) The effect of different water contents (u%) on c_p

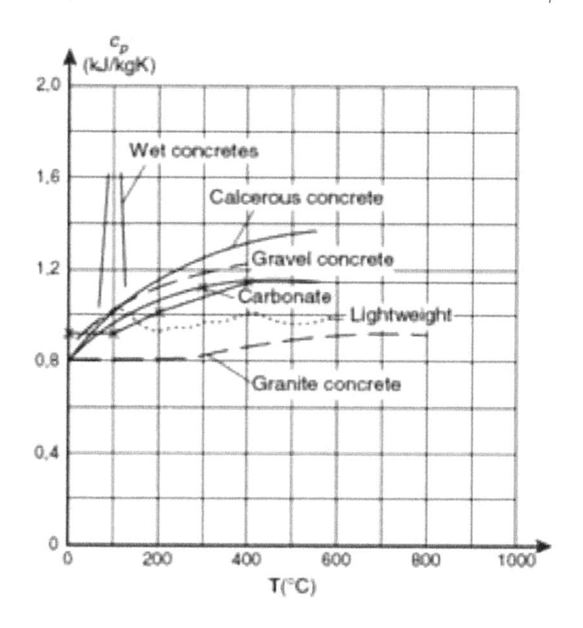

(b) Variation of c_p for different types of aggregate

Fig. 4.11: Specific heat in concrete [159, 251].

spalling [258]. Spalling is a hydro-thermal-mechanical phenomenon and can occur when the pore pressure inside the concrete components goes beyond the concrete's tensile strength [259]. It is believed that when the rate of heating is between 20-32°C, the possibility of thermal spalling increases significantly [260]. Spalling of concrete is categorized into aggregate spalling, explosive spalling, corner spalling, and surface spalling, with the first three of these occurring during the first 30 minutes of fire exposure. Surface spalling, however, generally occurs after 30-60 minutes, when the fire is normally at its maximum temperature [260]. Figure 4.12 shows the different types of spalling in a fire. Several studies have been performed to find the most important factors causing spalling, some of which are the heating rate, the section size and shape, the moisture content, the concrete strength, the quality, type and size of aggregate, and the lateral reinforcement [261-265]. In beam and column elements, thermal spalling is more important when the concrete cover is greater than 4.5 cm [266, 267], or when the beams and columns are made of high-strength concrete (HSC) [261] with particles smaller than the cement grains (micro silica, for instance) and a moisture content of more than 3-4% [268-270]. Experimental studies have also shown that HSC RC columns with 90° ties are more likely to sustain thermal spalling than those confined with a 135° tie [261] (Fig. 4.13).

In addition, tests have shown that concrete slabs are more sensitive to a fire's severity than columns and beams, in such a way that thermal spalling would occur after 2 minutes and 15 minutes under the ISO fire curve (a moderate-intensity fire curve) and the Hydrocarbon fire curve (a high-intensity fire curve) respectively [271].

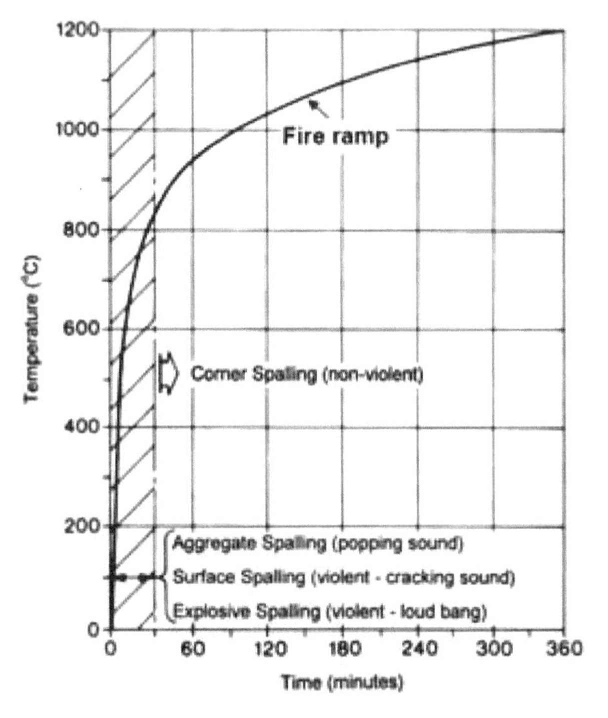

Fig. 4.12: Different types of spalling in a fire [260].

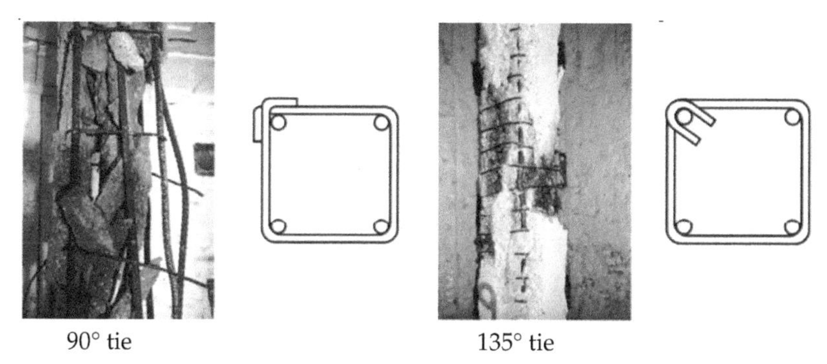

Fig. 4.13: Thermal spalling in HSC RC elements with 90° and 135° ties [261].

4.3.2.7 Thermal conductivity

Concrete has a very low thermal conductivity compared to steel. Nevertheless, its thermal conductivity varies and depends on numerous factors, the most important of which are the type of aggregate and the ratio of water to cement (W/C) [272]. Studies have shown that lightweight concrete (LWC), with a density between 1000-1900 kg/m^3, is more conductive than normal-weight concrete (NC), with a density between 2000-2900 kg/m^3 [273]. In addition, the water content has an important role in thermal conductivity, such that the thermal conductivity increases as the water content increases [274]. Figure 4.14 shows the thermal conductivities of LWC and NC. As is seen, for NC, two upper and lower boundaries are defined. These values have been gained from tests [159]. While the upper limit relates to concrete models, the lower limit pertains to composite steel/concrete models. It is worth mentioning that thermal conductivity reduces as the specific heat increases and vice versa.

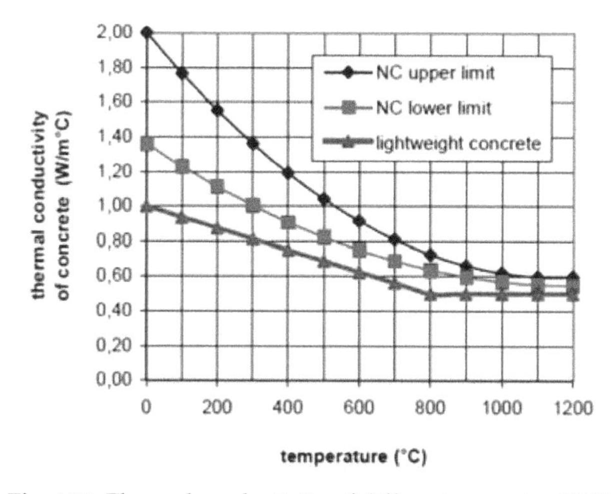

Fig. 4.14: Thermal conductivity of different concretes [159].

Chapter 5

Seismic Analysis

5.1 Introduction

It is well understood that in order to withstand the loads applied to a structure, the structural capacity must be greater than the structural demand. Given that structural behavior under seismic loads is much more complex than under gravity loads, more attention needs to be invested in seismic analysis than in gravitational analysis. Material nonlinearity is also an important point, which is further highlighted under seismic load. To include the material nonlinearity in the seismic analysis, the change from force-based seismic codes to performance-based seismic codes has provided a great help. This chapter presents a short review of different seismic structural analyses, both linear and nonlinear methods, and discusses their advantages and disadvantages.

5.2 Linear analyses

Using linear analyses, only the elastic behavior of structural components is considered. These methods are more applicable when most structural components remain in the elastic part, and there is almost no plasticity achieved. Often, when the demand-to-capacity ratio is less than two, the nonlinear behavior of a structure is not significant, and linear methods can be used [275]. Using linear methods, only primary elements are modeled and secondary elements are used to control the displacements produced by the seismic analysis. This is because the primary elements come under more cyclic loads, as the result of which, they are more susceptible to loss of strength and stiffness. In addition, the linear methods are based on the assumption that plastic hinges are created only at the ends of the elements – if they are created in a different position, the results of the linear analysis will no longer be accurate. There are some simplified assumptions for considering structural characteristics, such as the P-Δ effect and the cracking effect. The former can be applied as an additional lateral load to the structure, while the latter can be involved simply by reducing the cross-section dimensions.

5.2.1 Linear static analysis

Linear static analysis (LSA) is applicable when the first structural mode is the dominant mode, and thus the effects of the upper modes can be ignored. This clearly implies that LSA cannot be used for high-rise and irregular structures, as in both of these the first mode is not often dominant. LSA is based on two firm assumptions: 1) the material behavior is linear; and 2) although seismic loads have a dynamic nature, their effects on structures are determined statically and as a portion of the structural mass. The mass considered plays a role in the elastic spectrum acceleration. Using LSA, the mass value is determined in such a way that it can provide the maximum structural displacement, equal to its value predicted by the design earthquake. If, under that load, the structure behaves linearly, then the forces produced will be close to the predicted values. Otherwise, if the structure behaves nonlinearly, the forces' values will be greater than the yield forces.

5.2.2 Linear dynamic analysis

Linear dynamic analysis (LDA) is used to determine forces and displacements resulting from seismic loads, using the dynamic equilibriums of the elastic model. It is clear that as dynamic parameters are involved in the analysis, the results are thus more accurate than those from LSA, but at the same time, no material nonlinearity is considered. There are two assumptions with LDA: 1) the structural behavior can be controlled using linear combinations of different vibrational modes, but separately; and 2) the period of every mode is constant during the earthquake event. The LDA is performed through two methods, response spectrum analysis, and time history analysis [276].

5.2.2.1 Response spectrum analysis

The spectrum used in the response spectrum analysis is linear elastic, with no modification for nonlinear displacement. The results of the analysis are thus more accurate for those structures which will behave linearly during a seismic event. As the maximum modes do not happen simultaneously, and do not have identical direction, they cannot be added together [277]. The best way to address this issue is to use probabilistic theories in which different possibilities are considered, such as the square root of the sum of squares (SRSS) [278], or complete quadratic combination (CQC) [279]. The CQC is mostly employed for irregular structures. On the other hand, since most of an earthquake's energy is normally absorbed over the first few modes, for those structures with a large number of degrees of freedom, it is often enough to combine the first three to six of the modes. This can provide a reduction in the computation time. However, spectrum analysis has a number of deficiencies. For example, the assumption that the periods are independent is not always correct. This particularly relates to those structures that experience torsion during an earthquake event.

5.2.2.2 Time history analysis

Using time history analysis, the structural response is investigated as per the dynamic equilibriums and considering short time intervals. To do that, the structural response is calculated under at least three ground accelerations. Overall, for regular structures, time history analysis can be used for each of the directions separately, and it is not necessary to consider the interactions between the directions. For irregular structures, on the other hand, it is highly advised to perform a three-dimensional analysis. Using time history analysis, the structural responses in the time intervals are determined while ground accelerations at the structural base – often at the foundation level – are taken into consideration.

5.3 Nonlinear analyses

The use of linear analyses has some limitations. To address closely the structural response under seismic loading, nonlinear analyses are used, in such a way that the nonlinearity of stiffness and damping are considered. The nonlinearity of the stiffness encompasses the geometric nonlinearity and the material nonlinearity. While the material nonlinearity shows a hysteretic behavior under cyclic loading, there is no hysteretic behavior with the geometric nonlinearity. In addition, the damping nonlinearity is importantly of a nonlinear type. As most structures experience yielding under seismic excitations, it is therefore of importance to consider the material's nonlinearity during the analysis.

On the other hand, earthquake-induced forces have a dynamic nature and, therefore, only a dynamic analysis (being the most rigorous approach) can meet a realistic assessment [280]. Nevertheless, dynamic analyses are not routinely employed in everyday designs, mostly because of the complex nature of the analysis [281], in particular with respect to inelastic behavior. In addition, different ground motions may bring about different nonlinear responses, so choosing the right excitation is an important, and sometimes determining, consideration [282].

In view of the abovementioned difficulties, there is a more practical method currently used to determine the design forces. This method is known as static pushover analysis, and it was developed over the past two decades in accordance with the concept of design based on performance [283]. It is becoming a popular tool for the seismic performance evaluation of existing and new buildings [284], because the procedure is not only simple to use but also covers nonlinear behavior. Nevertheless, as certain simplifications are made during the analysis, the method becomes approximate only. Pushover method is based on the fundamental assumption that the structural response can be approximated by the response of an equivalent single-degree-of-freedom (SDOF) system, taken as a pre-assumed mode shape that remains constant throughout the analysis. This is obviously not a completely correct assumption, yet there exists a consensus on the use of this method as a rational tool for estimating the maximum seismic response of multi-degree-of-freedom (MDOF) structures, if the response of the structure is mostly governed by a single mode [285.291]. For those structures that cannot meet these limitations, advanced pushover-based methods can be employed, such as adaptive pushover analysis [292, 293] and modal pushover analysis [290, 294], in order to consider the effects of higher modes in the process.

5.3.1 Description of conventional pushover analysis

Pushover analysis was first described in FEMA273 [295] and then in FEMA356 [74]. Conventional pushover analysis is the nonlinear incremental-iterative solution of the equilibrium equation $KU = P$ in a finite element formulation, in which K, U and P stand for the nonlinear stiffness matrix, the displacement vector and a predefined load vector (which is laterally applied to the structure), respectively [296]. The lateral load is applied over the height of the structure incrementally, as shown in Fig. 5.1.

The analysis is carried out in two steps: the load-controlled step (controlled by gravity loads); and the displacement-controlled step (controlled by earthquake loads). While the former step normally happens prior to notable yielding, the latter step occurs after significant yielding [297]. It is also worth noting that the lateral load may be a set of displacements or forces, but it should have a constant ratio and a constant shape during the analysis. In this way, at the end of each iteration, the reaction vector (P_e) of the structure is assembled from the contribution of all finite elements. Then,

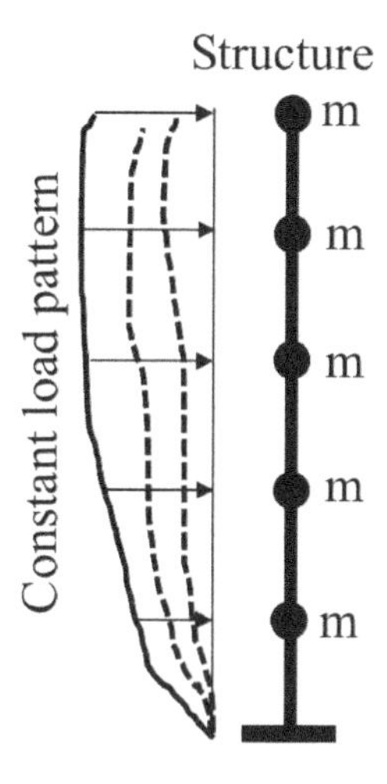

Fig. 5.1: Load pattern in the pushover analysis.

the out-of-balance forces are iteratively re-applied to the structure, until the specific predefined tolerance is reached, as represented by Equation 5.1.

$$\Delta U = [K_T]^{-1}(\lambda P_0 - P_e) \tag{5.1}$$

where ΔU is the calculated displacement increment within an iteration, K_T is the current nonlinear stiffness matrix, λ is the load factor within the corresponding load increment, P_0 is the initial load, and P_e is the reaction of the previous iteration. The reaction load vector (P_e) is then accounted for, using Equation 5.2.

$$P_e = \sum \int_v B_T . \sigma_{NL} . dV \tag{5.2}$$

where B is the strain-displacement matrix of each element, and σ_{NL} is the vector of the element's nonlinear stress, which is determined by its material constitutive law.

The process terminates when either a predefined limit state is reached or structural collapse is identified. The plotted roof displacement versus base shear is then interpreted as the capacity curve. Using this process, the structural behavior from elastic state to collapse state can be traced, as shown in Fig. 5.2.

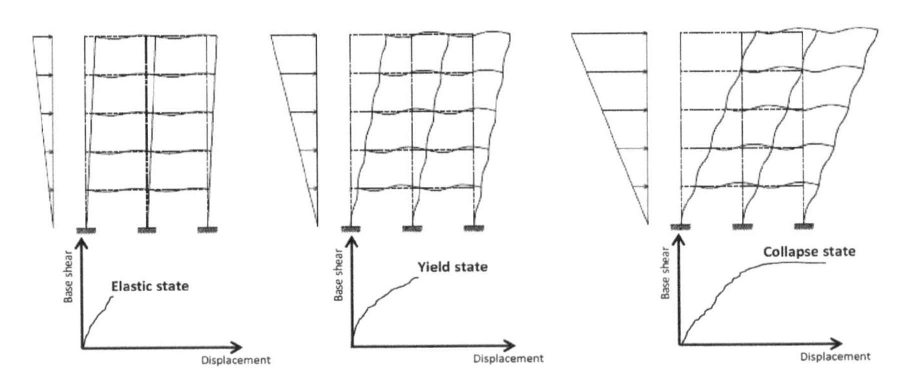

Fig. 5.2: Structural behavior through conventional pushover analysis [296].

In conventional pushover analysis, it is assumed that the target displacement of an SDOF structure corresponds to an MDOF structure, provided that the first mode of the structure is the predominant mode. In addition, studies have shown that, for structures with a fundamental period smaller than 2 seconds, this assumption is relatively accurate, specifically for elastic systems. For inelastic systems, however, the results are mostly conservative [284]. On the other hand and for the sake of simplification, the nonlinear base shear versus displacement curve can be replaced with an idealized bilinear diagram, incorporating the effective lateral stiffness and the effective yield strength, as shown in Fig. 5.3. In the figure, K_i, K_e and V_y

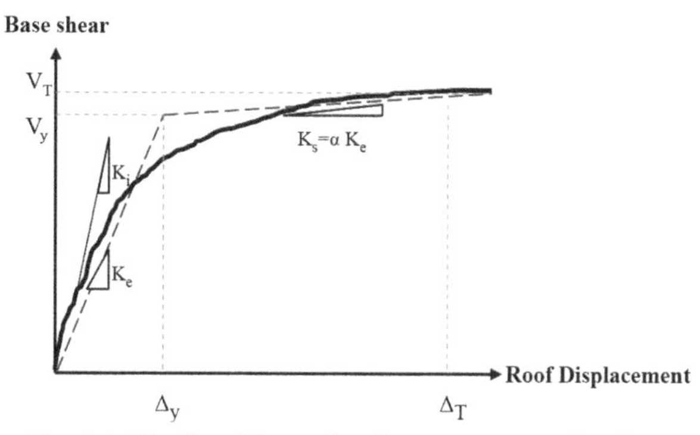

Fig. 5.3: Idealized force-displacement curve [284].

stand for elastic lateral stiffness, effective lateral stiffness and effective yield strength of the structure in the direction under consideration, respectively. The post-yield slope of the figure ($K_s = \alpha K_e$) then, may be positive or negative. In addition, V_y and V_T are the yield base-shear and the total base-shear, corresponding to the yield displacement (Δ_y) and the target displacement (Δ_T).

For the pushover curve, the lateral load shape (force or displacement), the way it is distributed over the height of the structure [298, 299], the number of load steps, and the convergence criteria are required, all of which play important roles in the level of accuracy of the analysis [296]. In some codes, such as FEMA356, it is strongly recommended that at least two different load distributions be applied to the structure, so that the most critical situation can be found. Furthermore, as there are several possibilities for failure modes, it is also recommended to consider the most critical failure mode [74, 300]. Figure 5.4 shows some commonly used force distributions in a pushover analysis.

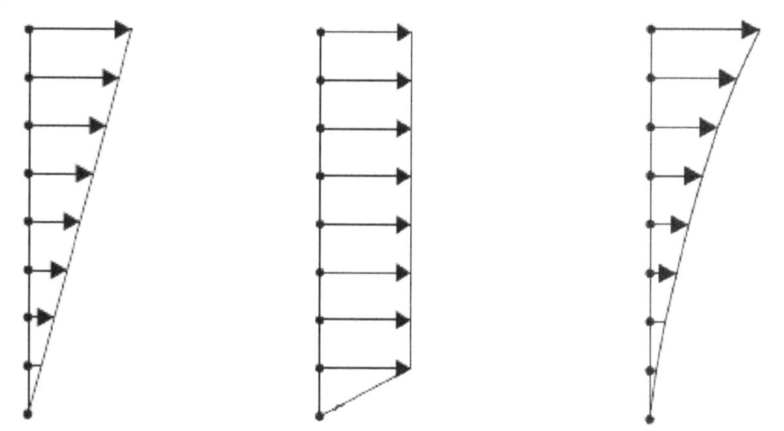

(a) Triangular distribution (b) Uniform distribution (c) Higher-mode distribution

Fig. 5.4: Different load distributions.

Several methodologies for estimating the target displacement have been proposed, some of the more well-known of which are the N2 method [285], the ATC40 method [301] the FEMA356 method [74] and the NERHP [302] method. These proposed methods are mostly based on the statistical relationships of the lateral displacement estimated by linear and nonlinear analyses. As an example, the target displacement according to the FEMA356 procedure is determined using Equation 5.3.

$$\Delta_T = C_0 C_1 C_2 C_3 S_a (T_e/2\pi)^2 g \tag{5.3}$$

The equation shows that the spectral displacement ($S_a(T_e/2\pi)^2 g$) is multiplied by the coefficients of C_1, where S_a is the spectral response acceleration and is determined according to the procedure mentioned in section 1.6.1.5 of FEMA356. T_e is the effective fundamental period, which is determined using Equation 5.4.

$$T_e = T_1 \sqrt{\frac{V_1/\delta_1}{V_y/\delta_y}} \tag{5.4}$$

where V_1, δ_1 and T_1 are calculated for the first increment of the lateral load.

The backgrounds of the coefficients C_1 in Equation 5.3 are explained below.

C_0 is the coefficient to modify the difference between the roof displacement of an MDOF and the equivalent SDOF system, which is determined using Equation 5.5.

$$C_0 = \frac{\sum_{i=1}^{n} w_i \phi_i}{\sum_{i=1}^{n} w_i {\phi_i}^2} \tag{5.5}$$

In this equation, w_i is the portion of the seismic weight (W) at level i, and ϕ_i is the amplitude of the shape vector at level i.

Clearly, when only first mode shape with elastic behavior is assumed, the value of C_0 is equal to 1.0. The C_1 coefficient is then applied to Equation 4.3, in order to modify the expected maximum inelastic displacement to the displacement computed for a linear elastic response. The coefficient is particularly important for structures with relatively short initial vibration periods and full hysteretic loops. In addition, the coefficient is mainly based on experimental observations and analytical investigations of the earthquake responses of yielding structures.

When the hysteretic loops demonstrate considerable degradation in stiffness or pinching, the energy absorption and dissipation capacities decline. In that case, the C_2 coefficient is applied to Equation 4.3. The mentioned effects are particularly important for structures that have short fundamental periods, low strength, and very pinched hysteretic loops. Nevertheless, studies have shown that the displacement response of MDOF systems is not considerably influenced by the pinched hysteretic behavior of every element and, as a consequence, the coefficient can be equal to 1.0.

On the other hand, it is well accepted that P-Δ effects as the result of gravity loads can often increase the lateral displacements. If P-Δ effects lead to a negative post-yield stiffness, then a significant inter-story drift and large displacement may be observed. In that case, the C_3 coefficient, which is larger than 1.0, is applied to Equation 4.3. Static P-Δ effects can be determined using a relatively simple method, proposed in section 3.4.5.2 of FEMA356. However, in the case of dynamic P-Δ effects, it is necessary to consider several factors, such as the slope α in Fig. 5.3, the fundamental period of the structure, the strength ratio, the relationship of the hysteretic load-deformation, and the frequency and duration of a strong earthquake. As several factors are involved in accounting for the dynamic P-Δ effects, it is difficult to propose a single modification factor. The C_3 coefficient suggested by FEMA356 considers only the effect of negative post-yield stiffness, which results in a significant simplification.

5.3.2 Plastic hinges and acceptance criteria

Performing a pushover analysis (as explained above) results in the computation of inertia forces generated at places of significant masses and, in a structure, which are displaced to the target displacement. The analysis may lead to some of the structural elements exhibiting yield in their sections, resulting in the formation of plastic hinges. It is worth mentioning that appropriately defining the nonlinear (plastic) characterization of the members is important. When an inelastic deformation is experienced (by, for example, an RC member), cracks tend to spread, resulting in the distribution of the curvature, as schematically shown in Fig. 5.5 [303]. It is worth mentioning that rebar is usually modeled as an elastic-perfectly-plastic material with parabolic strain hardening [304].

In order to simulate the inelastic behavior of the members, two main models have been developed: concentrated plasticity (lumped plasticity) models, and spread plasticity (distributed plasticity) models, the parameters of which are quantified using the members' strengths and deformation

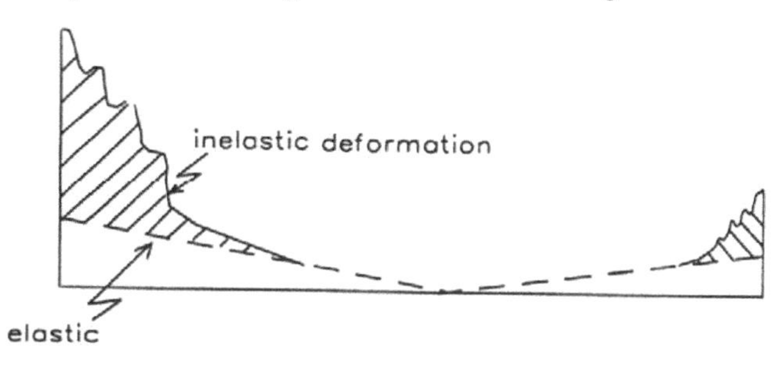

Fig. 5.5: Distribution of curvature in a member [303].

capacities [305]. The distributed plasticity models provide a skeleton for nonlinear structural analysis by defining the moment curvature at different locations of each member, which require more computational effort than the lumped plasticity models [303]. The lumped plasticity models, which are based on defining concentrated plastic hinges, however, are usually assigned at the expected locations of plasticity spread [306]. This is shown schematically, using idealized curvature distribution along a cantilever, in Fig. 5.6. As is seen in the figure, assuming a ductile behavior for the RC element, a typical moment-curvature curve can be idealized to a trilinear curve, with three stages – elastic stage, cracking stage and steel yielding stage [307]. Before the cracking moment (M_{cr}), the distribution of the stress across the cross-section is linear and elastic. Assuming a complete bond between concrete and rebar in the cross-section, the concrete and the rebar experience similar strain and similar modular stress ratios [308].

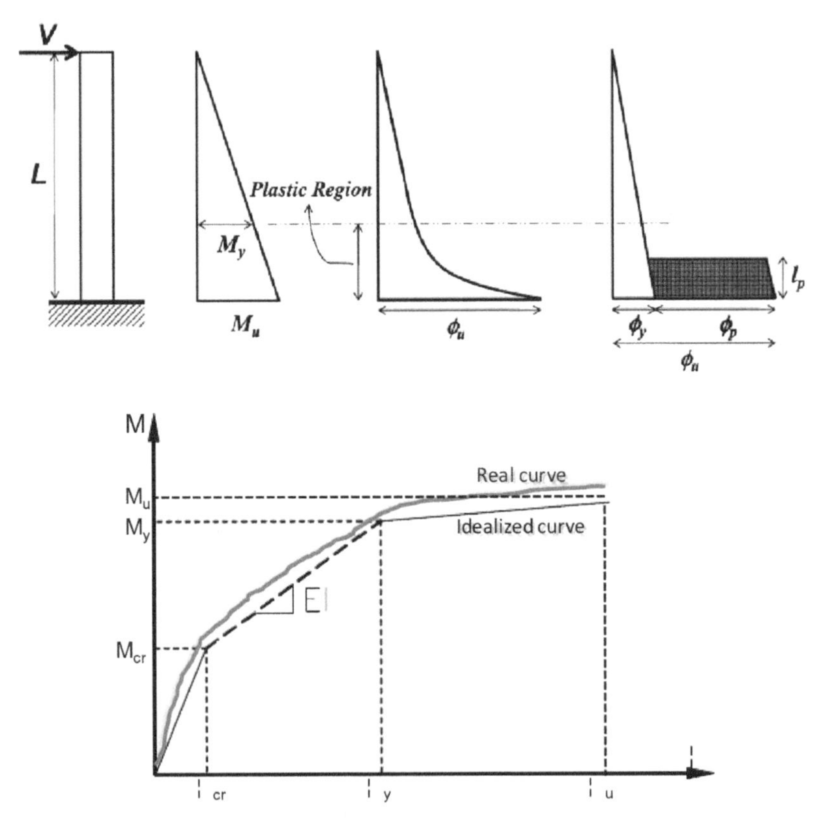

M_{cr} = cracking moment; M_y = yield moment; M_u = ultimate moment; l_p = plastic hinge length; Φ_{cr} = cracking curvature; Φ_y = yield curvature; Φ_p = plastic curvature; Φ_u = ultimate curvature

Fig. 5.6: Actual and idealized curvature distribution in a cantilever [80].

The modular stress ratio is defined as the ratio of the modulus of elasticity of the rebar to that of the concrete. However, when the maximum tensile stress in the cross-section increases to the modulus of rupture (defined as the cracking stress of concrete), the concrete cracks. Assuming that the plane cross-section remains in plane even after cracking, a linear strain distribution throughout the depth will be developed. Then, alongside the increasing of the bending moment to the yield moment (M_y), the neutral axis shifts towards the compressive face, as shown in Fig. 5.7.

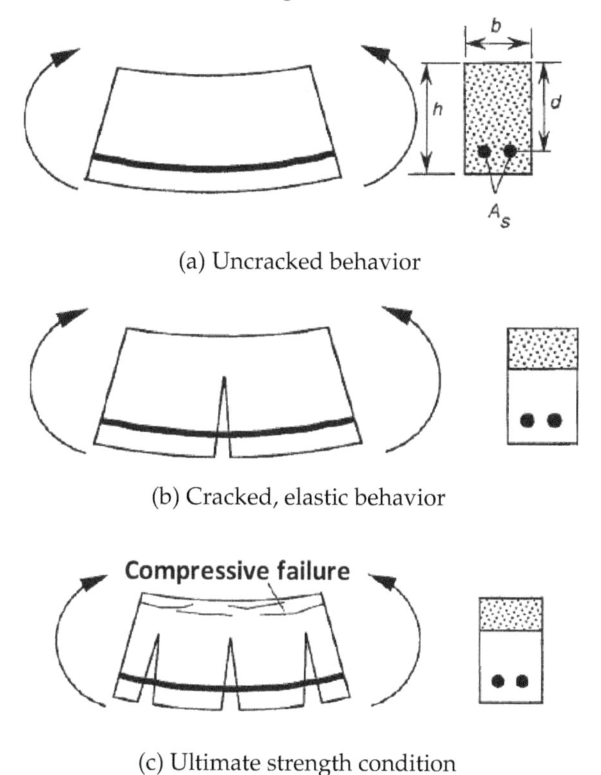

(a) Uncracked behavior

(b) Cracked, elastic behavior

(c) Ultimate strength condition

Fig. 5.7: Behavior of reinforced concrete member [308].

At the last stage when the rebar yields, the curvature increases significantly, while the moment ascends gradually to its maximum value. At that stage, the bending moment can still increase until the ultimate moment (M_u) reaches the ultimate condition (corresponding to the ultimate curvature (Φ_u). Several ultimate conditions can then be considered: a) when the moment capacity of the member drops by 20% [309]; b) when the rebar's tensile strain reaches ultimate [80]; and c) when the ultimate compressive strain in concrete reaches a limiting value. In most design procedures, limiting values of 0.003 or 0.0035 are conservatively assumed for unconfined reinforcement, after which concrete crushes in compression [80]. The stress-strain relation of both unconfined and confined concrete has been investigated in numerous studies [310-312], some of the more well-known of which are those by Scott *et al.* [310], Mander *et al.* [304, 313] and Priestley *et al.* [314]. As an example, Fig. 5.8 shows the concrete stress-strain relation proposed by Scott *et al.* [310].

The member will then experience plastic rotation as a result of plasticity developing over a certain length, called the plastic hinge length. This plastic rotation is defined as the difference between the ultimate curvature and the yield curvature multiplied by the plastic hinge length. Several empirical and numerical investigations have been performed to evaluate the equivalent length of the plastic hinge (l_p) [315, 316]. Of these, Park and Paulay's [80] formula is the simplest, as represented by Equation 5.6, in which H is the section height.

$$l_p = 0.5H \tag{5.6}$$

After deciding on the method of accounting for the ultimate condition of the concrete and the plastic hinge length, the acceptance criteria can be defined. For example, based on the FEMA356 code, structures are designed to meet different performance levels (such as IO, LS, and CP), as explained in Chapter 2. Therefore, if these performance targets were known, it would be possible to control whether the defined target is met. Figure 5.9 shows the idealized moment-curvature curve for different performance levels.

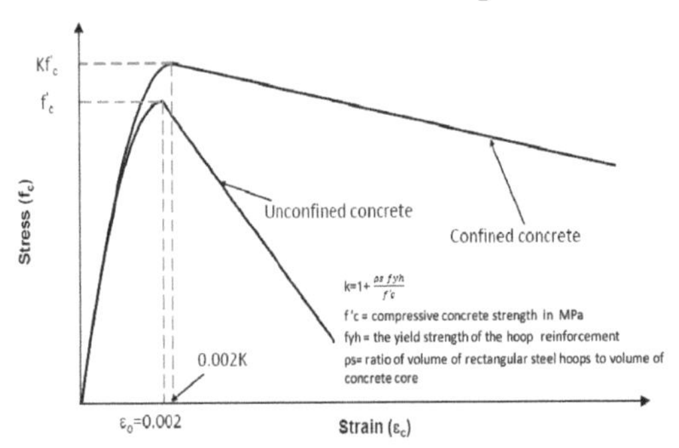

Fig. 5.8: Stress-strain relation for monotonic loading of confined and unconfined concrete [310].

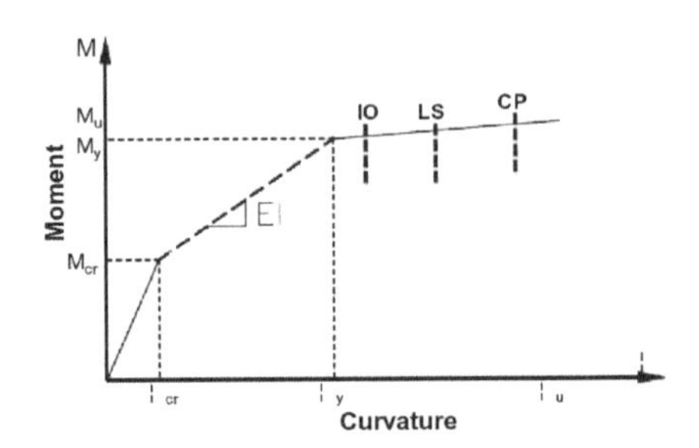

IO – immediate occupancy, LS – life safety, CP – collapse prevention

Fig. 5.9: Idealized moment-curvature curve for various performance levels.

Chapter 6

Post-earthquake Fire Analysis in Structures

6.1 Introduction

Although the performance of buildings subjected to PEF had been of some interest in the past, it received much more attention following the terrorist attack of '9/11'. In this respect, more attention has been paid to steel structures than to RC structures, possibility due to the belief that steel structures (particularly those that have not been fireproofed) are more vulnerable to high temperatures than RC structures. In the previous chapters, some fundamental information was provided, without which it would not be possible to talk about PEF analysis as a structural model.

At present, there have been few efforts to evaluate experimentally the performance of urban structures under PEF loading, mostly due to the tests being costly and the fact that their results cannot simply be generalized from one type of structure to other types. Computerized modeling can thus be a suitable substitute for experiment. Nevertheless, very few computer tools are currently available that are capable of performing a coupled analysis for all types of structures, i.e. considering seismic analysis and subsequent fire, allowing for the variation of material properties at elevated temperature. The reason for this limitation mostly pertains to the high complexity of PEF modeling.

The modeling commences with the application of gravity loads, which is followed by seismic loads. The seismic loads can be applied linearly or nonlinearly, each with its own limits (as pointed out in the previous chapter). After the earthquake load has finished, the structure might have sustained damage, ranging from very minor to major. The structure is then exposed to flames. The indispensable consideration is to properly transfer the damage arising from the previous steps, i.e. the gravity loads and earthquake loads, to the next step, which is the fire analysis. This consideration is important, as the extent of damage sustained can considerably change the fire resistance of the structure. The damage resulting from the gravity loading and the earthquake loading might be structurally dependent or non-structurally dependent. The structural damage might be as a result of incurring inter-story drift, or from the rotation of beam-to-column connections, etc., such that they may cause some degradation in the strength and stiffness. The non-structural damage might be as a result of internal partitions falling down, damage to active fire fighting facilities (such as sprinklers and smoke detectors), or damage to passive facilities (such as fireproofing materials sprayed on the surface of structural components). In steel structures, the damage to fireproofing materials can lead to direct exposure of structural components to fire. In RC structures, while the concrete cover often has no structural role in the computations, and while it is considered to protect the reinforcements against the environment, any damage or spalling incurred might substantially influence the fire resistance of the structure, since the reinforcement might be directly exposed to fire. On the other hand, although the seismic analysis can be done linearly, under fire condition the structural behavior is not linear, since almost all materials show a nonlinear behavior under elevated temperatures. As shown in Chapter 4, steel materials (as a specific case) are often too weak under fire conditions and they can lose their strength at a very rapid rate. It is evident that involving all of the abovementioned forms of damage, including their occurrence probability, in the PEF modeling will be a very complicated and time-consuming process. This chapter uses some functional tools to simulate the PEF loading in moment-resisting steel and RC structures.

6.2 PEF modeling of conventional structures

The method used in this chapter is called sequential analysis. Sequential analysis is a functional tool that considers the loads and corresponding changes in geometry in a number of steps. It allows the effects of residual deformations, as well as degradations in stiffness and strength resulting from the earthquake, to be used in the fire analysis. Employing commonly-used structural analysis software packages, and following the simple application of gravity loads, for which the structure mostly remains in the elastic region, the seismic analysis is performed and the equivalent seismic loads are derived. These loads are transferred into the second software package, which is capable of performing the structural fire analysis. There is often a degree of programming involved in connecting the two software packages with each other so that the process can become automatic. To perform the fire analysis, there are usually two steps: a sectional analysis in order to account for the distribution of the temperature inside the cross-section, and a structural analysis to account for the stresses and forces in the elements. There are some packages capable of performing structural fire analysis, such as Diana, Vulcan, InfoGraph and SAFIR. Here, the author's intention is not to compare the capability of these packages, each of which uses its own algorithm. The

SAFIR [317] program is a special nonlinear finite element program that allows simulation of the structural behavior under ambient and elevated temperature, and it is the program used for performing the fire analysis in this book. The results of the simulator have been validated in numerous tests conducted for conventional buildings, in both two and three dimensions [318, 319]. The stress-strain material laws for steel and concrete are linear-elliptic and nonlinear, respectively, and are embedded in the program according to Eurocodes. In addition, the material laws have been written considering monotonic loading cases, but not cyclic loading cases. Therefore, there is no energy dissipation during cyclic loading in SAFIR, which is in turn a limit. It is worth noting that in Eurocode 2, the total strain (ε_{tot}) is implicitly considered as the sum of free thermal strain (FTS, ε_{th}), mechanical strain (ε_{m}), and possible basic creep strain (ε_{cr}), as shown in Equation 6.1.

$$\varepsilon_{tot} = \varepsilon_{th} + \varepsilon_{m} \ (+ \ \varepsilon_{cr}) \tag{6.1}$$

In SAFIR, however, the total strain is explicitly considered by splitting into FTS, instantaneous stress-related strain (ε_{σ}), transient creep strain (ε_{tr}) and basic creep strain (ε_{cr}), as shown in Equation 6.2.

$$\varepsilon_{tot} = \varepsilon_{th} + \varepsilon_{\sigma} + \varepsilon_{tr} \ (+ \ \varepsilon_{cr}) \tag{6.2}$$

Overall, the structural fire analysis is performed in steps, which are thermal cross-sectional analysis, torsional analysis (if required) and structural analysis. The first step is thermal analysis, which is performed in order to predict the temperature distribution inside the cross-section. To do this, the thermal characteristics of the materials are either manually introduced, or chosen from default values already available within the program. The cross-sections are then discretized into several fibers (Fig. 6.1). The fibers are simulated as two-dimensional solid elements. In order to reduce the computational effort, fibers are modeled using isoperimetric triangle or quadrilateral solids. For two-dimensional models, it is assumed that no heat is transferred along the longitudinal axis of the beam. Conduction is considered the main mode of heat transfer in plane and solid sections. However, for those sections with internal cavities, radiation is also considered. This is a good feature of the program when hollow-core concrete slabs are to be modeled [320]. In addition, various fire curves can be defined for the thermal analysis –standard fire curves, natural fire curves or localized fire curves. While some of the well-known fire curves, such as ISO834 and ASTM E119, have already been embedded into the program, it is possible to introduce new fire curves as required. If a fully developed fire is to be introduced, the cooling phase can also be considered, which is a useful capability of the program.

The elastic torsional analysis may then be performed for three-dimensional structures, with beam elements subjected to warping at ambient temperature. The analysis is performed based on the geometry of the members, while any cracking in the concrete caused by the structural loading is ignored. Hence, the torsional results are more conservative than reality. On the other hand, the elastic torsional stiffness is related to the modulus of elasticity and Poisson's ratio. As both of these characteristics decrease as the temperature increases, the elastic torsional stiffness also decreases. Nevertheless, the reduction is normally ignored, mostly to reduce the complexity of the problem [247].

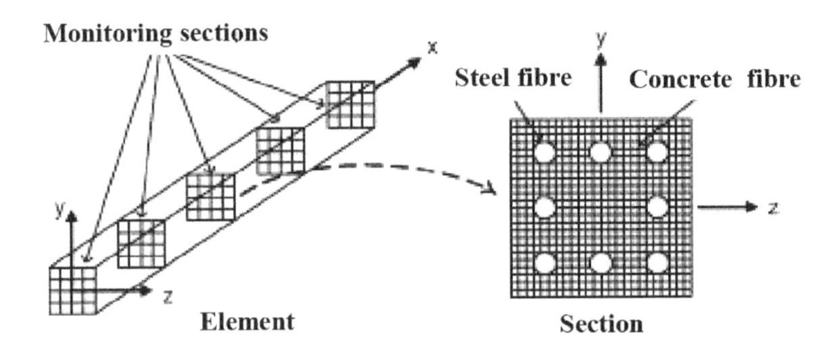

Fig. 6.1: Element and section discretization [321].

As all results in SAFIR are provided in time-temperature fashion, the torsional results are thus enclosed in the time-temperature profile acquired from the performed thermal analysis. The results are consequently used for the three-dimensional structural analysis.

Finally, the structural analysis is performed in either two or three-dimensional fashion, in order to determine the structural performance subjected to the applied loads (gravity, lateral, and thermal loads). For this, the generated time-temperature profile resulting from the thermal and/or torsional analysis with consideration of the nonlinear temperature-dependent mechanical properties is utilized in the structural analysis. Then, using an iterative technique based on the Newton-Raphson method, the equilibria between external loads and internal stress at each time step are computed, as shown in Fig. 6.2. The iteration will run as long as convergence is achieved. Consequently, the required information, such as displacements at the nodes, axial and bending moment in the elements, stress, strain and tangent modulus of the fibers, can be generated at every time step. The following assumptions are made during the modeling:

- Based on Bernoulli theory, plane sections remain plane, even after bending. Hence, a linear distribution of strain throughout the cross-section of the concrete sections can be assumed [322].
- For the elements that have experienced plastic deformations under compression, the unloading path is linear, from the maximum compressive strain in the loading curve parallel to the tangent at the origin [323].

- Plastification is considered only in the longitudinal direction of the beam elements.
- In the beam elements and three-dimensional structures, non-uniform torsion is considered.
- In the beam elements with steel fibers, local buckling is not considered.

It is worth mentioning that since SAFIR employs a fiber model, spread of plasticity can be considered appropriately. Indeed, unlike the lumped plasticity model, in the fiber element model, the plasticity is spatially distributed both in cross-section and along the member. Another important point to mention is that SAFIR uses the arc-length technique, which ensures that local failure of a single element does not result in the overall failure of the structure. Thus, the failure of the structure in fire becomes a global failure, which is more realistic [320]. In addition, as co-rotational procedure is used to solve the large displacements, the swift increase in the displacement can then be tolerated [324].

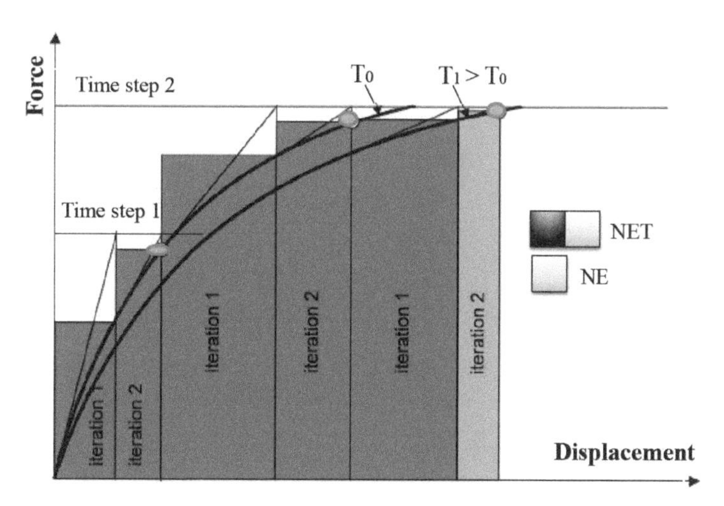

NE: The norm of the energy computed at each iteration
NET: The norm of the total energy computed as the summation of all the NEs

Fig. 6.2: Convergence iterations for the SAFIR structural analysis [325].

6.3 Nonlinear sequential analysis

As mentioned earlier, investigating PEF responses involves modeling a structure's response under gravity loads, seismic loads, and thermal loads. The states of the structure (displacements and stress states) under gravity and seismic loads have to be transferred for the fire modeling. Sequential analysis is a nonlinear process in terms of both earthquake and fire, and thus it can provide a good estimate for the response of a structure under multi-hazard loading. Figure 6.3 shows the schematic stages of nonlinear sequential analysis. As shown, the first stage of loading is the application of gravity loads, consisting of dead and live loads, which are assumed to be static and uniform throughout the analysis. Using a time-load fashion, the gravity loads are applied to the structure from $t = 0$, when the analysis initiates, until the analysis terminates ($t = \infty$).

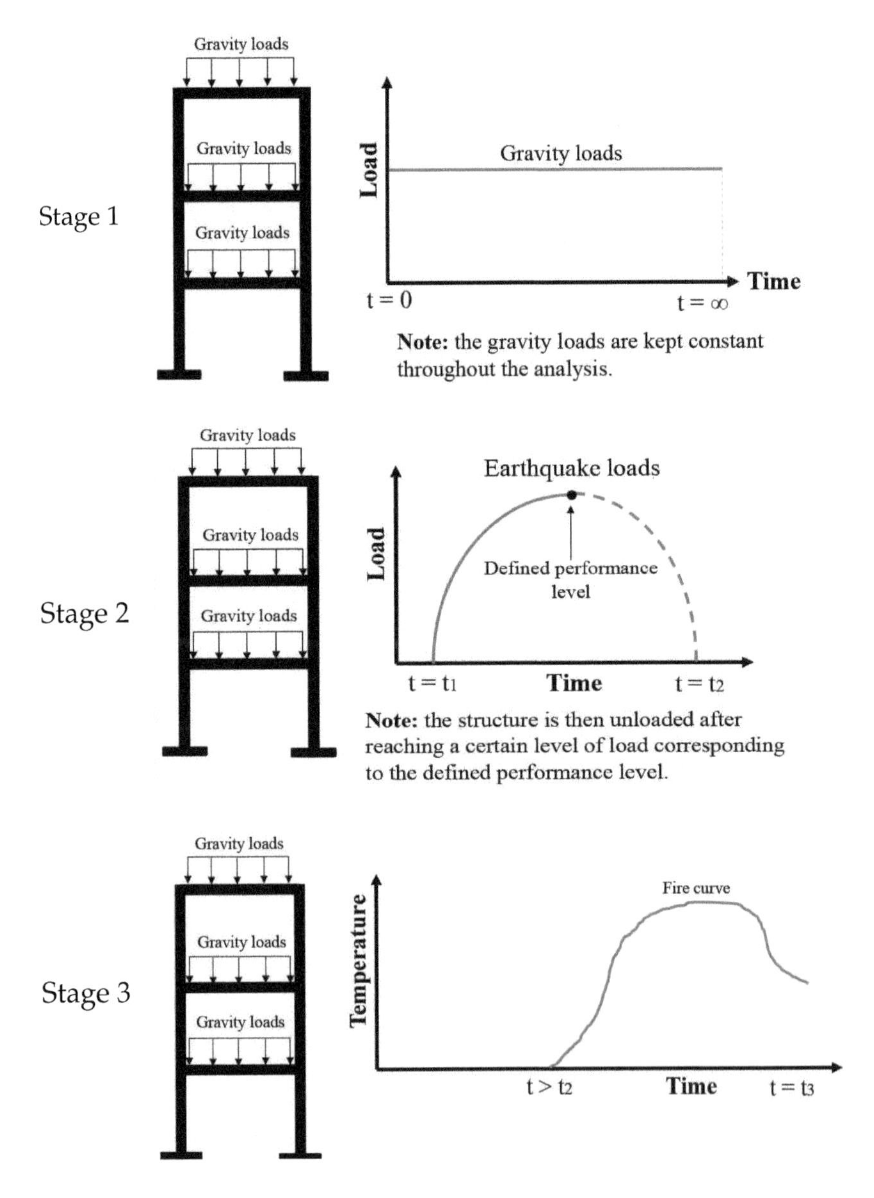

Fig. 6.3: Stages of the sequential analysis.

Simulation of the seismic loads is then performed using a pseudo earthquake load, which can be applied in a dynamic nonlinear fashion, or in a static nonlinear (pushover) style (the merits and limits of these methods were discussed earlier). It is evident that when pushover analysis is used, no dynamic effects are considered. As also shown in Stage 2 of Fig. 6.3, the structure is unloaded after reaching a certain level of load corresponding to the defined performance level, such as IO, LS, or CP. This assumption is in line with the seismic design philosophy, in which the performance level of structures shall not exceed the assumed level when subjected to the design earthquake. Therefore, an assumed structure is pushed to these levels and then unloaded. This stage takes place between t_1 and t_2. Given that long-term effects such as creep and shrinkage are not included, load duration is not important for either gravity or earthquake loads. Finally, the fire loads are applied to the structure (from $t > t_2$ to $t = t_3$), using one of the fire curves explained in Chapter 3. Prior to fire loading, the properties of the structures are set to the reference temperature, but evidently, during fire, the mechanical properties will vary with temperature.

When performing the structural analysis, including gravity and earthquake loads, almost any finite element program that is capable of handling linear and nonlinear seismic analysis can be employed. Nevertheless, the more appropriate programs would be those that are compatible with the most well-established seismic codes (such as FEMA 356), in such a way that the formation and state of hinges in the seismic analysis can be found throughout the wider analysis (because the seismic acceptance criteria can be controlled much more easily). The case studies explained hereafter have been seismically analyzed using SAP2000 [326], a well-known finite-element-based program, which uses both lumped (concentrated) plasticity and spread plasticity models. For models created using concentrated plasticity, the hinges are defined at the expected locations of plasticity spread, and based on the idealized curvature distribution along the structural member. The results of the seismic analysis, including the lateral forces and the state of the created plastic hinges, can therefore be stored in a file. On the other hand, although SAFIR is a finite-element tool specifically written for fire analysis, and considering some of its limitations, it can potentially be employed for performing an analysis that includes gravity, lateral and thermal loads. Nevertheless, the program has no compatibility with the well-established seismic codes, such as FEMA356. Therefore, no acceptance or failure criteria can be defined for the structure, if a seismic analysis is to be performed. Moreover, as cross-sections in SAFIR are discretized into fibers, the distributed type plasticity is considered in the analysis, which might make it different to SAP2000 when the lumped plasticity is used.

To ensure that the possible differences can be considered, a comparison is made here between the results of a three-story RC model seismically analyzed, once by SAP2000 (using lumped plasticity), and once by SAFIR (using spread plasticity). The model's details and characteristics are shown in Fig. 6.4. The structure is first created in the SAP2000 environment using the lumped plastic hinges, and then pushed using a pushover fashion to arrive at 2% drift (which corresponds to the LS level of performance according to the FEMA356 code). An identical model is also created in the SAFIR environment, which inherently uses distributed plasticity. The model created in SAFIR is then pushed monotonically to arrive at 2% drift, in order to provide an identical condition to the model created in the SAP2000. Results of the pushover analysis from both programs are shown in Fig. 6.5, showing that they are broadly similar to each other.

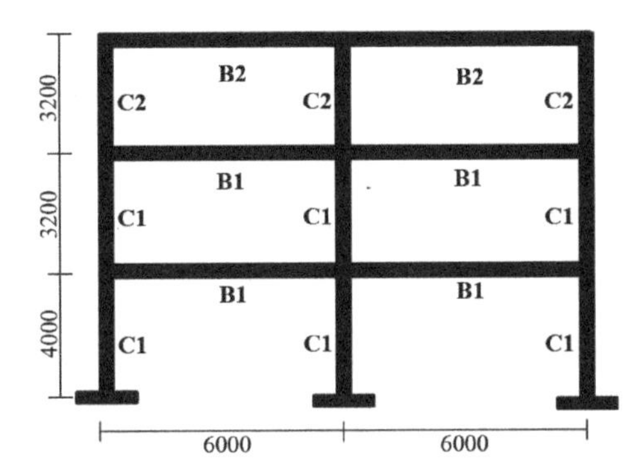

B1: 400×400- 5T20 (top) and 4T20 (bot) ,

B2: 400×400- 3T20 (top) and 3T20 (bot)

C1: 400×400- 12T20 , C2: 400×400- 8T20

Fig. 6.4: Geometric properties of the model.

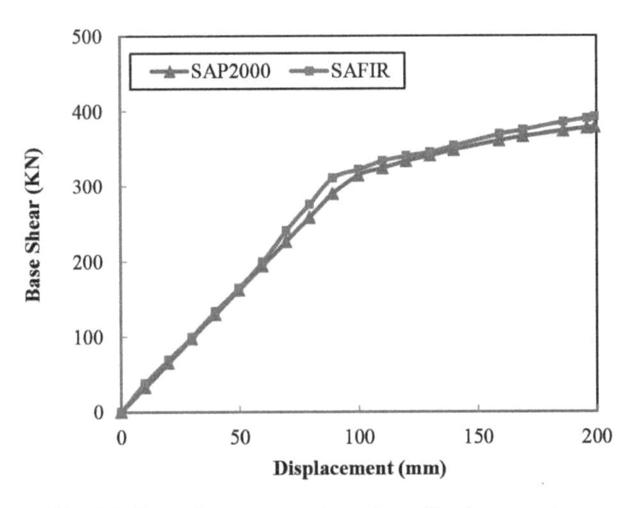

Fig. 6.5: Base shear versus top story displacement.

In addition to the above, another control is also performed for the plastic hinge states. To do that, the plastic hinges' locations and states in SAP2000 are compared with those in SAFIR. Using Equation 5.6 (introduced in Chapter 5), the plastic hinge length can be determined. Figure 6.6a shows the plastic hinge locations and states after the pushover analysis has been performed by SAP2000, and Fig. 6.6b shows the results from the same procedure when performed by SAFIR. As seen, the hinge states in both analyses are roughly identical. In addition, the plastic hinge length as per the SAFIR analysis is close to that from Equation 5.6, showing a good conformity. This reveals that, in SAFIR, only those parts of the elements located adjacent to the joints sustained plastic deformations, while the rest remained elastic throughout the loading history. The locations matched those of the lumped plasticity under SAP2000. The results of the comparisons show that although the two programs use different algorithms, the results of the analyses are not necessarily different.

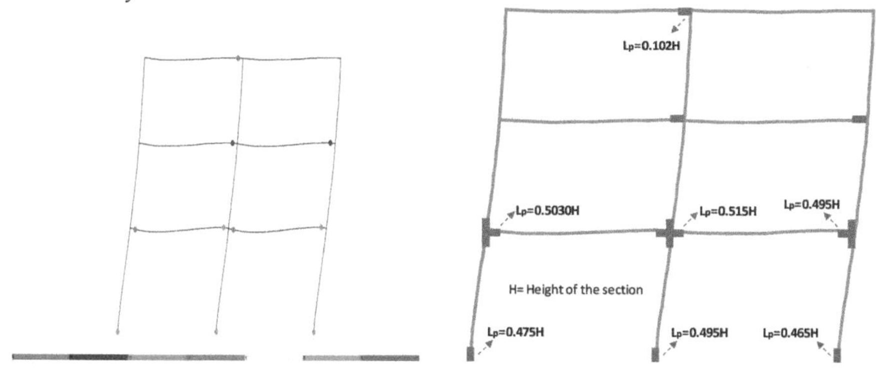

(a) Hinge states resulting from SAP2000

(b) Hinge states and lengths resulting from SAFIR

Fig. 6.6: Comparison between the hinge states and locations from SAP2000 and SAFIR [327, 328].

To perform the sequential analysis, as pointed out, the first step is to store the results of the performed seismic analysis performed by SAP2000, and then to import those results into SAFIR. For doing this, a function has to be developed as a linkage between SAP2000 and SAFIR, which means there is a need to perform some programming. The state of the created plastic hinges can also be known when the seismic analysis is performed using SAP2000. While for bare steel structures, extracting only the pushover forces resulting from SAP2000 and casting them into SAFIR is adequate for the subsequent fire analysis, the states of the potential hinges becomes important in fireproofed-steel and RC structures. In bare steel structures, there is no difference between the fire frontiers of cross-sections before or after earthquake, as shown in Fig. 6.7. In fireproofed-steel and RC structures, however, the fire frontiers of the cross-sections may vary before or after earthquake, which, in turn, may speed up degradation of the strength and stiffness of the members. The vulnerability of fireproofing materials under seismic loading in steel structures was discussed in previous chapters, and this information can be used when performing the sequential analysis of fireproofed steel structures under PEF loading.

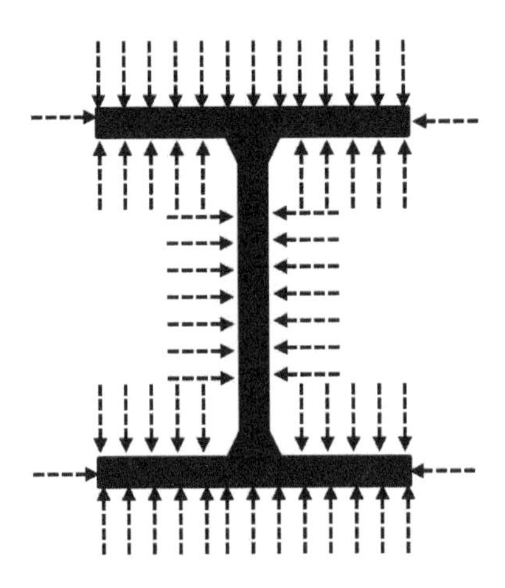

Note: the arrows show fire frontiers

Fig. 6.7: Schematically applied fire frontiers on a steel cross-section, before and after earthquake.

For RC structures, there have been a number of investigations to find out the extent of the damage while they are under loading. The well-known seismic code FEMA356 has some prescriptive definitions of damage states, where various levels of structural performance are discussed, as explained in previous chapters. These definitions can provide an estimate for how the fire frontiers should be applied to the cross-sections' surfaces. There has also been some research conducted to assess the behavior of intact and damaged RC structures exposed to fire, in order to show how the temperature penetration inside the members is affected by any previous damage. Beeby and Scott [329] performed a numerical investigation into the behavior of an RC specimen subjected to pure tension, in order to formulate and predict the crack width and its development under loading. They then carried out an experimental and theoretical study into the behavior of RC members in relation to tensile stresses inside the reinforcement bars [330].

A similar study was performed by Kong *et al.* [331] to clarify the different cracking phases of RC structures in the vicinity of tension zones. Assuming ductile behavior of RC elements, a typical moment-curvature relation is idealized to separate stages [307]. The elastic behavior of concrete is mainly limited to the pre-cracking stage. After the cracking of concrete, and upon increased loading, the number of cracks increases, while the crack width reduces until the tensile stress in the concrete reaches tensile strength. At this stage, no further cracks appear, and the tensile forces are carried by the rebars. This stage continues until concrete crushing occurs. Based on the above explanation and assuming both very small crack widths in concrete and the elastic behavior of the reinforcement, Kong [332] proposed an empirical equation to find out the heat transfer length. Nevertheless, there were some deficiencies with the equation, as the rebars' plasticity was not considered.

Shi *et al.* [333] investigated different RC flexural specimens under loading to find the influence of the cover thickness on the behavior of the specimen when exposed to fire. While the top side of the specimens was not subjected to fire, the results showed that the cover at the bottom has an important role on the ultimate loading capacity. No influence, however, was observed from increasing the cover thickness.

Another study was conducted by Vejmelkov *et al.* [334] to find the effects of cracks on the thermal characteristics of concrete when exposed to fire. Their investigation showed that the conductivity of cracked members decreases as the crack width increases. This was attributed to the assumption that the air within the cracks acts as an insulation, which hampers the quick development of heat. However, no information was provided in Vejmelkov *et al.*'s work about the crack dimensions, and therefore no relationship could be found between the crack width and the conductivity.

An experimental study was performed by Ervine *et al.* [335] on an RC element subjected to static loads followed by fire loads. Applying two concentrated vertical loads to the specimen and recording the subsequent deflection, nucleation and propagation of cracks were observed through the element. The model was then subjected to fire loads (simulated by a panel heater) to find the effect of cracking on the heat propagation inside the section. The results showed that minor tensile cracking would not considerably change the heat penetration inside the section. They concluded that the fire resistances of intact specimens and of the specimens with minor damage (crack width up to 1 mm at the surface of the specimen and 0.5 mm at the rebar level) were roughly identical. However, exposing the rebars directly to fire, i.e. in the case of cover removal or with values of crack widths larger than 1 mm, considerably changes both the thermal and the structural behavior of the specimens.

The effect of cracking on the temperature distributions inside the RC components was also investigated by Wu *et al.* [336]. They considered three different parameters: the width, projected length, and inclination angle of the crack. Through both experiment and numerical investigations, they found that cracks of limited width would hamper the heat transfer in the concrete, repeating the results of Vejmelkov *et al.*'s work, pointed out above. They also found that cracking with widths less than 3 mm and projected lengths of less than 100 mm had almost no influence on the temperature distribution inside the cross-section. However, as the projected length increased to 160 mm, some differences were observed. On the other hand, when the cracks were inclined, the heat penetration inside the cross-section was slightly greater than for perpendicular cracks.

This author collected the information from a number of tests conducted on RC joints subjected to cyclic loading at the laboratory of the School of Civil Engineering at the University of Queensland [78]. The damage states of the specimens were monitored at various drifts, corresponding to various performance levels. The arrangements for these tests were explained in Chapter 2 (see Figs. 2.15 and 2.16), where the seismic performance levels were introduced. As shown in Fig. 6.8a, at IO level of performance, the fire frontiers should be applied on the surface of the cross-section, as schematically shown in Fig. 6.8b, meaning that there is almost no difference between the heat propagation in a cross-section before and after earthquake.

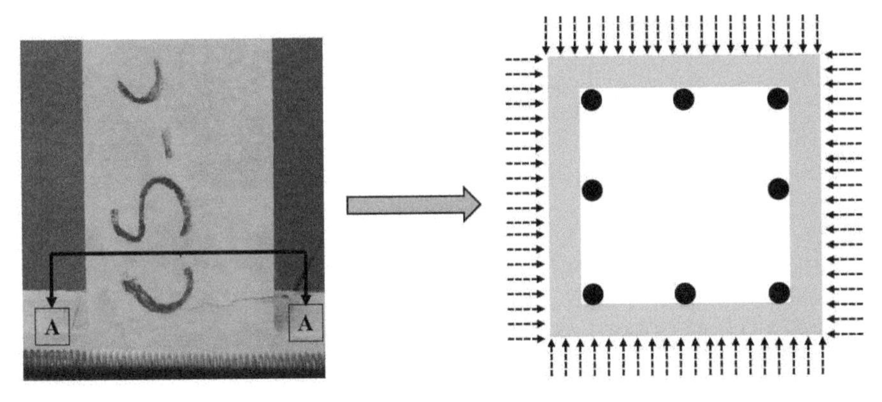

(a) The almost intact member, though some minor cracking has occurred

(b) In section A-A, the fire frontiers are thus applied to the surface of the cross-sections

Fig. 6.8: The damage state at the IO level of performance and the corresponding fire frontiers.

On the other hand, as seen in Fig. 6.9a, the concrete cover at the LS level of performance has been almost spalled off, or has at least incurred serious damage. The length of the spalled-off region was almost 82 mm, which corresponds to the plastic hinge length, and which is greatly similar to Park and Paulay's equation as pointed out earlier. The concrete cover for the tests was 25 mm. It is therefore reasonable to apply the fire frontiers to the reinforcement rebars directly, as shown in Fig. 6.9b.

Analogously, at the CP level of performance, when the member has sustained much more damage compared to the LS performance level, along with the spalling of cover, a part of the concrete core (about 28 mm) has also spalled off, with a length of about 89 mm as shown in Fig. 6.10a. This means that the fire frontiers should be applied to the concrete core directly, as shown in Fig. 6.10b.

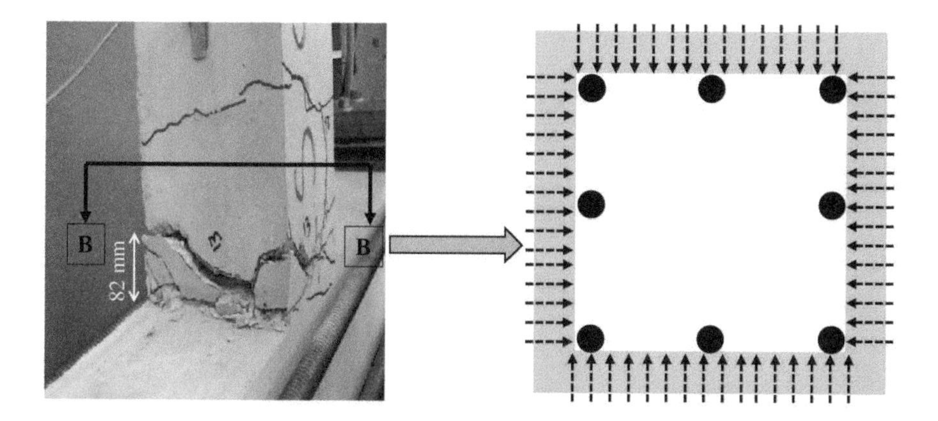

(a) The cover has been extensively damaged or spalled

(b) In section B-B, the fire frontiers are thus applied to the rebars directly

Fig. 6.9: The damage state at LS level of performance and the corresponding fire frontiers.

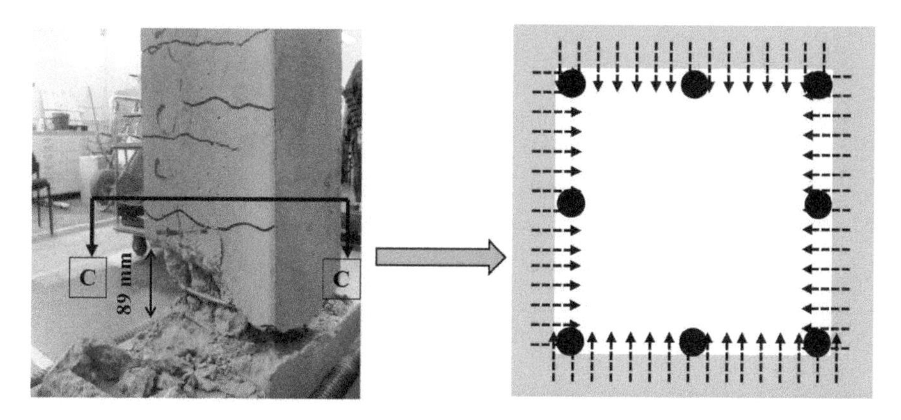

(a) The cover and some parts of the concrete core have sustained extensive damage

(b) In section C-C, the fire frontiers are thus applied to the rebars and the concrete core directly

Fig. 6.10: The damage state at CP level of performance and the corresponding fire frontiers.

To provide an understanding of the effects of the damage state to the temperature distribution, thermal sectional analyses are performed on the test discussed above. To do that, a generalized exponential fire curve (introduced by Equation 6.3) is used, where T_0 is the ambient temperature, T_{max} is the maximum temperature and α is an optional 'rate of heating'. Assuming an ambient temperature of 20°C, a maximum temperature of 900°C, a rate of heating of 0.01 and a fire duration of 60 minutes, the fire curve is applied to the cross-sections at various performance levels, i.e. IO, LS, and CP.

$$T(t) = T_0 + (T_{max} - T_0)\,(1 - e^{-\alpha t}) \tag{6.3}$$

For the thermal analysis, it is assumed that the concrete's moisture level is 40 kg/m^3 and that the emissivity is 0.7. Moreover, the thermal expansion coefficient of rebar and concrete are assumed to be 12×10^{-6}/°C and 10×10^{-6}/°C, respectively. Poisson's ratio of 0.2 is considered for the concrete. As it is believed that the damage sustained at IO level of performance has almost no significant effect on the thermal propagation inside the cross-section, the results of the thermal analysis are assumed to be similar to those for the intact cross-section, i.e. prior to the seismic loading.

Figure 6.11 shows the temperature distribution at the IO performance level. As seen, the bottom layer (Node 1) heats up rapidly, such that its temperature reaches around 900°C after one hour. Other layers, however, show a very slow rate of temperature rise. Given that the critical temperature of 593°C is introduced as a preliminary failure criterion for reinforcement, the figure thus shows that no failure has occurred after a one-hour fire exposure.

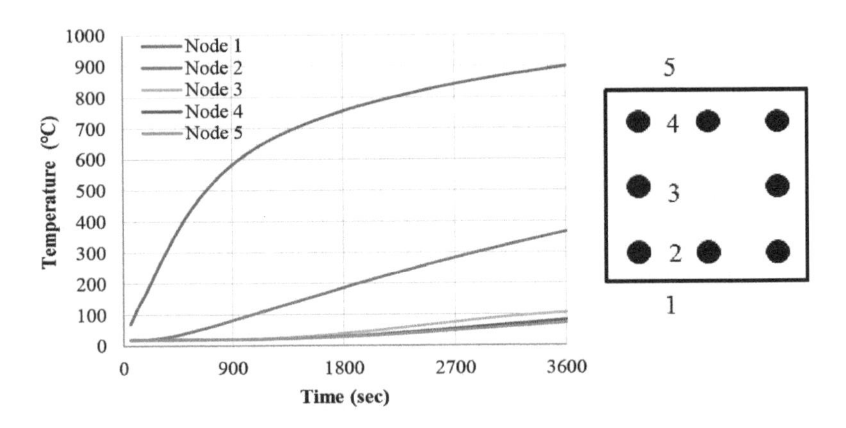

Fig. 6.11: The distribution of temperature inside the column section at the IO performance level.

Figure 6.12 shows the temperature distribution at the LS level of performance. As shown, the fire frontiers are applied to the rebar directly, causing a rapid increase in the temperature of the outer layers. The temperature in the inner layers, however, increases at a slow rate. Using the critical temperature criterion, the thermal analysis shows that the rebar is no longer able to perform as a structural material.

Figure 6.13 shows the temperature distribution at the CP level of performance, when the fire frontiers have been applied to the core directly. The results of the thermal analysis show that all of the layers experience a very rapid rise in temperature, in a way that both rebar and concrete are not able to perform as structural materials, and a very rapid failure should be expected.

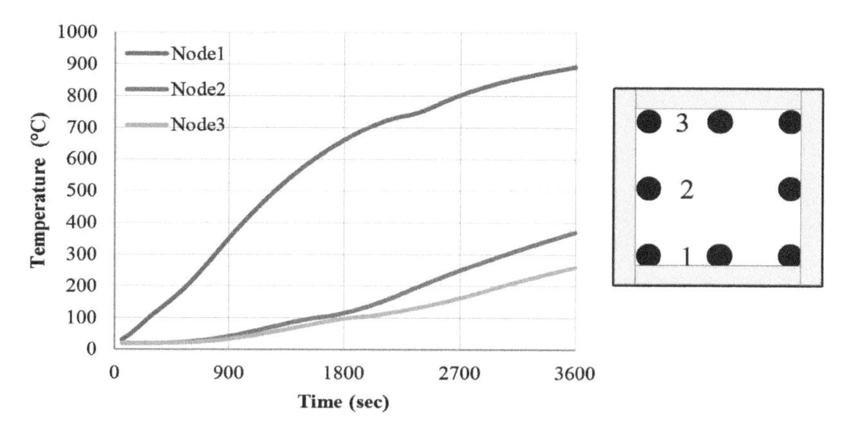

Fig. 6.12: The distribution of temperature inside the column section at the LS performance level.

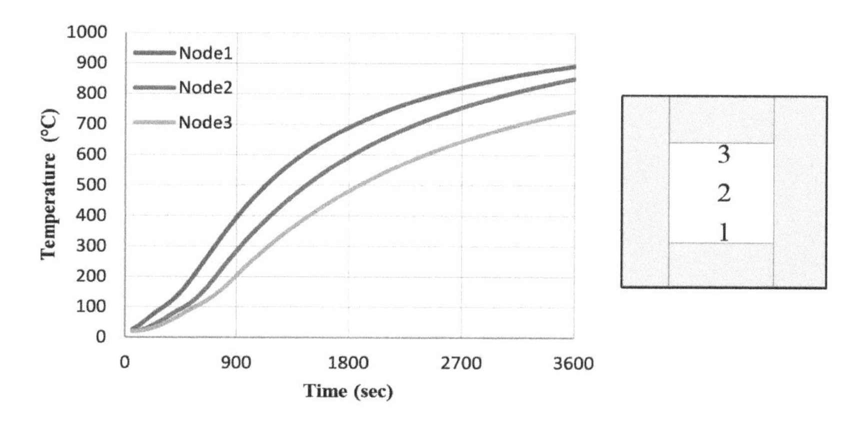

Fig. 6.13: The distribution of temperature inside the column section at the CP performance level.

These explanations can be summarized in a systematic application as shown in Fig. 6.14. As is seen, two functions are written as inputs to the structural fire analysis: a function for the seismic loads; and a function for the physical damage state.

The former function is developed in a way it can provide both the loading and unloading processes of the seismic loads. After the earthquake phase, the structure might sustain some permanent (residual) displacement, dependent on the performance level. This residual displacement thus acts as a starting point for the PEF analysis. The latter function provides a condition through which the sectional analysis can be performed more accurately. This function for RC components is written based on the tests explained in Figs. 6.8 to 6.10. For fireproofed steel components, the results of the analytical

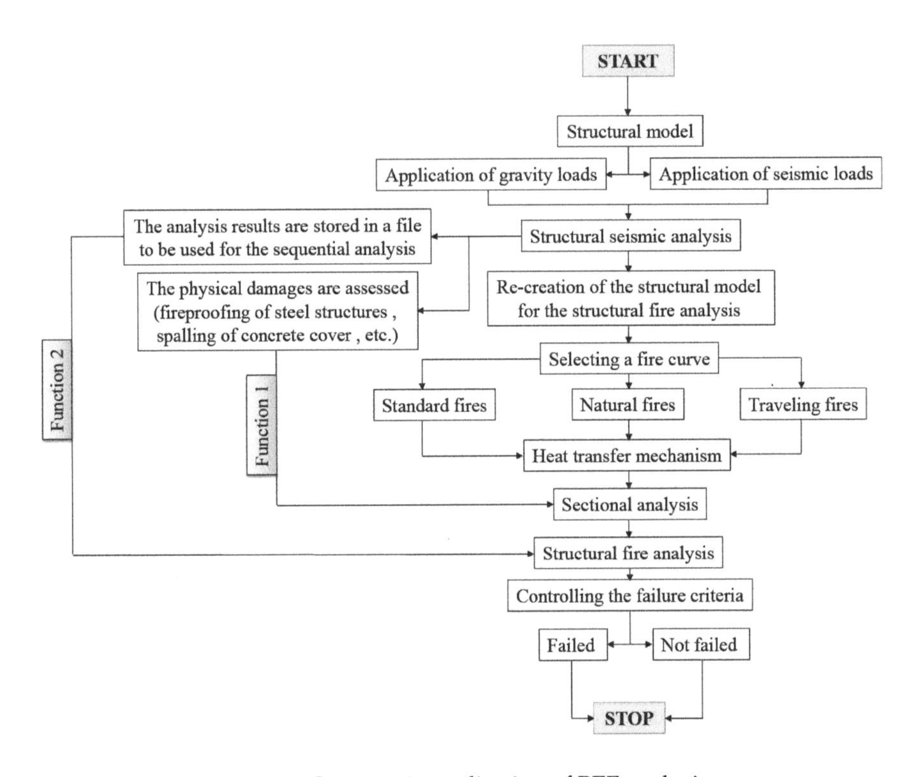

Fig. 6.14: Systematic application of PEF analysis.

investigations performed by the author are used as explained in Chapter 2. As mentioned there, fireproof materials are often not reliable when subjected to earthquake loads. The areas prone to sustaining earthquake damage are mainly concentrated in the plastic hinge length, while the rest of the fireproofed areas remain almost intact. It is evident that when no earthquake damage has been sustained by the components, whether RC or steel, the problem reduces to analyzing the structural model at a normal condition. The fire resistance of the structure is thus determined as for the undamaged structure. It is also worth mentioning that when the natural fire curves are employed, the availability or otherwise of fire fighting facilities established either inside or outside of buildings can also be a factor.

As an example of using the application explained in Fig. 6.14, the fire resistance of the RC joint, tested as described above, is determined. For the structural fire analysis, the mechanical characteristics are also defined. A compressive strength of 28 MPa for the concrete, longitudinal and transverse reinforcing bars with a yield stress of 400 MPa, and a concrete density of 2400 Kg/m^3 are used for the analysis. The constant load shown in the test configuration is maintained throughout the analysis, representing the gravity loading. The analysis is conducted for various performance levels by applying various drifts. The cyclic loading was shown in Fig. 2.15 of

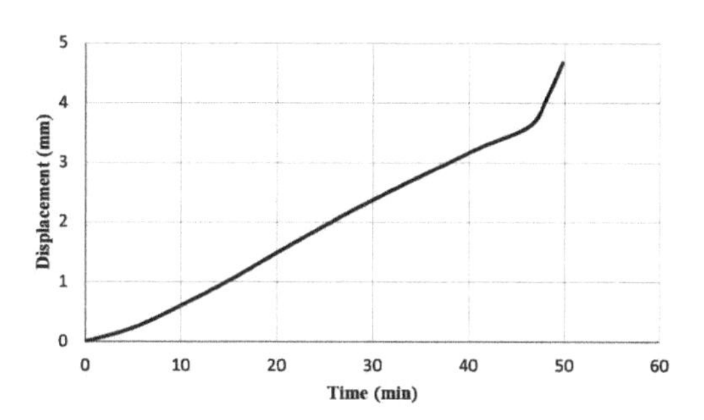

(a) Prior to the application of cyclic loading

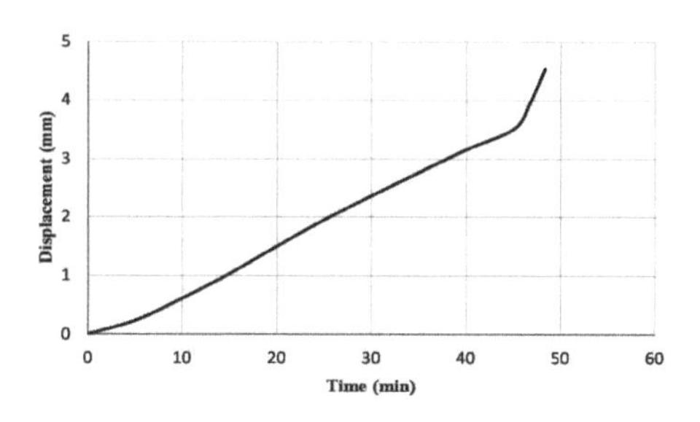

(b) At the IO level of performance

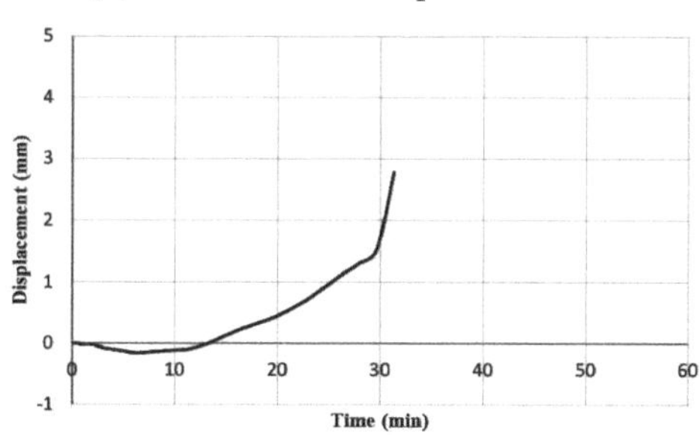

(c) At the LS level of performance

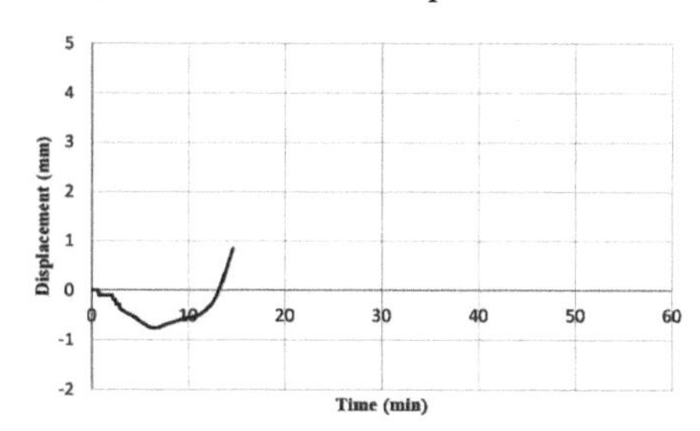

(d) At the CP level of performance

Fig. 6.15: Fire resistance of the specimen for various damage levels [78].

Chapter 2. Using the envelope curve of the cyclic loading, and pushing the specimen to different lateral drifts representing different performance levels, the structural fire analysis is performed. To do so, the SAFIR program is used, where the fire resistance is defined as the time at which the displacements exceed a threshold. The threshold is defined by the curve for displacement versus time step merging toward the vertical asymptote by a 1% error.

Figure 6.15 shows the fire resistance of the specimen before and after applying the cyclic loading. The fire resistance of the specimen prior to applying the lateral loading is almost identical with that at the IO performance level, which is about 48 minutes. At the LS and CP levels of performance, however, the fire resistance decreases considerably, to about 32 minutes and 15 minutes, respectively. This significant reduction can be ascribed to degradation in the strength and stiffness and to the direct exposure to fire of the rebars. The axes of these figures are scaled identically, to provide a better understanding of the analysis results.

Based on the application explained in Fig. 6.14 and the information given above, the behavior of moment-resisting steel and RC structures, including low- and high-rise structures under fire and PEF loading, are investigated in the following sections. Standard fires, parametric fires, and traveling fires are used, in order to provide a better understanding about the structural response under elevated temperature.

6.4 The behavior of RC portal frames under PEF loading

Two seismically designed portal RC frames with fixed supports and differing geometry are monotonically pushed to arrive at different lateral drifts, corresponding to the IO and LS levels of performance. Structural design of the frames is based on the ACI 318-02 code, with moderate ductility, and the design base shear is accounted for with a PGA of 0.3g, representing a high seismic hazard. The soil class D of FEMA356 is assumed to account for the spectral acceleration. Figure 6.16 shows the properties of the designed frames and Fig. 6.17 shows the schematic illustration of the pushover analysis for the mentioned performance levels. The frames are made of normal-strength concrete with a compressive strength of 25 MPa and longitudinal and transverse reinforcing bars with yield stress of 400 MPa. Concrete cover of 40 mm is considered for both beams and columns. The concrete type is assumed to be siliceous, with a density of 2400 kg/m^3. For the thermal analysis, it is assumed that the concrete moisture is 40 kg/m^3. The emissivity is assumed to be 0.7, and the coefficient of convection between concrete and the air is assumed to be 35 W/m^2K. Moreover, the thermal expansion coefficients of rebar and concrete are assumed to be $12\times10^{-6}/°C$ and $10\times 10^{-6}/°C$, respectively. Poisson's ratio of 0.2 is considered for the concrete. As no high-strength concrete is used here, and the concrete cover is not more than 50 mm, the possibility of thermal spalling would be ignored. The frames

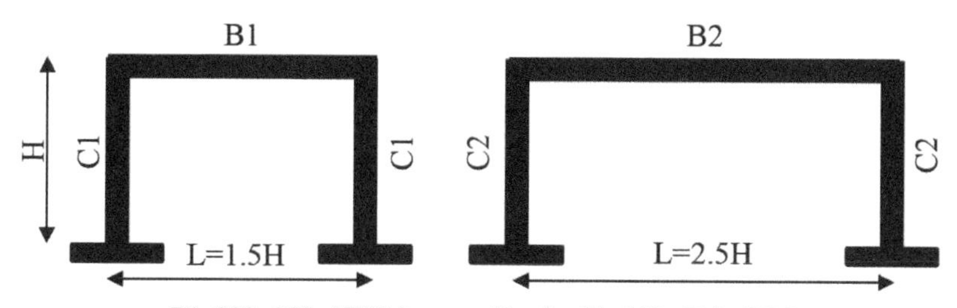

B1: 300×300- 3T20 (top and bot) , C1: 300×300- 8T20
B2: 350×350- 5T20 (top) and 4T20 (bot) , C2: 350×350- 8T20

Fig. 6.16: Geometric properties of the frames, H = 3000 mm.

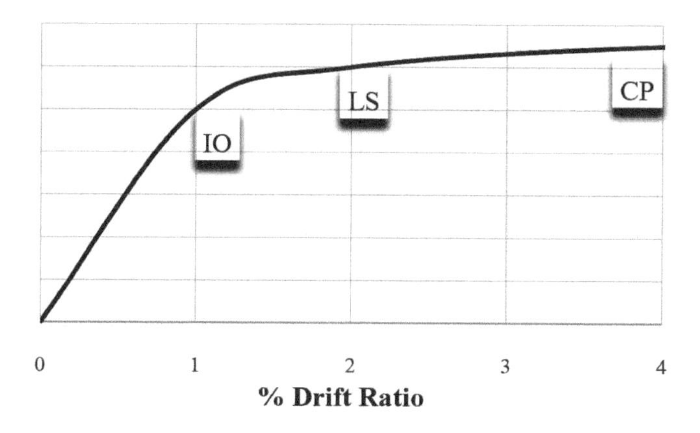

Fig. 6.17: Pushover curve.

are dimensioned for a height (H) of 3000 mm and load combinations of 8.0 kPa for dead load and 2.5 kPa for live load. The combination of 100% dead load and 20% live load is used to find the required mass for calculating the earthquake load [337]. Furthermore, the frames are exposed to the standard ISO834 fire curve in three fire scenarios. In the first scenario, only the beam elements are exposed to fire; in the second scenario, only the columns are exposed to fire; and in the third scenario, both beams and columns are subjected to fire.

In modeling the columns, an imperfection of H/300 is considered over the unfavorable direction. The external side of the columns is not subjected to fire, because it is assumed the fire occurs only inside the frame. The top side of the beam is also not subjected to fire, as it is assumed that it is protected by the concrete slab. In addition to the PEF analysis, the fire-alone analysis is also used prior to the earthquake loading, in order to compare the results and to provide more understanding about the structural behavior under various scenarios. Figure 6.18 shows the distribution of temperature inside the column C1 after four hours of fire exposure and at different damage states. It should be noted that the four-hour fire exposure seems to be unrealistic, but the time is considered here only as an example. As seen from Figs. 6.18a and b, as more damage is sustained by the members, more heat penetrates inside the cross-sections. This in turn causes a rapid reduction in the strength and stiffness of the concrete and rebars, and thus load-bearing capacity decreases. Accepting that the temperature of 593°C in the rebars is a simple criterion for failure [338], the possible failure in members can be detected. This failure criterion, however, has its root in the prescriptive-based approach and can be seen as very superficial. At the IO level of performance and as shown in Fig. 6.18d, the temperature of the rebar number 1 has passed 593°C at 93 minutes. At the LS level of performance, as the rebar number 1 is directly exposed to fire, its temperature increases at a rapid rate, and reaches 593°C at 23 minutes, as shown in Fig. 6.18e. While the maximum temperature of rebar as a failure criterion is simple to use, it cannot appropriately reflect the structural response under fire loading. To address this deficiency, performance-based failure criteria, such as deflection and the rate of deflection, can be used (as discussed previously).

Figure 6.19 shows displacement against time for the frame with L = 1.5H, under the scenarios of beam and columns exposed to fire, but separately. As is seen, there is a correlation between the fire resistance and the performance levels. While the lateral displacement in the frame increases, the fire resistance decreases, such that the fire resistance of the frame under the LS level of performance is much lower than that under the IO level. Fig. 6.19a shows that under the beam-exposure scenario and under the fire-alone or IO level of performance, the frame fails at 148 minutes. However, it fails at 75 minutes when it is pushed to arrive at the LS level of performance. By contrast, Fig. 6.19b shows that the frame collapses at around 238 minutes under the fire-alone and IO scenarios, and it collapses at around 152 minutes under the LS scenario. The figures also show a minor difference between the fire resistance levels at the IO level and fire alone – at the IO level, only minor damage occurs, resulting in insignificant residual displacement and degradation in strength and stiffness. From a different aspect, the fire resistance declines considerably when only the beam is exposed to fire, compared to exposing only the columns to fire. This means the beam is more vulnerable to fire than the columns, the reason for which can be correlated to the 'strong column-weak beam' concept, which is a condition that it is strongly advised to meet for seismically designed structures.

Interestingly, Fig. 6.20a shows that there is a close similarity in the fire resistance when all members are exposed to fire and when only the beam is exposed to fire. This implies that the fire resistance of the frame is mostly dependent on the fire resistance of the beam. In Figs. 6.19 and 6.20, the shape of the failure is shown, and two types of failure are observed: local, and global. The local collapse is correlated to the collapse of beams, and the global collapse is mainly governed by considerable lateral displacement of the columns. It is evident from the figures that the frame fails locally in the

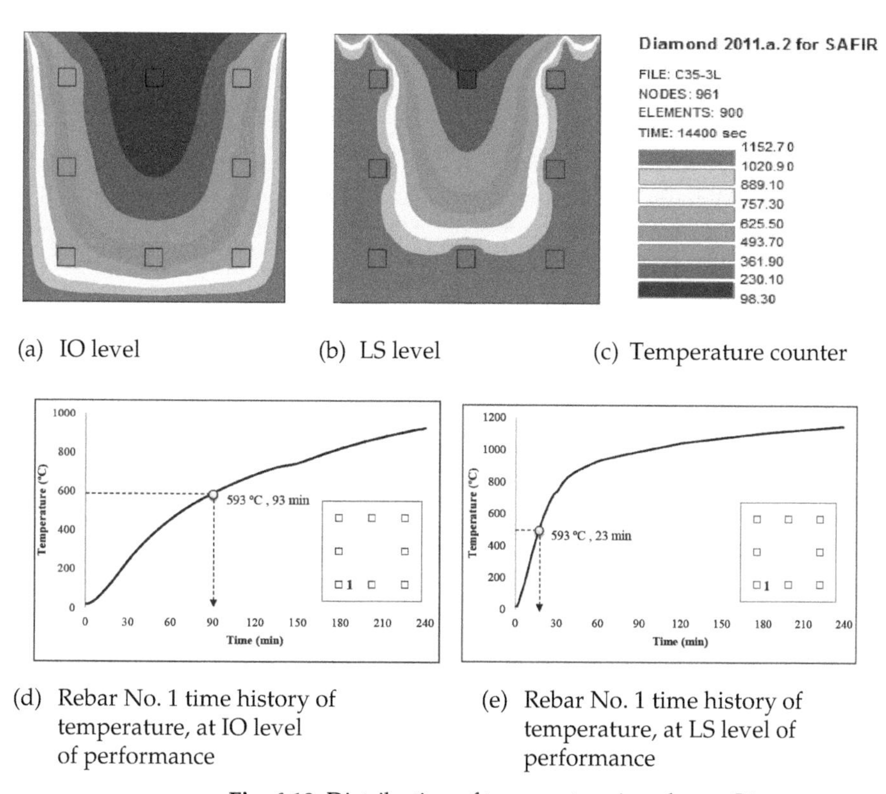

(a) IO level (b) LS level (c) Temperature counter

(d) Rebar No. 1 time history of temperature, at IO level of performance

(e) Rebar No. 1 time history of temperature, at LS level of performance

Fig. 6.18: Distribution of temperature in column C1 after four-hour fire exposure.

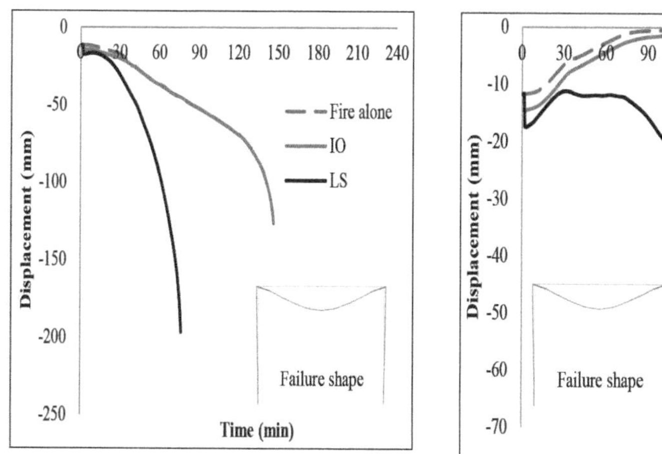

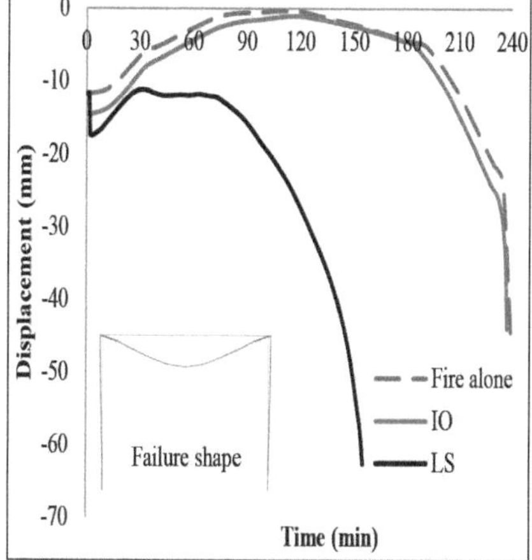

(a) Only the beam exposed to fire (b) Only the columns exposed to fire

Fig. 6.19: Fire resistance of the case "L = 1.5H", members separately exposed to fire [339].

case of fire alone and the IO level of performance, and fails globally at the LS level. The sharp increase and then decrease in Fig. 6.20b represents the displacement resulting from lateral pushover, both loading and unloading.

The above results were achieved based on the threshold defined by the SAFIR analysis, in which the analysis terminates when the curve for displacement versus time step merges toward the vertical asymptote by a 1% error. The results, however, need to be additionally controlled by performance-based failure criteria, such as the mid-span deflection and the rate of deflection. Regarding the deflection criterion, and for Fig. 6.19a, the maximum deflection is limited to $L/20$, where L is the length span in mm. For the case here, $L = 4500$ mm, if the deflection exceeds 225 mm, the frame is considered to have failed. As the maximum deflection is about 200 mm (i.e. lower than 225 mm), the frame has thus not failed under all the performance levels. As for the rate of deflection, it is limited to $L^2/9000d$ (mm/min), where d is the effective depth of beam in mm. For the case here, $d = 250$ mm; thus the maximum rate of deflection allowed is 9 mm/min. Based on this criterion, Fig. 6.19a shows that the frame has failed at 142 minutes for the fire-alone and IO scenarios, and has failed at 62 minutes for the LS scenario. Controlling the frame under the scenarios shown in Fig. 6.19b, confirms that it has not failed under the mid-span deflection criterion. Based on the rate of deflection criterion, however, the frame has failed at 230 minutes in the fire-alone and IO scenarios, and at 130 minutes in the case of the LS scenario. Therefore, there are considerable differences between the fire resistances determined based on different failure criteria. These differences, however, have their source in the performance-based attitude. Hence, the objective of the design dictates which one of the above results should be considered as the criterion for action.

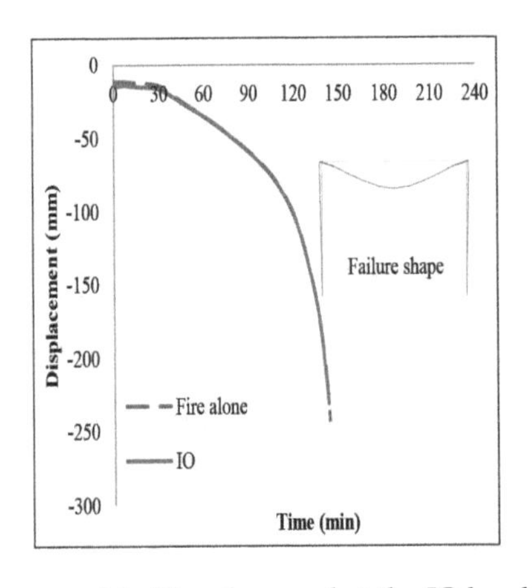

(a) Fire alone and at the IO level

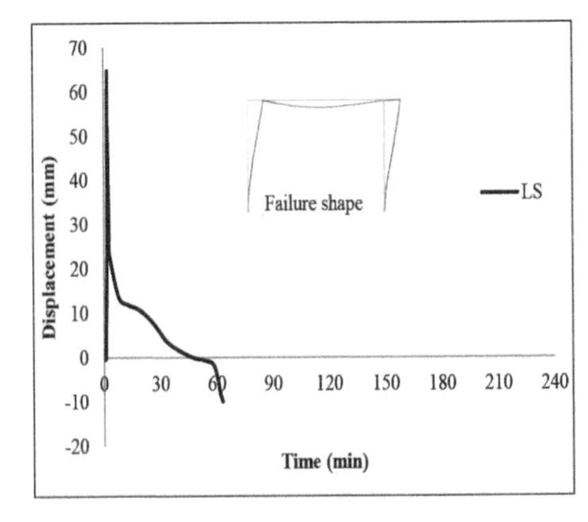

(b) The LS level

Fig. 6.20: Fire resistance of the case "L = 1.5H", all members exposed to fire [339].

Figures 6.21 and 6.22 show the fire resistance of the frame with L = 2.5H, for different fire scenarios. Similar to the frame with the L = 1.5H configuration, as the frame is pushed more from the IO to LS level of performance, the fire resistance of the frame reduces. As seen, there is a similarity between the fire resistance when all members are exposed to fire and that when only the beam is exposed to fire, which indicates that the fire resistance of the frame depends mostly on the fire resistance of the beam. Fig. 6.21a shows that, based on the threshold defined by the SAFIR analysis here, the frame has failed at 230 minutes under the fire-alone and IO scenarios, but it has failed at 118 minutes under the LS scenario. As a specific note with regard to Fig. 6.21b, it is understood that no failure is observed in the fire alone and IO scenarios, but the frame fails at around 225 minutes under the LS scenario when the threshold defined by the SAFIR analysis is considered. On the other hand, when the frame is controlled under other performance-based failure criteria, the results might differ from those found by the criterion defined above. Based on the mid-span deflection criterion and given that L = 7500 mm, the deflection allowed is 375 mm, and as d = 300 mm, the rate of deflection allowed is 20.8 mm/min. Based on the mid-span deflection, Fig. 6.21a confirms that the frame has not failed in the fire-alone and IO scenarios, but it has failed at 100 minutes in the LS scenario. If the rate of deflection is considered, the figure shows that even in the fire-alone and IO scenarios, the frame has failed at 225 minutes, and for the LS scenario, it has failed at 98 minutes. Figure 6.21b shows that no failure has occurred when the mid-span deflection limit is considered. The frame is, however, considered to have failed when the rate of deflection is considered, but only for the LS scenario, which occurs at 215 minutes. Besides, as is seen from Fig. 6.22b, the frame

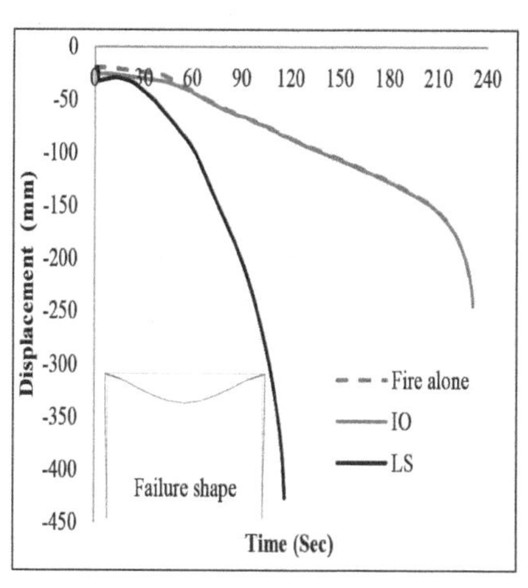

(a) Only the beam exposed to fire

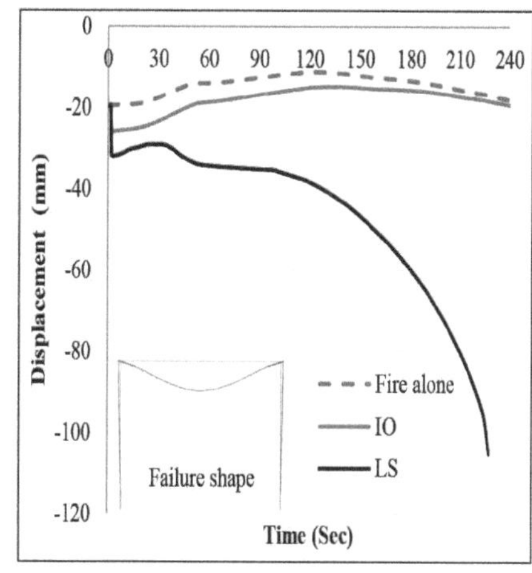

(b) Only the columns exposed to fire

Fig. 6.21: Fire resistance of the case "L = 2.5H", members separately exposed to fire [339].

fails globally at the LS level, but locally at other performance levels. As a note, it is mentionable that while the fire resistance of the frame with $L = 2.5H$ is greater than that of the frame with $L = 1.5H$, the PEF resistances of both frames are closely similar [339]. This shows the significant effect of seismic damage on the post-earthquake behavior of RC portal frames.

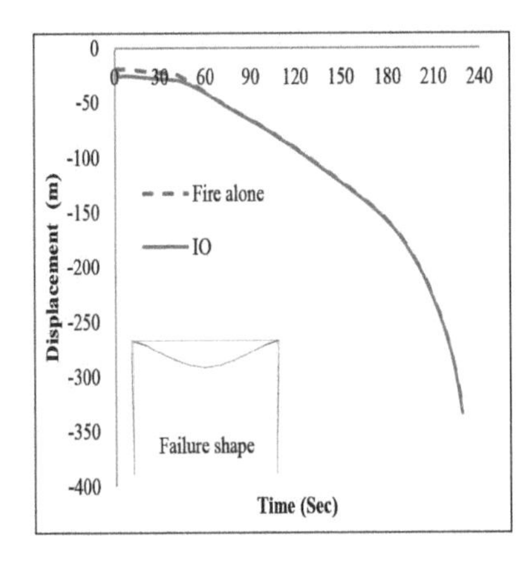

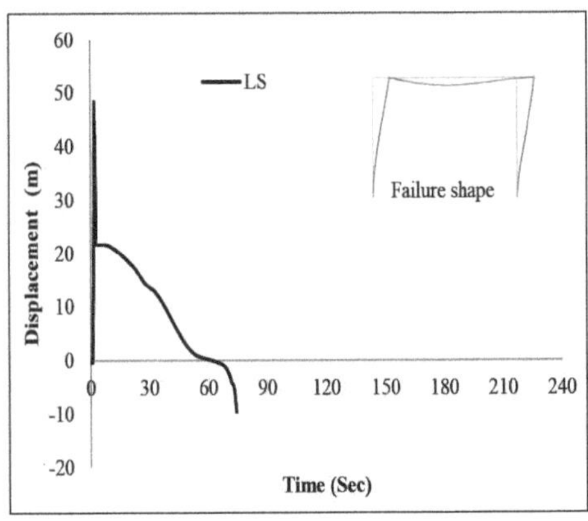

(a) Fire alone and IO level (b) LS level

Fig. 6.22: Fire resistance of the case "L = 2.5H", all the members exposed to fire [339].

6.5 The behavior of multi-story RC structures under natural fires

The example explained in Chapter 3 for the parametric fire curve is used here to investigate the behavior of multi-story RC structures when faced with different fire scenarios, before and after earthquake. Figure 6.23 shows the plan view of the structure. The fire curves were plotted in Chapter 3, Fig. 3.5. The structure was assumed appropriate for two different occupancies: an educational facility, corresponding to the IO level of performance; and a residential facility, corresponding to the LS level of performance. The structure is also intended to work for two different performance levels: IO level and LS level. The structure is designed for a PGA of 0.3g and uses normal-strength concrete with a compressive strength of 25 MPa, and longitudinal and transverse reinforcing bars with a yield stress of 400 MPa. The structure is dimensioned for the load combinations of 8.0 kPa for dead load and 2.5 kPa for live load. Combinations of 100% dead load and 20% live load for the designed frame in the LS level of performance, and 40% live load for the designed frame in the IO level of performance are used to find the required mass for calculating the earthquake load.

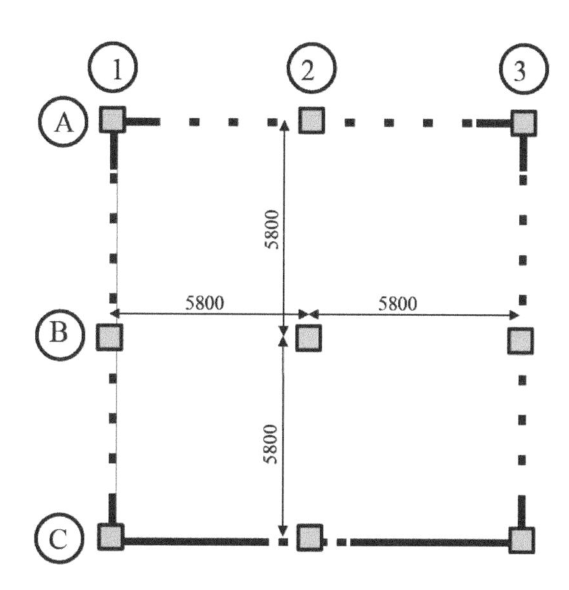

Fig. 6.23: The plan view of the case study.

The structural details are shown in Fig. 6.24, along with details of the fire scenario used. For the thermal analysis, it is assumed that the concrete's moisture content is 2%. Moreover, the thermal expansion coefficients of the rebar and concrete are assumed to be $12\times10^{-6}/°C$ and $10\times10^{-6}/°C$, respectively. Poisson's ratio of 0.2 is considered for the concrete. The concrete cover is 40 mm. No thermal spalling is considered during the fire analysis. As a benchmark, the fire analysis is also performed at the normal situation, i.e. when no earthquake has occurred. To perform the fire and PEF analyses, the middle frame of the structure, Frame B, is selected, meaning that the analysis will be two-dimensional. This gives rise to the question as to whether two-dimensional analysis is adequate. In response, detailed comparisons already performed by researchers such as Usmani [340], Flint [341], Röben *et al.* [342], Quiel, and Garlock [343] have shown that there is a close agreement between the results of two-dimensional and three-dimensional models. Therefore, in the light of previous studies, a two-dimensional model is used here for ease of analysis.

Using the pushover analysis, the structure is pushed to arrive at two different target displacements, which are accounted for using Equation 5-3. The target displacements at the IO and LS performance levels are 162 mm and 175 mm, respectively. Figure 6.25 shows the pushover curves for both performance levels and the response spectrum used for the analysis. Figure 6.26 shows the structure after the pushover analysis and the plastic hinge states for the mentioned performance levels.

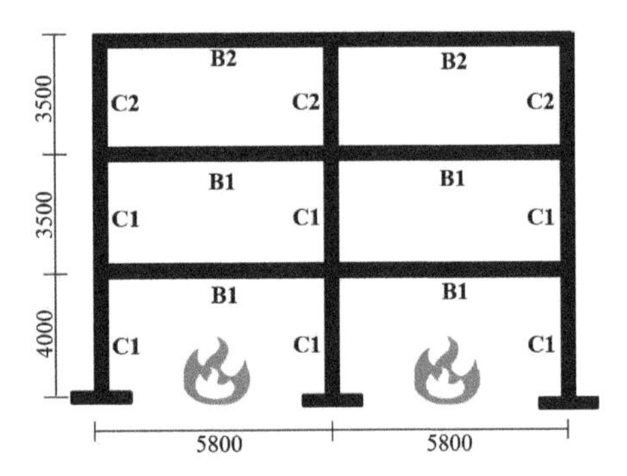

B1: 400×400- 5T20 (top and bot) , C1: 400×400- 12T20
B2: 400×400- 4T20 (top and bot) , C2: 400×400- 8T20

(a) Frame A (IO performance level)

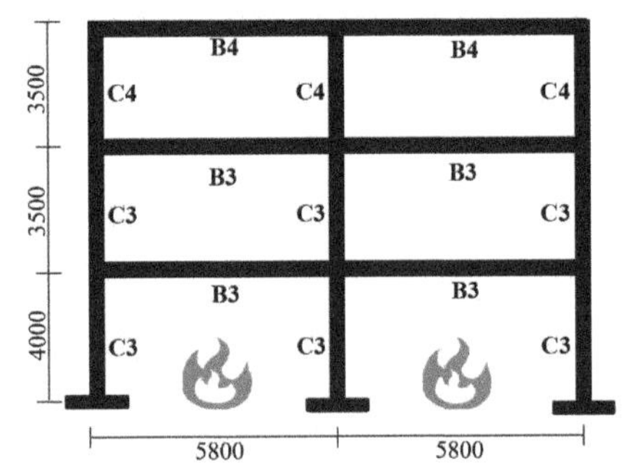

B3: 350×350- 4T20 (top and bot) , C3: 350×350- 12T20
B4: 350×350- 3T20 (top and bot) , C4: 350×350- 8T20

(b) Frame B (LS performance level)

Fig. 6.24: Designed frames and the considered fire scenario [344].

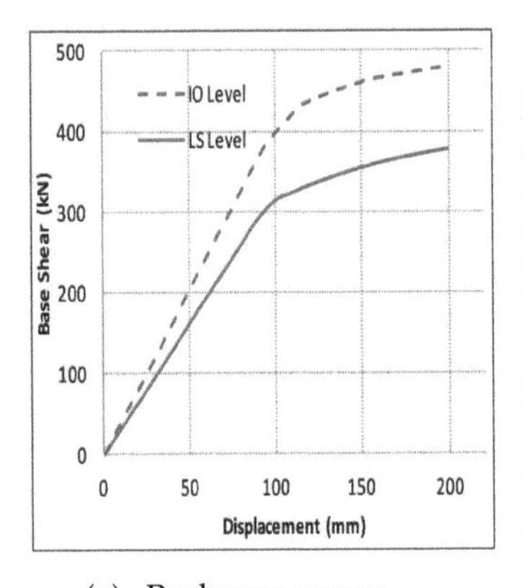

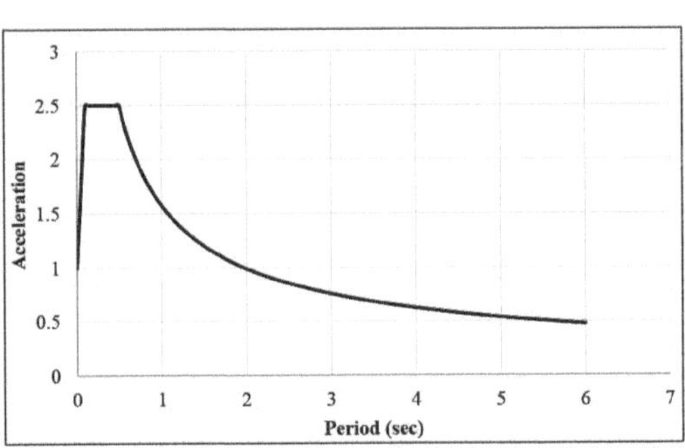

(a) Pushover curves (b) The response spectrum

Fig. 6.25: The pushover curves and the response spectrum.

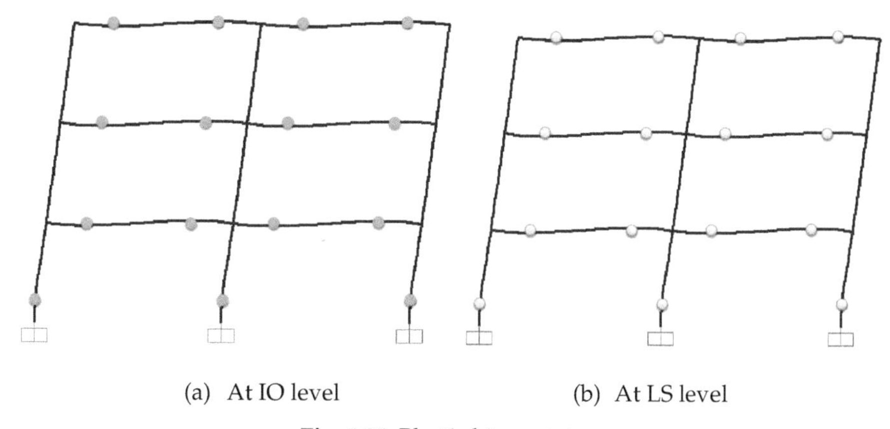

(a) At IO level (b) At LS level

Fig. 6.26: Plastic hinge states.

Using sequential analysis, the PEF resistance of each structure is determined. To do that, the pushover forces are introduced as a function written inside the SAFIR computing environment. The damage states of the structural elements are also defined in such a way that the fire frontiers can be properly applied to the cross-sections. The fire analysis is performed under several scenarios, as shown in Fig. 6.23b. Thus, a comparison can be made between different fire scenarios. The fire analysis is terminated when the displacements exceed the defined thresholds. The thresholds are similar to that used in the case study explained in section 6.4, where the displacements versus time step are merged toward the vertical asymptote by a 1% error. The fire resistance is also accounted for using performance-based failure criteria, i.e. mid-span deflection and the rate of deflection, as previously discussed.

Figure 6.27 shows the reaction versus time for column 2-B at the IO level of performance. As is seen in Fig. 6.27a, when the earthquake loads are yet to be applied, no failure occurs during the analysis. After the application of earthquake loads, however, the structure collapses at around 70 minutes, as shown in Fig. 6.27b. The initial sharp increase and decrease at the beginning of the figure are due to the loading and unloading of the pushover loads. It can be seen that the reaction of column 2-B increases while the thermal loads are applied to the frame. This increase continues to around 30 minutes, when the load-bearing capacity of the column starts to decrease. When the load-bearing capacity becomes lower than the demand value, the column is considered to have failed.

Using the mid-span deflection and rate of deflection criteria, the analysis results are plotted for the structure, as shown in Fig. 6.28. Figure 6.28a represents deflection versus time for the IO level of performance, before the application of earthquake loads. The figure shows that there is no failure over the analysis, confirming that the structure will remain stable for the fire-alone scenario. The PEF analysis is shown in Fig. 6.28b. The structure is

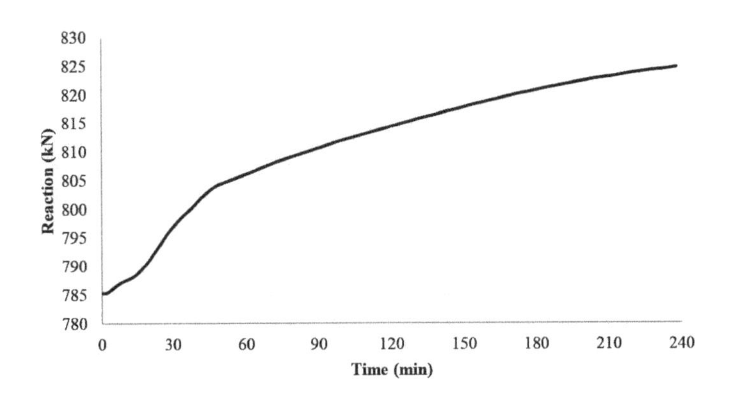

(a) Before earthquake

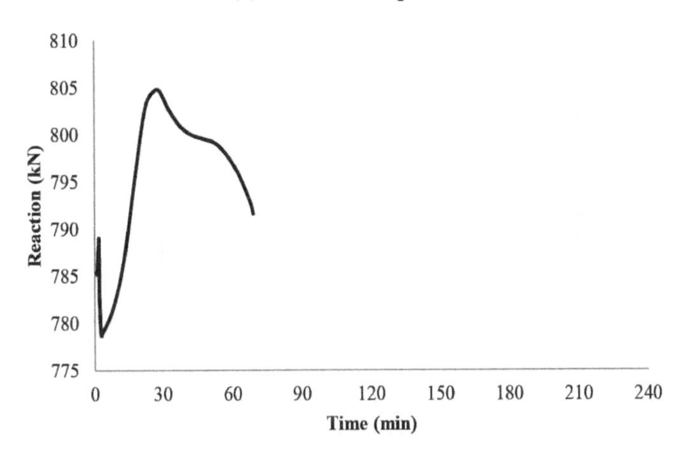

(b) After earthquake

Fig. 6.27: IO level – reaction.

considered to have failed if the mid-span deflection goes beyond 290 mm. The figure, however, shows that no failure has occurred, as the maximum deflection is around 120 mm. However, for the rate of deflection, the figure shows that the structure has failed at around 60 minutes, because the rate of deflection is larger than 10.7 mm/min (given that the effective depth of the beam is considered 350 mm). This means that the fire resistance of the structure has declined considerably after the earthquake. For the case here, this considerable reduction mostly relates to significant increase in the fire loads, and not necessarily to the structural damage. As mentioned earlier, it is expected that at the IO performance level, only minor damage will be sustained by structures. Nonetheless, as there is a high possibility that fire extinguishment facilities would not work properly after an earthquake, the fire load density will thus increase (see Equation 3.20). The presence or

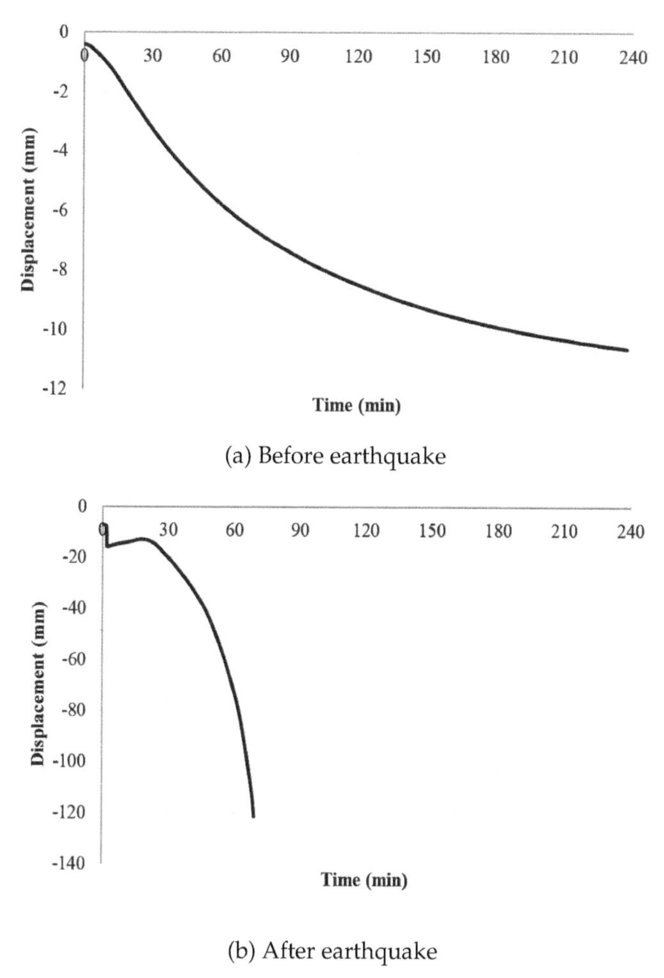

(a) Before earthquake

(b) After earthquake

Fig. 6.28: IO level – displacement.

absence of fire fighting facilities and their role in drawing fire curves were pointed out in Chapter 3 when the parametric fire curves were introduced.

Figure 6.29 shows the fire resistance of the structure designed for the LS performance level. As seen in Fig. 6.29a, no failure occurs at the normal condition, which means the structure will remain stable under application of the fire loads before earthquake. The response of the structure under application of the fire loads after earthquake is shown in Fig. 6.29b. The structure fails at about 45 minutes after application of the PEF loads, showing a considerable decline compared with the fire resistance at the initial condition. The sharp increase and then decrease at the first part of the figure is due to the loading and unloading of the pushover loads prior to application of the fire loads.

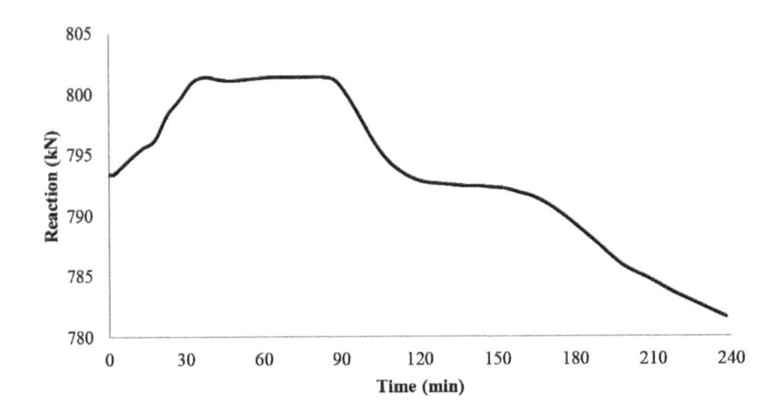

(a) Before earthquake

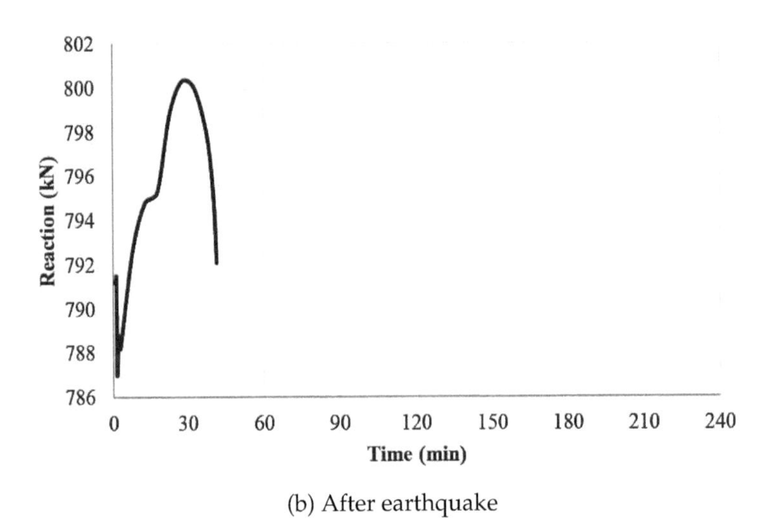

(b) After earthquake

Fig. 6.29: LS level – reaction.

Figure 6.30 shows the evaluation of the structural response under fire and PEF loading, using the mid-span deflection and rate of deflection criteria. As shown in Fig. 6.30a, no failure occurs at the LS performance level based on either of these criteria. The allowed values of these criteria for the case investigated here are 290 mm and 12.45 mm/min, respectively. The effective depth of the beam is considered 300 mm. The mid-span deflection criterion is also satisfied by the structure after the application of PEF loads, as shown in Fig. 6.30b. However, controlling the rate of deflection shows that the structure has failed after around 45 minutes. This rapid decline is strongly linked to the influence of the earthquake damage, equally as important as the increase to the fire loads. The reason for the latter was discussed earlier.

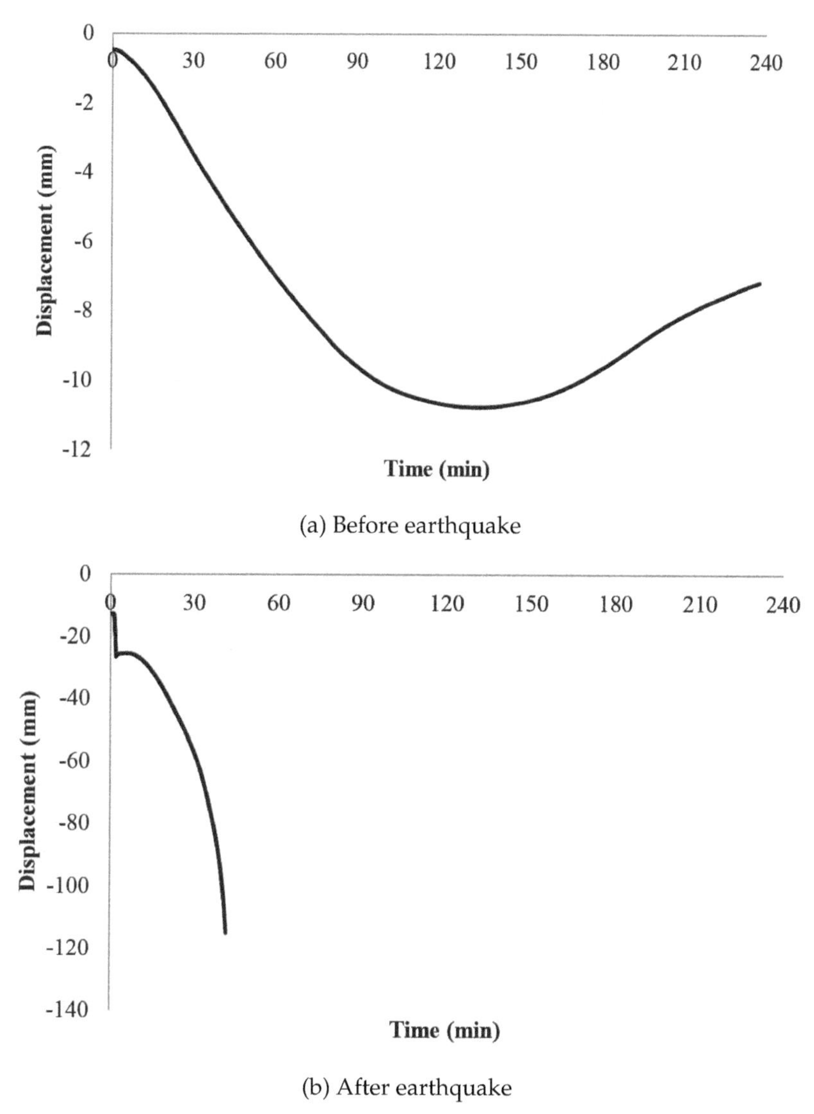

(a) Before earthquake

(b) After earthquake

Fig. 6.30: LS level – displacement.

It is also worth noting that two types of collapse were noted during the analyses: global failure, and local failure. Global collapse is defined as the collapse configuration in which the frame fails because of considerable lateral movement of columns. Local collapse involves mainly failure of the beams. In the studied frames, at both performance levels, i.e. IO and LS, and in the case of PEF, global collapse occurred. However, local collapse occurred when the frame was subjected to fire alone. These failure types are shown in Fig. 6.31.

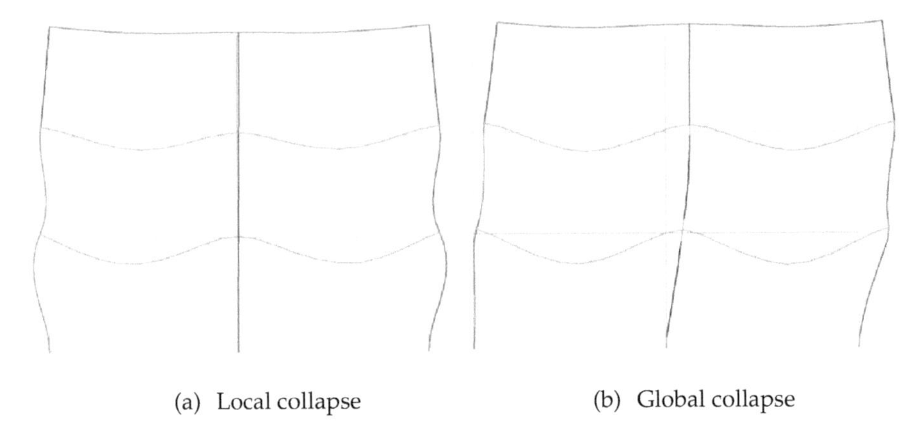

(a) Local collapse (b) Global collapse

Fig. 6.31: The collapse mechanism of the cases studied.

6.6 The effect of vertically traveling fire on tall RC Structures

A tall RC structure (as an office occupancy) is designed for a PGA of 0.30g and the LS level of performance, as shown in Fig. 6.32. The floor, the ceiling and the external walls are made of normal-weight concrete with a 100 mm thickness and standard bricks, respectively. The compressive strength of the concrete is 35 MPa and the yield stress of the longitudinal and transverse reinforcement bars is 400 MPa. The frame is dimensioned for the load combinations of 5.5 kPa for dead load and 3.0 kPa for live load. A combination of 100% dead load and 20% live load for the designed frame is used to find the required mass for calculating the earthquake load. The resulting sections for the designed structure are shown in Fig. 6.32b. Table 6.1 shows the thermal characteristics of the materials in the building. The characteristic fire load density of 511 MJ/m^2 is assumed to account for the thermal actions, which is modified using Equation 3.20 where effective fire fighting measures are involved.

Table 6.1: Thermal characteristics of the considered materials

Description	*Specific heat (J/kg K)*	*Material's density (kg/m³)*	*Thermal conductivity (W/mK)*
Floor and Roof	1000	2400	1.6
External walls	840	1600	0.7

The thermal properties of steel and concrete slab are taken into account using Eurocode 4. For the thermal analysis, it is assumed that the concrete's moisture content is 2.2%. Moreover, the thermal expansion coefficients of

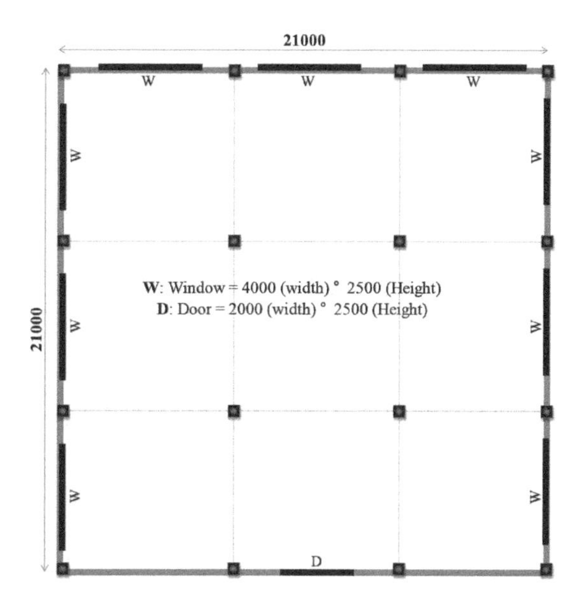

(a) Plan view of the building and the opening

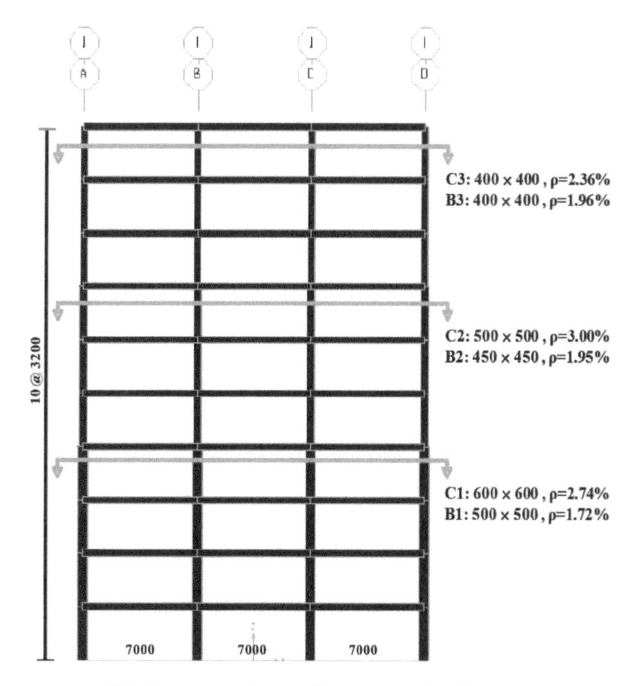

(b) Cross-sections of beams and columns

Fig. 6.32: The case study (dimensions are in mm) [345].

rebar and concrete are assumed to be $12\times10^{-6}/°C$ and $10\times10^{-6}/°C$, respectively. Poisson's ratio of 0.2 is considered for the concrete. It is worth noting that the specific heat of concrete depends on its percentage of water content. Experiments have shown that between 100°C and 200°C, which corresponds to the temperature needed for water evaporation, the peak value of specific heat is observed [159, 251]. As most of the moisture content migrates during the heating phase, no peak should re-occur during the cooling phase. This may result in a difference during the penetration of higher temperature inside the concrete. However, studies have shown that this difference will not be significant and can be ignored [346].

In high-rise structures, there are many more possible fire scenarios than in low- and mid-rise structures. This adds to the complexity of the analysis. As explained in Chapter 3, when one story of a building catches fire, there is always the possibility of vertically traveling fire. In low- and mid-rise buildings, this traveling might occur quickly, in such a way that it can reasonably be assumed that all of the floors will be on fire simultaneously. In high-rise structures, however, the time taken for the fire to travel from one floor to the next might vary, from a few minutes to much longer. Yet, as there are numerous uncertainties, it is very difficult to predict the rate of fire spread between floors. Possibly, using observations from real fires could provide an estimate to make an assumption. For example, it was said by eyewitnesses that during the fire in the Windsor Tower in Madrid in 2005, and in the Tamweel Tower in Dubai in 2012, the time taken for the fire to spread between the floors was between 6 and 30 minutes [347]. Categorizing a 6.minute delay as rapid spread, and a 30-minute delay as slow spread, various scenarios can be defined. For the case introduced here, the investigation is carried out for two fire-spread assumptions: a concurrent fire; and a fire with a 25-minute inter-story delay. A 5-minute delay in the spread of fire between the stories is not considered, as it is believed that the results of this would be largely similar to those for the concurrent fire. The fire is assumed to start from three different floors and then spread to the others, as shown in Fig. 6.33. Taking advantage of the delayed fire assumption can provide possible scenarios via which the structural responses during the cooling phase can be investigated.

The case study here is investigated under an iBMB fire curve. The advantages of the iBMB curve over other parametric fire curves were explained in Chapter 3. The fire load of the compartment, $q_{t,d}$ (MJ/m^2), is determined using Equation 3.19. Assuming two scenarios, the availability or non-availability of fire fighting measures, the fire load can be modified using Equation 3.20. These assumptions are shown in Table 6.2. In addition, δ_{q1} and δ_{q2} are set to 1.50 and 1.0, respectively. The combustion factor m is assumed to be 0.8, which is the value used for most cellulosic materials. For the adoption of the iBMB fire curve, a maximum fire load density needs to be selected as a reference, e.g. $q'' = 1300$ MJ/m^2 for ordinary buildings.

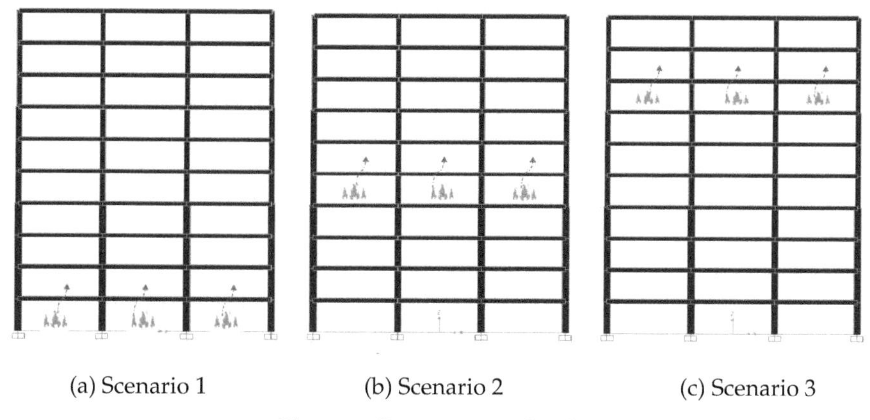

(a) Scenario 1 (b) Scenario 2 (c) Scenario 3

Fig. 6.33: Fire scenarios [345].

Table 6.2: Firefighting measures

Fire situation	*Automatic water extinguishing systems*	*Water supply*	*Automatic fire detection*	*Automatic fire alarm*	*Automatic transmission*	*An onsite fire brigade*	*An off-site fire brigade*	*Safe access routes*	*Normal fire fighting devices*	*Smoke exhaust systems*	*Sum*
	δ_1	δ_2	δ_3 or δ_4		δ_5	δ_6 or δ_7		δ_8	δ_9	δ_{10}	$\Pi\ \delta_{1\ to\ 10}$
Available	1.0	1.0	0.87	~~0.73~~	0.87	~~0.61~~	0.78	1.0	1.5	1.0	**0.89**
Not available	1.0	1.0	1.0	~~1.0~~	0.87	~~1.0~~	1.0	1.5	1.5	1.5	**2.94**

The analysis is performed only for PEF loading. For the seismic analysis, the structure is subjected to a monotonically-increasing lateral load, to arrive at a certain level of displacement corresponding to the LS level of performance. As mentioned in FEMA356, the target displacement should be calculated for controlling the deformed state of the structure for the assumed performance level. To do this, the widely accepted Equation 5.3 is used. Using that equation, the target displacement for the case study here is accounted for, and is 760 mm. Performing the pushover analysis, it is found that the first modal period is 1.54 seconds. This means that the conventional pushover analysis is adequately accurate, because the conventional pushover analysis is often proper when the first modal period is not more than 2.0 seconds. Figure 6.34 shows the pushover curve for the mentioned performance level and the response spectrum curve used in the pushover analysis.

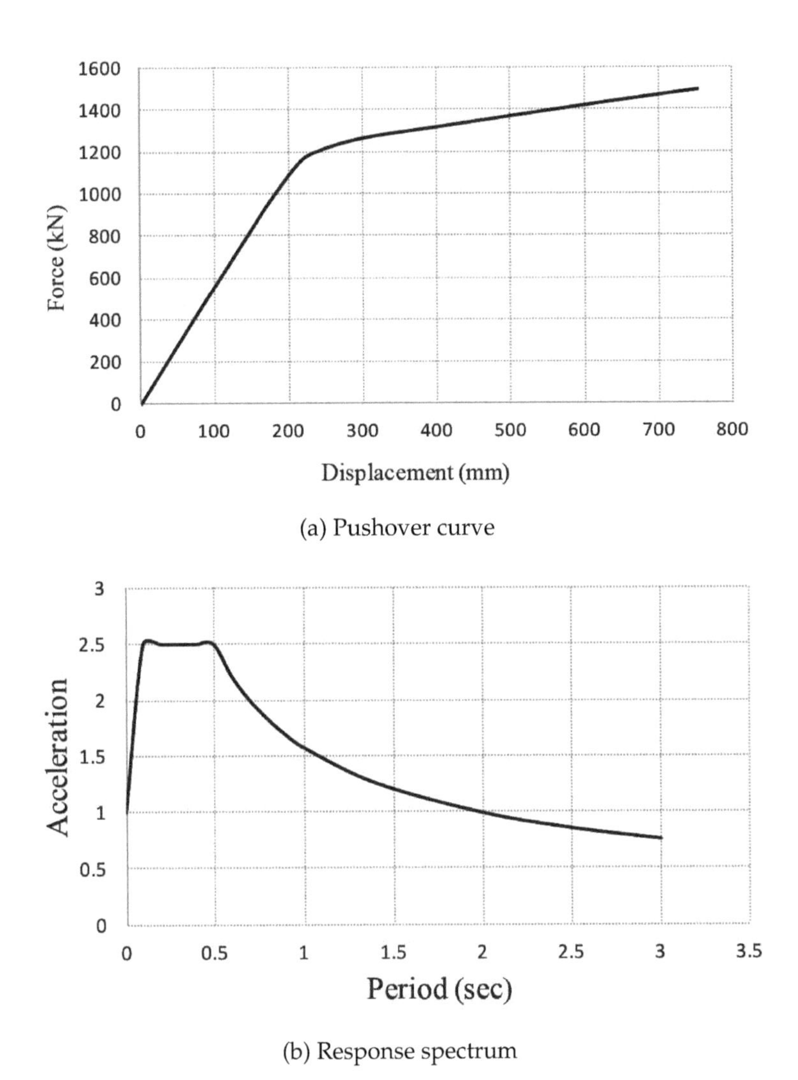

(a) Pushover curve

(b) Response spectrum

Fig. 6.34: The pushover curve at LS level and the response spectrum.

To plot the fire curve, equations introduced for the iBMB method are used. The summarized calculations of the fire curve are shown in Table 6.3, and the fire curve used for the PEF analysis is shown in Fig. 6.35. The fire curve is subjected to all sides of the interior columns. The exterior side of the external columns, however, is not exposed to fire. Meanwhile, only three sides of the beams are exposed to fire, because it can generally be assumed that the top side of the beams is protected by the concrete slab. To apply the fire frontiers to the cross-section of beams and columns, the damage states at the LS level are taken into consideration, as explained in detail in previous sections.

Table 6.3: Summarized calculations of the fire curve based on iBMB method

$A_f = 441.0$ m^2
$A_w = 95.0$ m^2
$A_t = 1150.8$ m^2
$A_T = 1055.8$ m^2
$h_w = 2.5$ m
Ventilation factor = 150.2 m$^{3/2}$
Opening factor = 0.142 m$^{1/2}$
q'' (increased by active measures) = 950 MJ/m^2
Total fire load $Q_{950} = 418068$ MJ
Averaged thermal properties $b = 1717$
$\dot{Q}_{max} = \text{Min}\,(\dot{Q}_{max,v}\,\&\,\dot{Q}_{max,f}) = \text{Min}\,(181.74\&110.25) = 110.25$: Thus **Fuel-controlled**
$Q_{1300\,(\text{reference})} = q\,A_f = 1300 \times 441.0 = 573300$ MJ
$t_1 = 10$ min; $Q_1 = 800$ MJ
$Q_{2,950} = 291847$ MJ; $t_2 \approx 72$ min
$Q_{3,950} = 125420$ MJ; $t_3 \approx 178$ min
$T_{2,950} = 1340$ °C
$T_{3,950} = 297$ °C

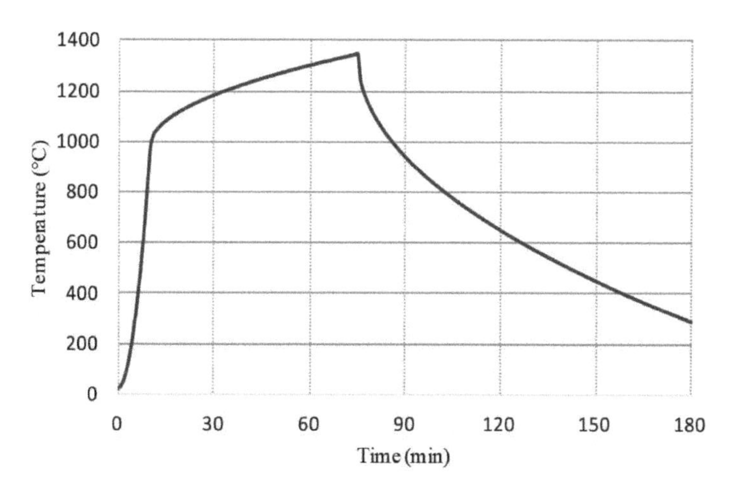

Fig. 6.35: Time-temperature based on iBMB curve.

As an example, Fig. 6.36 shows the temperature distribution inside beam B-C on the third floor during the heating and then cooling regime, before and after the application of seismic loads. The figure shows that exposing the section to fire during the heating phase results in a rapid increase in temperature in the outer layers and a gradual increase in the central inner layers. During the cooling phase, however, the surface of the concrete shows a drop in temperature at a fast pace while the temperature is still rising in the rest of the section. This relates to the low conductivity coefficient of concrete, which leads to a slower rate of gaining or losing heat during the analysis.

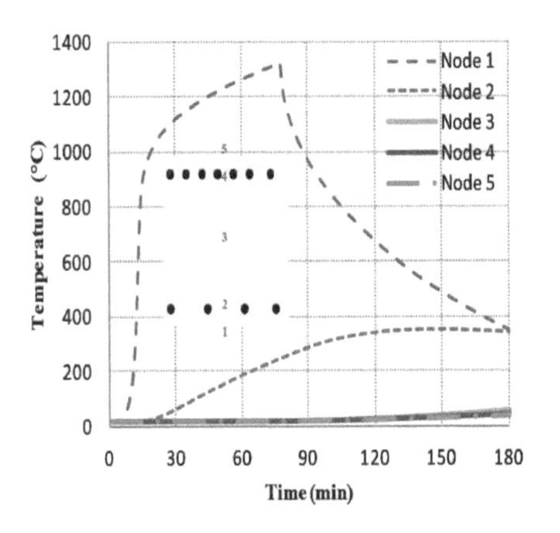

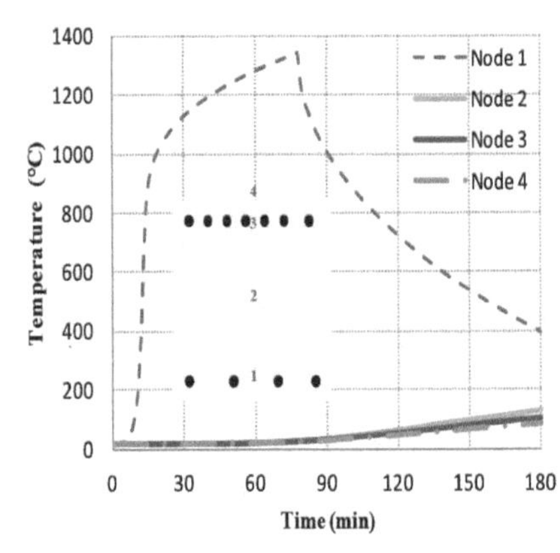

(a) Before earthquake (b) After earthquake at the LS performance level

Fig. 6.36: Distribution of temperature inside beam B-C on the third floor during the heating and cooling phases.

Figure 6.37 shows the displacement versus time for the mid-span of the beam, when the PEF is applied to the first four floors of the structure, with both a 25-minute delay in the fire traveling and concurrently, but as separate scenarios. Based on the thresholds defined for the SAFIR analysis, the fire resistance under concurrent fire is about 100 minutes and under delayed fire is about 160 minutes. Based on the mid-span deflection criterion, if the deflection exceeds 350 mm, the structure is considered to have failed. As the deflection is less than the limit, it has thus not failed in either scenario. Moreover, using the rate of deflection criteria, the frame is considered to have failed at 150 minutes under delayed fire. However, based on the rate of deflection criteria, no failure has occurred when a concurrent fire is applied to the frame. It is worth mentioning that applying a delayed fire to the frame leads to some fluctuations during the analysis. These fluctuations occur because while one floor is experiencing a cooling phase, the other floor might be under heating, resulting in a fluctuation in the deflection. It can also be understood from the figure that, in the delayed fire, failure occurs when fire reaches the fourth floor.

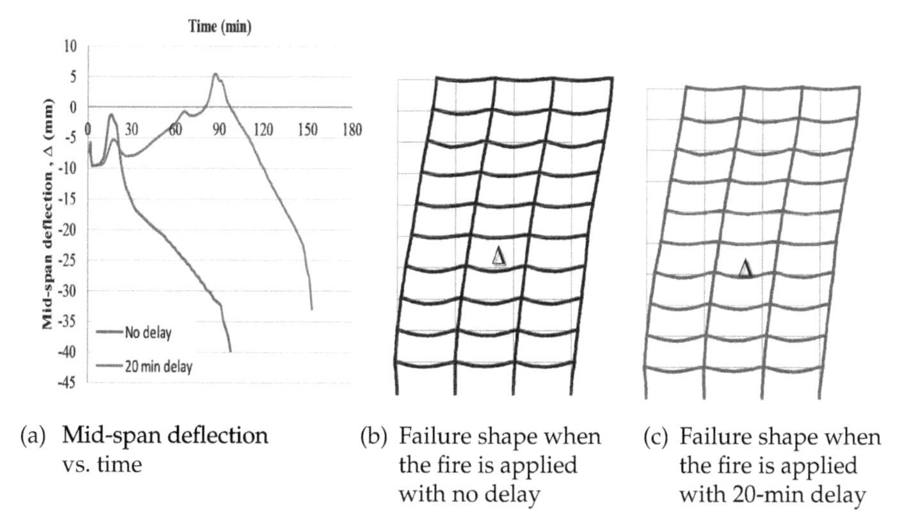

(a) Mid-span deflection vs. time

(b) Failure shape when the fire is applied with no delay

(c) Failure shape when the fire is applied with 20-min delay

Fig. 6.37: PEF resistance of the frame based on scenario 1.

Figure 6.38 shows the PEF resistance of the frame when the fire is applied to the fifth, sixth and seventh floors, either concurrently or with a 25-minute delay. As is seen, based on the thresholds defined for the SAFIR analysis, failure occurs at 120 minutes and 150 minutes when the fire is applied concurrently and with delay, respectively. No failure, however, is seen if the mid-span deflection criterion is considered, either in the concurrent fire or in the delayed fire. Moreover, based on the rate of deflection criterion, the frame collapses at around 100 minutes if the fire is applied concurrently, and at around 135 minutes if the fire is applied with a 25-minute delay.

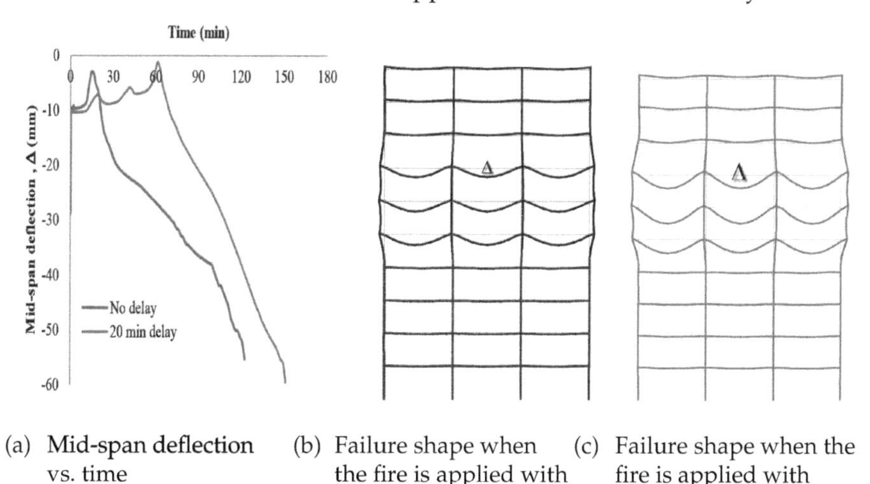

(a) Mid-span deflection vs. time

(b) Failure shape when the fire is applied with no delay

(c) Failure shape when the fire is applied with 20-min delay

Fig. 6.38: PEF resistance of the frame based on scenario 2.

Figure 6.39 shows the PEF resistance of the frame when the fire is concurrently applied to the eighth to tenth floors concurrently and with a 25-minute delay. Using the thresholds defined for the SAFIR analysis, failure occurs at 50 minutes and 75 minutes when the fire is applied concurrently or with delay, respectively. While no failure is observed during the analysis when using the deflection criterion, using the rate of deflection criterion shows that the frame fails at about 40 minutes and 60 minutes when the concurrent fire and the delayed fire are assumed, respectively. The notable difference between the PEF resistances in the last fire scenario and the first two fire scenarios is mainly due to the marked difference between the stiffness of the structural members, which changes alongside the height of the structure.

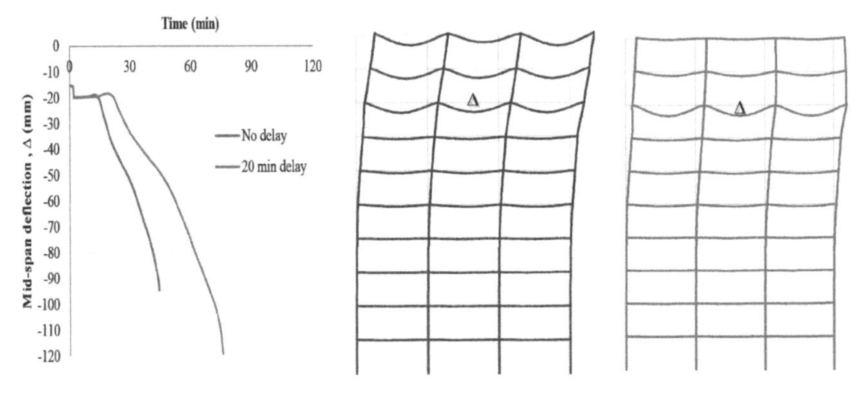

(a) Mid-span deflection vs. time (b) Failure shape when the fire is applied with no delay (c) Failure shape when the fire is applied with 20-min delay

Fig. 6.39: PEF resistance of the frame based on scenario 3.

It is also mentionable that the failure mode in the first fire scenario is global, meaning that the frame fails as a result of notable lateral movement of the columns, as shown in Fig. 6.37b and c. In the second scenario, the frame fails locally, such that the failure mainly pertains to the deflection of the beams, as shown in Fig. 6.38b and c. In the third scenario, the failure fashion seems to be more global than local. Overall, it can be said that the method of application of fire to the floors can considerably influence the fire resistance as well as the failure shape. This is one possible explanation for the complexity of fire when applied vertically to structures.

6.7 The effect of horizontally traveling fire on multi-story RC structures

As discussed in Chapter 3, the application of parametric fires is limited to small- and medium-sized compartments up to 500 m^2, with heights up to 4 m, and with no openings in the roof. These conditions apply because, although the temperature inside small or medium compartments can be assumed to be uniform, this assumption is not correct in large compartments. Instead, tests and observations have shown that in large compartments, fire travels from one spot to the next, resulting in a non-uniform temperature within the burning compartments.

In this section, the application of horizontally traveling fire is used in a three-story RC structure with the plan view and the structural properties shown in Fig. 6.40. The building is designed based on the ACI 318 code and for LS level of performance, using FEMA356 code and a PGA of 0.35g. The structure is dimensioned for the load combinations of 7.5 kPa for dead load and 2.0 kPa for live load. A combination of 100% dead load and 20% live load for the designed frame is used to find the required mass for calculating the earthquake load. The slab is made of normal-weight concrete with a 100 mm thickness. The compressive strength of concrete is 25 MPa and the yield stress of longitudinal and transverse reinforcing bars is 300 MPa. The thermal properties of the steel and the concrete slab are taken into account using Eurocode 4. For the thermal analysis, it is assumed that the concrete's moisture content is 2.0%. Moreover, the thermal expansion coefficients of the rebar and concrete are assumed to be $12 \times 10^{-6}/°C$ and $10 \times 10^{-6}/°C$, respectively. Poisson's ratio of 0.2 is considered for the concrete.

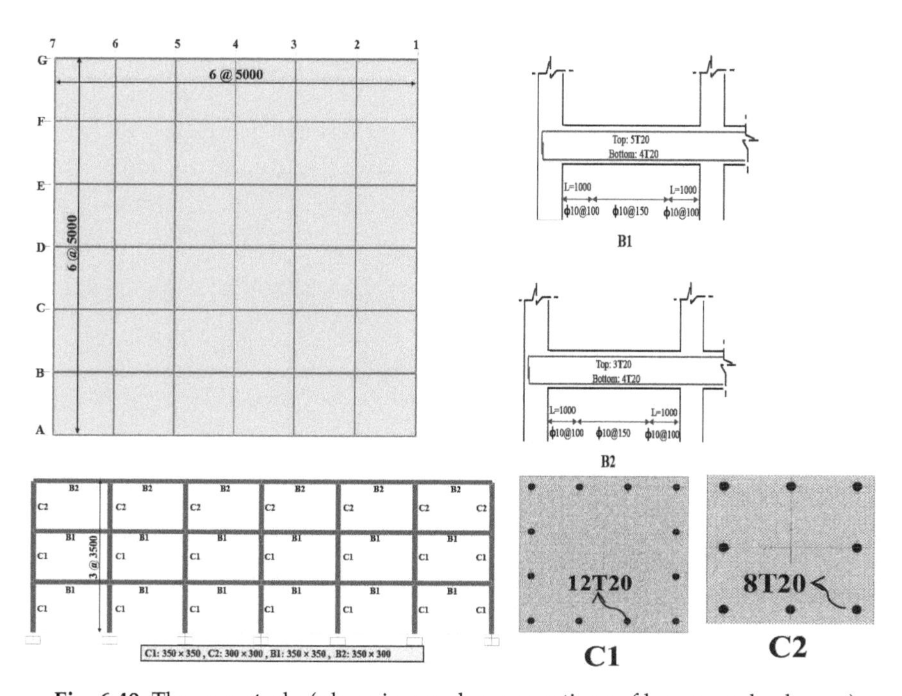

Fig. 6.40: The case study (plan view and cross-sections of beams and columns) (Dimensions in mm) [229].

The PEF analysis is performed using the sequential analysis method. Fig. 6.41 shows the pushover curve for the LS performance level. The traveling fire methodology is applied to the first floor of the case study, as shown in Fig. 6.42. As is seen, there are six bays in both directions, each of which is 25 m^2. The fire is assumed to initiate from axis 7 toward the other axes. Several fire sizes can be assumed, from 1% to 100%, some of which are shown in Table 6.4. This table includes information about the HRR, the maximum total burning duration, the spread rate, the near temperature, and the far field temperature. The near field temperature is assumed to be 1200°C.

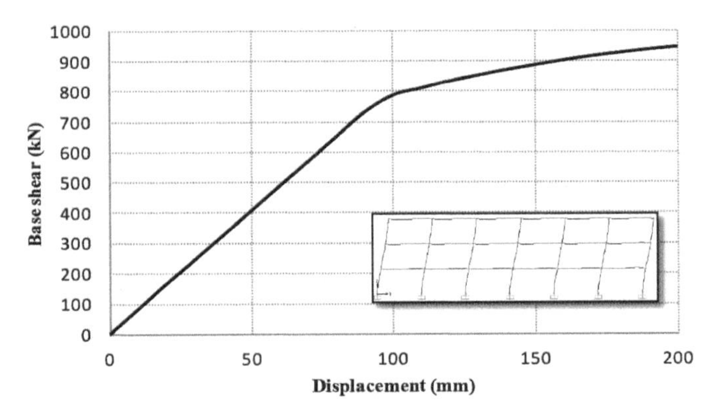

Fig. 6.41: The pushover curve at LS level.

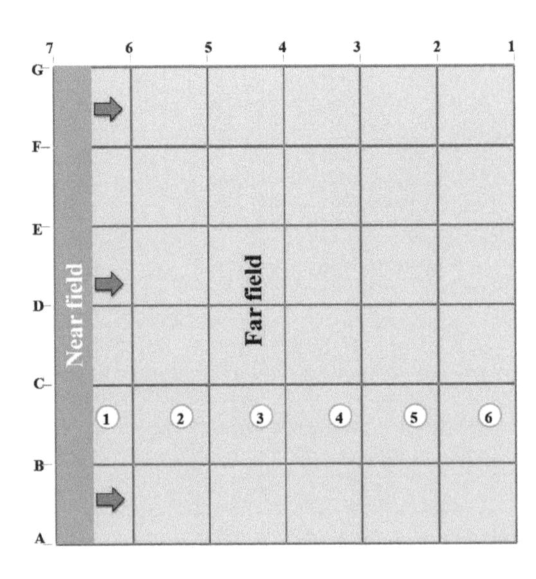

Fig. 6.42: Application of the traveling fire to the case study.

Table 6.4: The size range of the fire

Fire size (%)	$A_f\,(m^2)$	$r_{nf}\,(m)$	$\dot{Q}(MW)$	$t^*_{total}\,(min)$	$s\,(m/min)$	$T_{nf}\,(°C)$	$T_{ff}(°C)$
1	9	1.69	4.50	3232	0.01	1200	192
2.5	22.5	2.67	11.2	1312	0.02	1200	282
5	45	3.78	22.5	672	0.04	1200	352
10	90	5.35	45.0	352	0.09	1200	447
17.0	153	6.98	76.5	220	0.15	1200	502
25	225	8.46	112.5	160	0.23	1200	558
33	297	9.72	148.5	129	0.31	1200	642
50	450	11.97	225.0	96	0.47	1200	726
67	603	13.85	301.5	80	0.63	1200	828
75	675	14.66	337.5	75	0.70	1200	929
83	747	15.42	373.5	71	0.78	1200	1065
100 (uniform fire)	900	16.92	450.0	64	0.94	1200	1200

Assuming a fire size of 17%, which is almost equal to the size of each bay, and referring to Table 6.4, the fire curves are plotted, some of which are shown in Fig. 6.43a. The curves show that while some bays are in the heating phase, others are experiencing the cooling phase (Fig. 6.43b). To highlight the difference between a standard fire and a traveling fire, the ISO834 curve is also shown in Fig. 6.43a.

Performing a sequential analysis where the earthquake-damaged structure is subjected to fire, the fire resistance is calculated. It is evident that the fire is not applied to cross-sections of all the bays simultaneously;

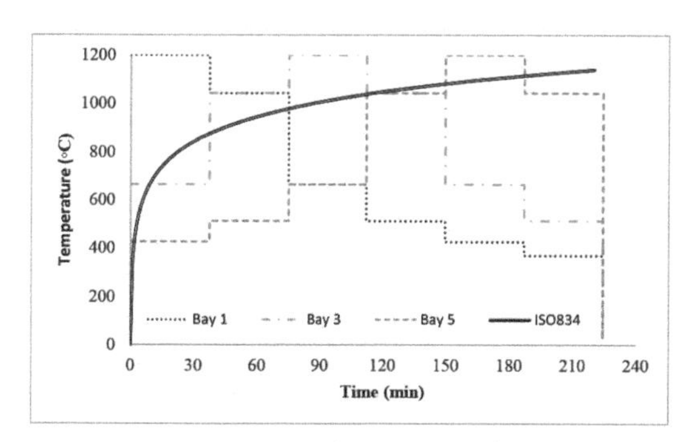

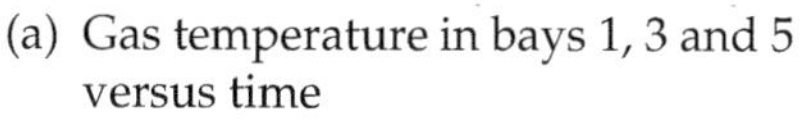
(a) Gas temperature in bays 1, 3 and 5 versus time

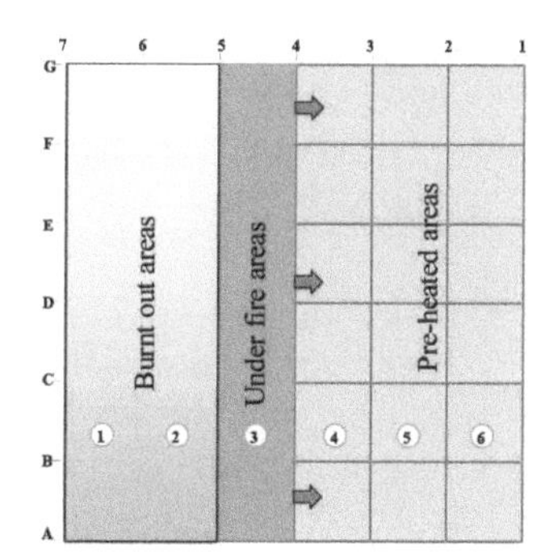

(b) Progression of fire along the length

Fig. 6.43: The gas temperature in bays, assuming a fire size of 17% and progression of the fire along the length of the case study.

hence, the temperature distributions inside the identical cross-sections (but in different locations) are not identical. As an example, Fig. 6.44a and b show the temperature distribution inside column C1 at axes 7 and 1 under the traveling fire. To draw a comparison, the temperature distribution inside the column C1 at axis 7 under the ISO834 fire curve and 100% fire size is also illustrated, as shown in Fig. 6.44c and d. As seen in Fig. 6.44a, b, and d, the external layers gain and lose heat at a rapid rate during both heating and cooling phases, while the rise and drop in temperature in the internal layers occur at a gradual rate. This relates to the low conductivity coefficient of concrete, as explained earlier.

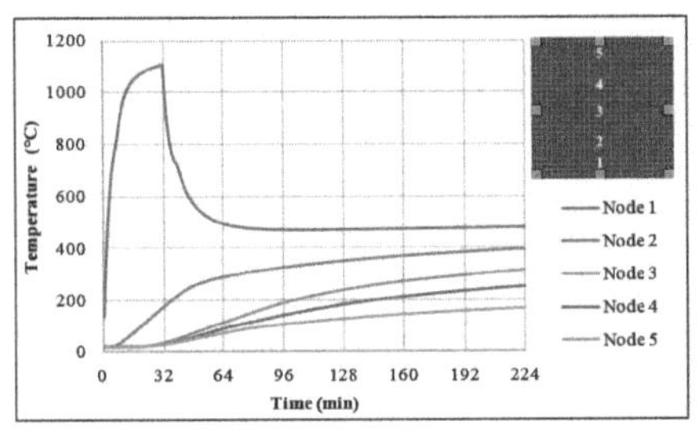

(a) At axis 7 under the traveling fire

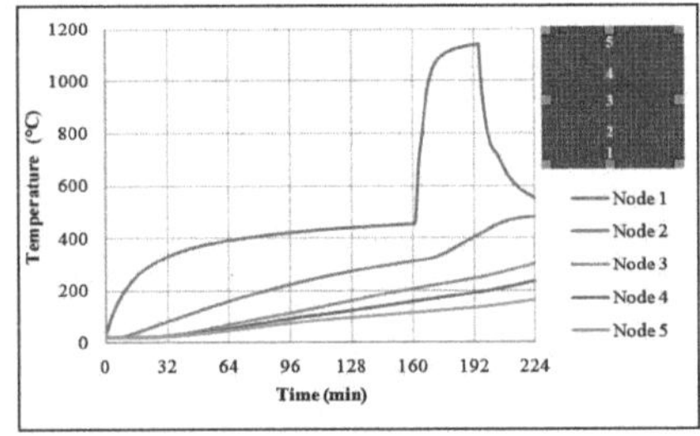

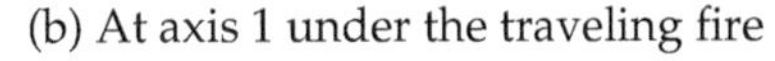
(b) At axis 1 under the traveling fire

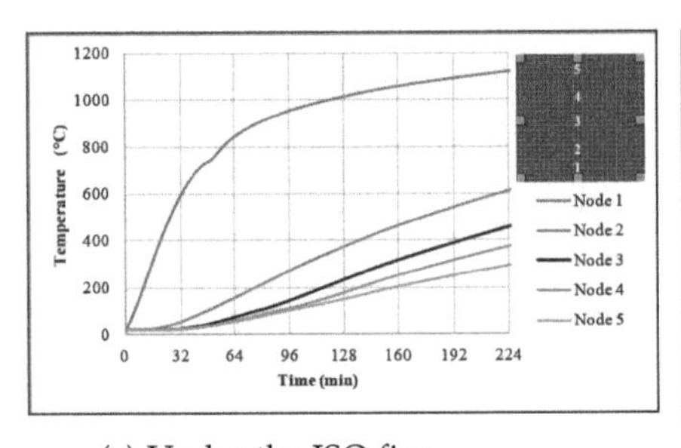

(c) Under the ISO fire

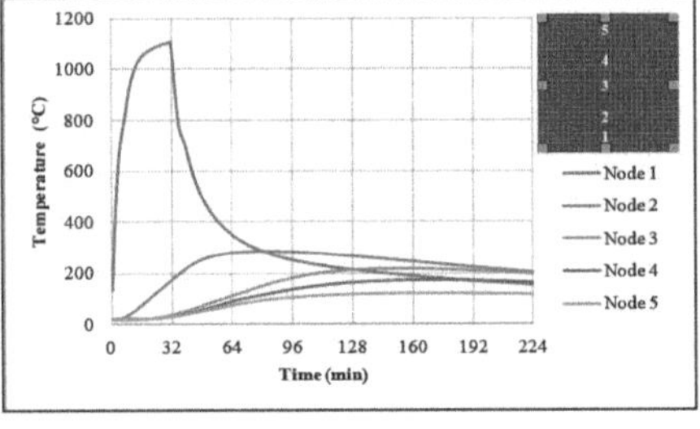

(d) Under the100% fire

Fig. 6.44: Distribution of temperature inside column C1 at axis 7 under the traveling fire, the ISO fire and the 100% fire.

Figure 6.45 shows the PEF resistance of the frame based on traveling fire. As per the thresholds previously defined in the SAFIR analysis, the fire resistance is determined by the time at which the displacements, either globally (i.e. the drift of a certain point) or locally (i.e. the deformations at the middle of a beam) merge toward the vertical asymptote by a 1% error. The sharp increase seen at the beginning in Fig. 6.45 is due to the structure being first laterally pushed to arrive at a certain level of displacement and then being unloaded.

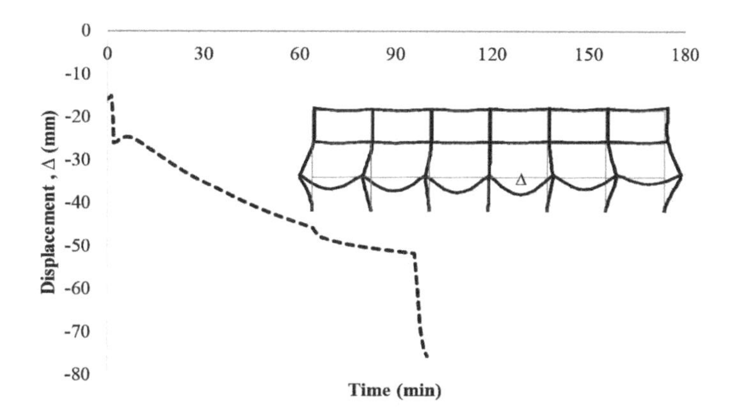

Fig. 6.45: PEF resistance of the frame based on the traveling fire [229].

The damaged structure is then loaded with fire as a sequential load, which arrives at the structure in its residually deformed state. Based on the analysis results, it takes about 100 minutes for the frame to fail. The fire resistance, however, is different when it is determined by using the mid-span deflection and the rate of deflection criteria. Since the mid-span deflection for the case here is limited to 250 mm, the frame has thus not failed. As for the rate of deflection, it is limited to 9.25 mm/min, and thus, the frame must be considered to have failed at 91 minutes. It is worth mentioning that two types of collapse, local and global, may be observed during the analysis. On the other hand, regarding the failure type, it seems that local collapse is the predominant failure mode for the frame subjected to the traveling fire, as shown in Fig. 6.45. To make a comparison, the PEF resistance based on the uniform fires (i.e. ISO fire and 100% fire size) are also accounted for. Fig. 6.46 shows the PEF resistance of the frame under the ISO curve. The resistance

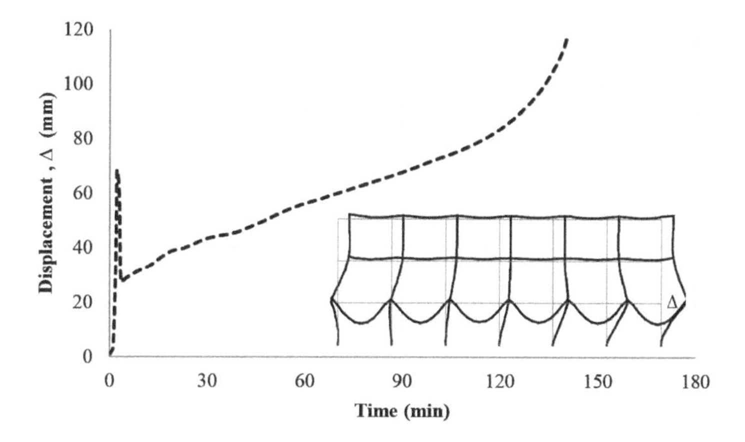

Fig. 6.46: PEF resistance of the frame based on the ISO834 curve [229].

of the frame is seen to be around 140 minutes for the defined thresholds. However, no failure is observed when the deflection and rate of deflection criteria are controlled. In terms of the collapse mode, the frame fails laterally, as shown in the figure.

Figure 6.47 shows time versus deflection of the frame under a 100% fire size. As is seen, no failure occurs during the analysis, showing that the frame remains stable under the fire.

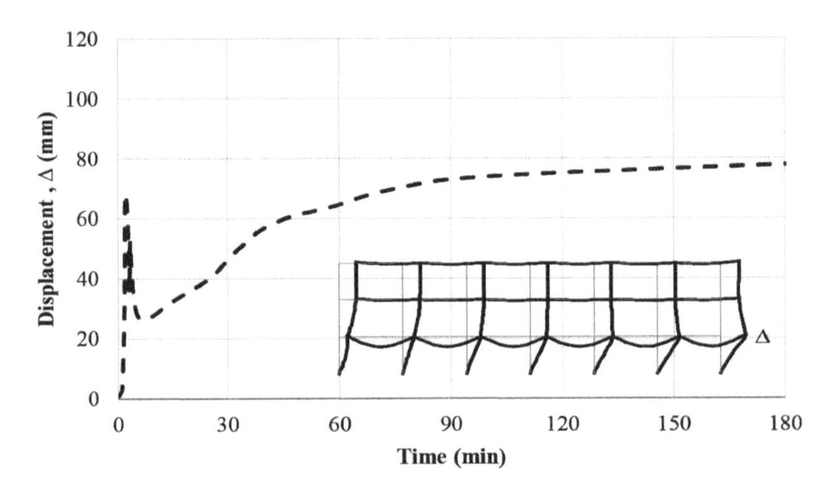

Fig. 6.47: PEF resistance of the frame based on the 100% fire size [229].

It is evident from the above results that the vulnerability of the structure is greater for a traveling fire than for uniform fires, such as ISO or 100% fire size. While no failure occurs under the standard fire, the frame fails under the non-uniform fire. This conclusion contradicts the usual belief that considering a uniform fire inside a compartment is a conservative assumption.

6.8 The behavior of moment-resisting steel structures under standard and natural fires

The application of parametric fires is used here to monitor the response of a five-story moment-resisting steel structure under pre- and post-earthquake fire. It is assumed that the structure has not been fireproofed and thus, the fire frontiers are applied directly to the bare cross-sections, either before or after the earthquake. The structure's specifications are shown in Fig. 6.48 and Table 6.5; it has been designed for the IO performance level as if it were an educational building. This assumption is needed for calculating the fire load density. The floor and the ceiling are made of normal-weight concrete and the compartment partitions are built with standard bricks.

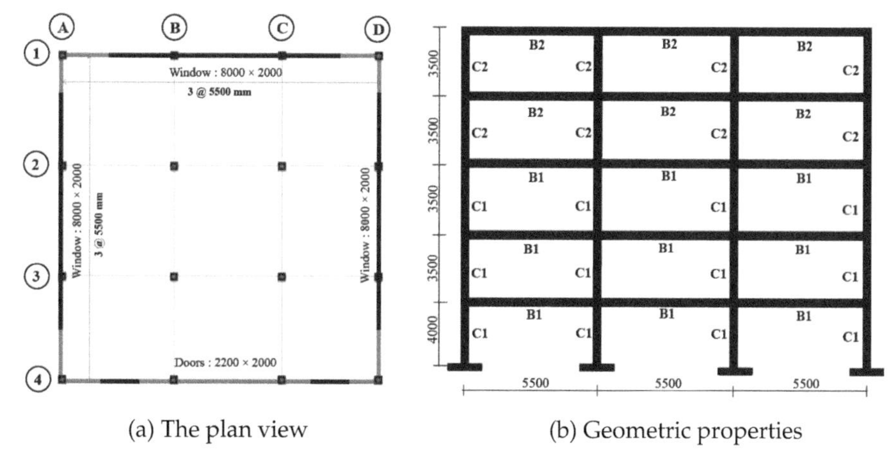

(a) The plan view (b) Geometric properties

Fig. 6.48: The case study (dimensions are in mm) [348].

Table 6.5: Sections dimensions of the case study

Dimensions are in mm	*C1*	*C2*	*B1*	*B2*
	H = 400 B = 300 $t_f = 20$ $t_w = 15$	H = 350 B = 300 $t_f = 15$ $t_w = 10$	H = 300 B = 300 $t_f = 20$ $t_w = 15$	H = 250 B = 250 $t_f = 15$ $t_w = 10$

The structure is loaded, and then designed for a PGA of 0.35g using steel plates (welded compact sections) with the yield stress of 240 MPa. The structure is dimensioned for the load combinations of 8.0 kPa for dead load and 2.5 kPa for live load. A combination of 100% dead load and 40% of live load for the designed frame at the IO level of performance is used to find the required mass for calculating the earthquake load. In order to improve the understanding of structural behavior, the fire-alone analyses are also performed. It is assumed that only the first story is on fire, and that no vertically traveling fire occurs. The exterior side of the external columns is not exposed to fire. Meanwhile, only three sides of the beams are exposed to fire, because it is assumed that the top side is well protected by the concrete slab. Table 6.6 shows the thermal characteristics of the materials in the building.

Using Equation 5.3, the target displacement is determined, which brings the structure to arrive at its corresponding performance level. The pushover loads that are used to perform the sequential analysis are shown in Fig. 6.49. The temperature evolution is determined based on the parametric fire methodology, as explained in Chapter 3. The presence or absence of

Table 6.6: Thermal characteristics of the considered materials

	Specific heat (J/kgK)	*Material's density (kg/m³)*	*Thermal conductivity (W/mK)*
Floor and Roof	1000	2400	1.6
Walls	840	1600	0.7

fire fighting measures are also considered with regard to the fire load of the compartment, as per the assumptions made in Table 3.8. The fire load density of 347 MJ/m^2 for an educational occupancy is used. The fire curves that apply when the fire fighting measures are available (say before earthquake), and when they are not available (say after earthquake), are shown in Fig. 6.50. The cross-sections of the structure are subjected to each fire curve, in order to record the temperature evolution over time.

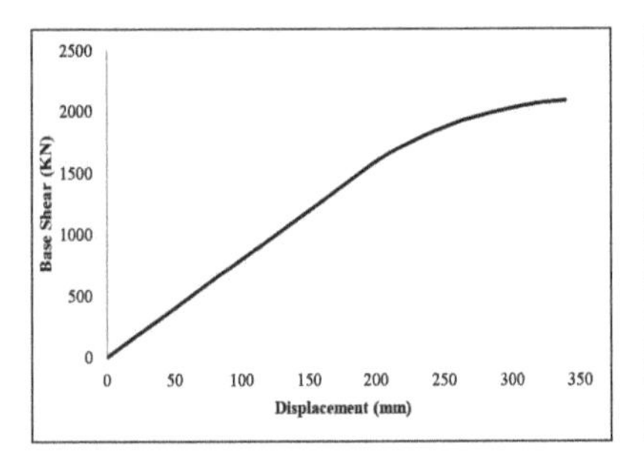

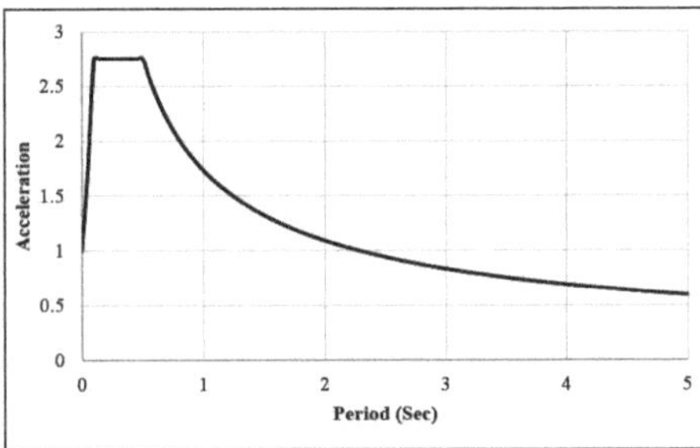

(a) The pushover curve (b) The response spectrum

Fig. 6.49: The pushover analysis and the response spectrum.

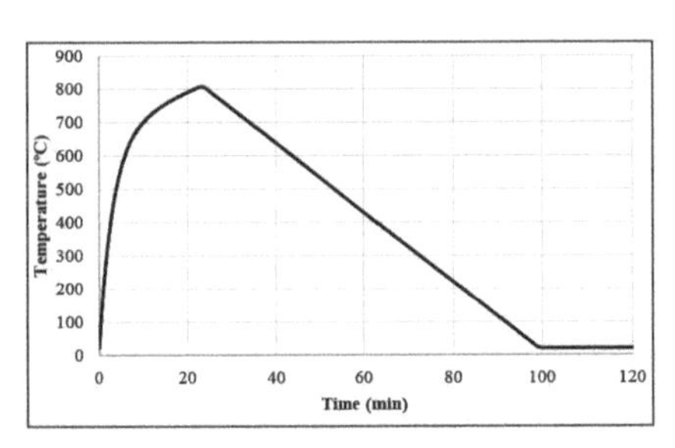

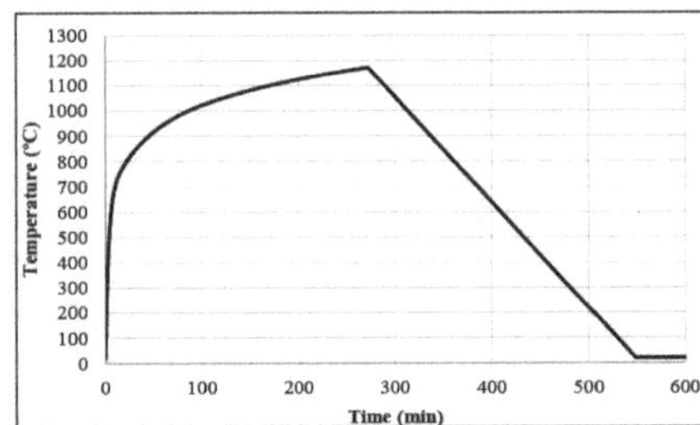

(a) When fire fighting facilities are available (b) When fire fighting facilities are not available

Fig. 6.50: Temperature changes over time, when the presence and absence of fire fighting facilities are considered.

Figure 6.51 shows the fire resistance of the structure under the ISO834 fire, before and after the earthquake. The time versus displacements are plotted for beam B-C and column D separately, in order to provide more understanding about the analysis results. As is seen, based on the threshold defined, the structure fails at 32 minutes and 28 minutes, before and after the earthquake, respectively. This shows an earthquake-induced reduction of 12.5% in the PEF resistance. If the rate of deflection is considered as the failure criterion, the frame fails at 28 minutes under the fire-alone scenario and 25 minutes under the PEF scenario.

Figure 6.52 shows the fire resistance of the structure under the parametric fire curves, when the presence or absence of fire fighting measures have been involved in determining the fire curves. As seen, the structure does not fail prior to earthquake, as it does not exceed the threshold defined by either

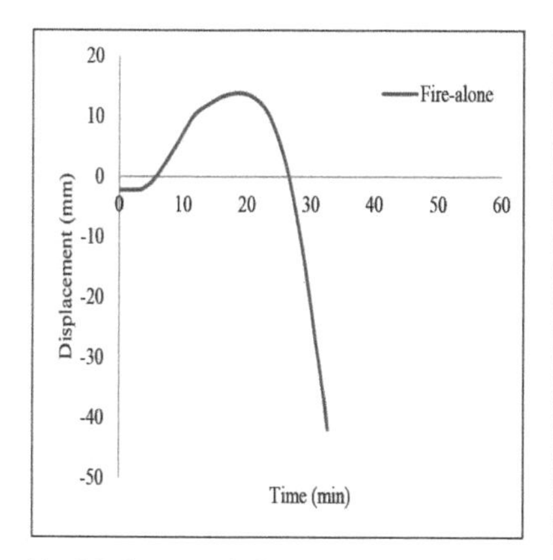

(a) Mid-span deflection of beam B-C in the first floor

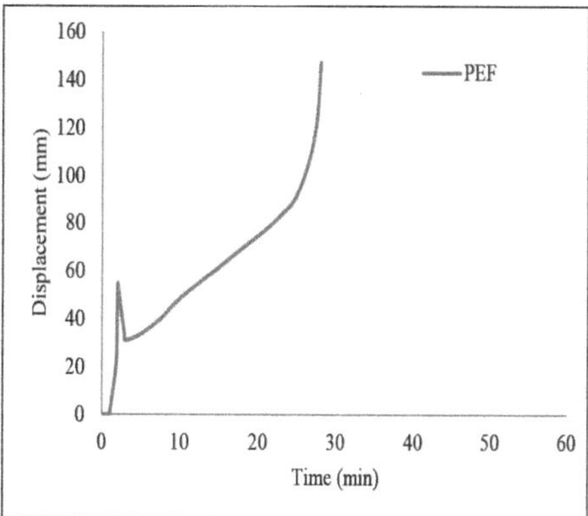

(b) Lateral displacement of column D in the first floor

Fig. 6.51: Fire resistance of the structure under ISO834 fire curve.

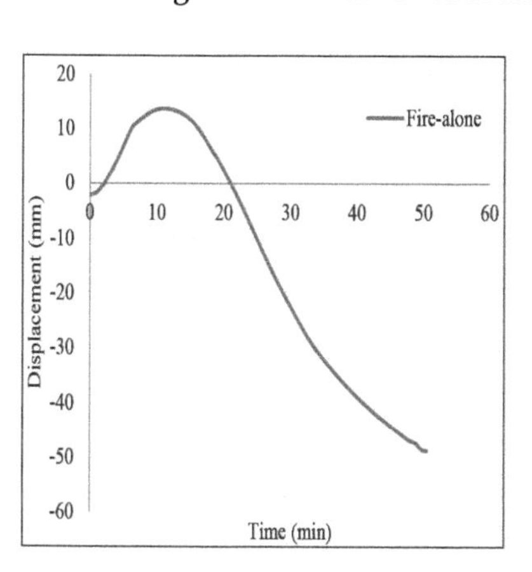

(a) Mid-span deflection of beam B-C in the first floor

(b) Lateral displacement of column D in the first floor

Fig. 6.52: Fire resistance of the structure under natural fire curve.

the SAFIR analysis or the performance-based criteria. After the earthquake, however, it fails at 22 minutes under the defined threshold and at 21 minutes under the rate of deflection criterion, showing that both earthquake damage and the non-availability of fire fighting measures can significantly reduce the fire resistance.

It is also evident from Figs. 6.51 and 6.52 that the PEF resistance based on the parametric fire is much lower than that of the ISO834 fire. In addition, the frame fails in different fashions under the fire-alone and PEF scenarios, such that it fails locally under the fire-alone scenario and laterally under PEF scenario, as shown in Fig. 6.53.

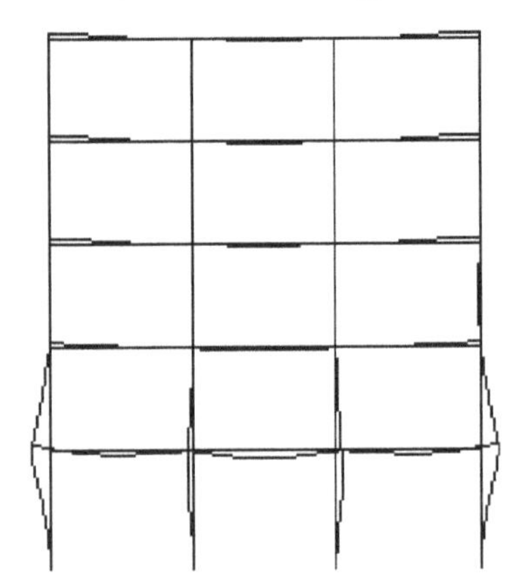

(a) The structure deformed before the earthquake

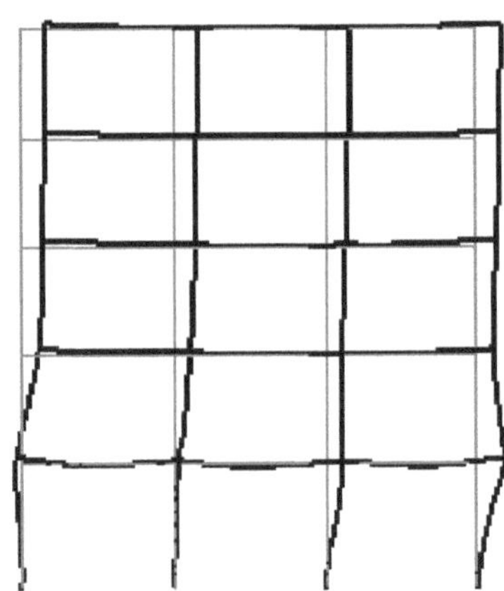

(b) The structure deformed after the earthquake

Fig. 6.53: Collapse mechanism of the structure.

6.9 Behavior of tall steel structures under vertically traveling fire

The behavior of tall RC structures under a severe iBMB fire was investigated in Section 6.6. Investigating the response of tall steel structures subject to PEF loading when the fire travels vertically is also of interest. Here, a ten-story moment-resisting steel structure with office occupancy type is considered (its plan view and properties are shown in Fig. 6.54). The structure is loaded and then designed, using steel profiles with a yield stress of 240 MPa and a PGA of 0.35g. The frame is dimensioned for load combinations of 6.5 kPa for dead load and 2.0 kPa for live load. A combination of 100% dead load and 20% live load for the designed frame is used to find the required mass for calculating the earthquake load [349]. The resulting sections for the designed frame are shown in Fig. 6.54b. The materials used in the building and their thermal characteristics are shown in Table 6.7. The slab depth is 100 mm. As the building is used as an office, a fire load density of 511 MJ/m^2 is used to determine the thermal actions. The fire load density is then modified by involving the presence or absence of fire fighting measures and by using Equation 3.20. The thermal properties of the steel and slab are taken into account using Eurocode 4. For the thermal analysis, it is assumed that the concrete's moisture content is 2.2%. Additionally, the thermal expansion coefficients of the rebar and concrete are assumed as $12\times10^{-6}/°C$ and $10\times 10^{-6}/°C$, respectively. Poisson's ratio of 0.2 is considered for the concrete.

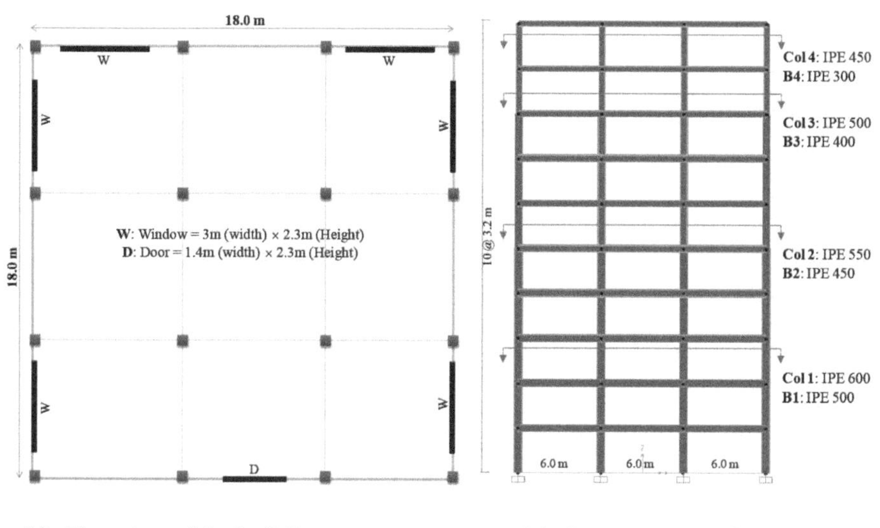

(a) Plan view of the building and the opening

(b) Cross-sections of beams and columns

Fig. 6.54: The case study [350].

Table 6.7: Thermal characteristics of the considered materials

	Specific heat (J/kgK)	*Material's density (kg/m³)*	*Thermal conductivity (W/mK)*
Floor and Roof	1000	2400	1.6
External walls	840	1600	0.7

To perform the PEF analysis, it is assumed that bare profiles are exposed to the fire. This assumption might seem very conservative, as tall steel structures are often protected by a layer of fireproofing materials in order to meet the minimum fire resistance at the normal condition. For example, according to the French and the British fire codes, conventional buildings (such as offices and dwellings) of up to 11 stories and with a total height up to 33 meters have to have a fire resistance of at least 120 minutes [351]. This required time is independent of the nature of the structural skeleton, i.e. either RC or steel. In RC structures, meeting the minimum fire resistance rating is simple, as the concrete cover can provide adequate resistance. In steel structures, as they are very vulnerable to heat, they are often protected by fireproofing materials. While it is expected that fireproofing materials will work properly in a fire event, their workability after an earthquake cannot be guaranteed, as discussed in previous chapters. As shown in Chapter 2, the fireproofing materials are particularly vulnerable in the vicinity of beam-to-column connections, where the plastic hinges are often created. Therefore, even though the fireproofing materials may remain intact throughout the rest of the length, their presence might not be effective in hampering fire attack.

To monitor the structural response under a vertically traveling fire, three scenarios are assumed, as shown in Fig. 6.55. It is assumed that the fire initiates from one story and then travels to the story above, either with a delay or immediately. In addition, two different delay times for fire spread between the stories are assumed: 5 minutes (as rapid traveling) and 25 minutes (as slow traveling). These assumptions would provide a better understanding of the structural responses under different conditions. The fire curve used in this study is determined based on the parametric fire introduced in Chapter 3. To modify the fire load density, it is assumed that δ_{q1} is 1.54, and δ_{q2} is 1.0. The combustion factor is assumed to be 0.8, which is good enough for most cellulosic materials. It is also assumed that the value of $\delta_{q1}\,\delta_{q2}\,\delta_n\,m$ is 1.86. The fire curve and a summary of its calculation are shown in Fig. 6.56.

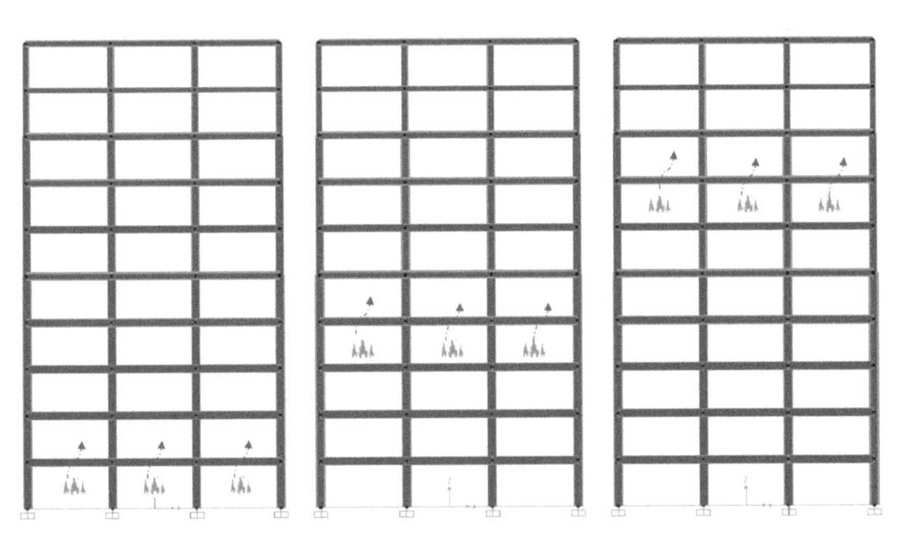

(a) Scenario 1: No delay

(b) Scenario 2: 5-min delay

(c) Scenario 3: 25-min delay

Fig. 6.55: Fire scenarios.

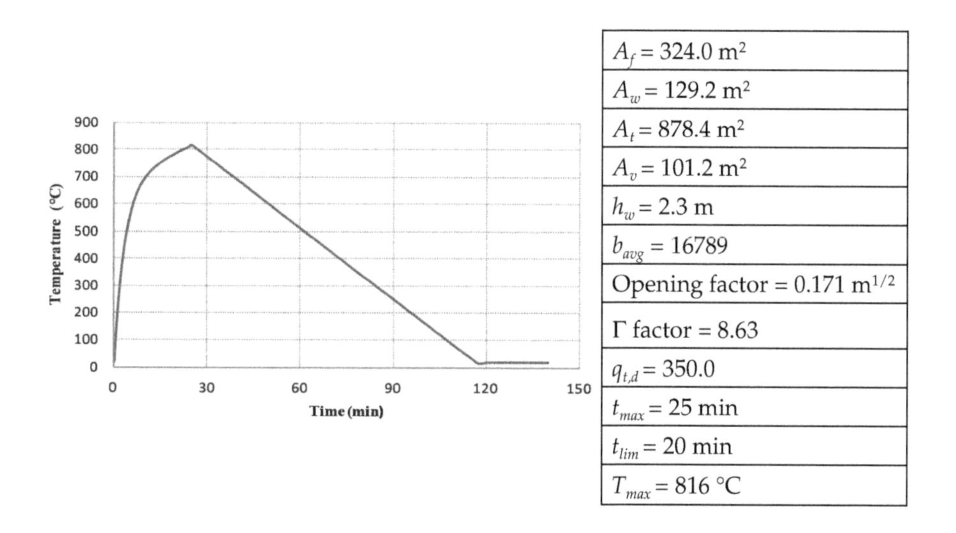

$A_f = 324.0$ m^2
$A_w = 129.2$ m^2
$A_t = 878.4$ m^2
$A_v = 101.2$ m^2
$h_w = 2.3$ m
$b_{avg} = 16789$
Opening factor = 0.171 m$^{1/2}$
Γ factor = 8.63
$q_{t,d} = 350.0$
$t_{max} = 25$ min
$t_{lim} = 20$ min
$T_{max} = 816$ °C

Fig. 6.56: Time-temperature based on Natural fire curve.

As an example, Fig. 6.57 shows the temperature distribution inside an IPE500 (used as a beam) and the concrete slab during heating and then cooling regimes. As is seen, exposing the steel profile and the slab above to fire during the heating phase results in a rapid increase in the temperature of the steel, but only a gradual increase in the slab's temperature. During the cooling phase, while the steel is losing its heat at a rapid rate, the temperature of the slab is still on the rise. From the figure, it is also evident that around 25 minutes after the fire's initiation, the temperature on the other side of the slab (Node 5) is about 50°C. Although this disparity seems not to be notable, it is also added to the ambient temperature in the subsequent floor, when a 25-minute delay is considered, in order to provide more accuracy.

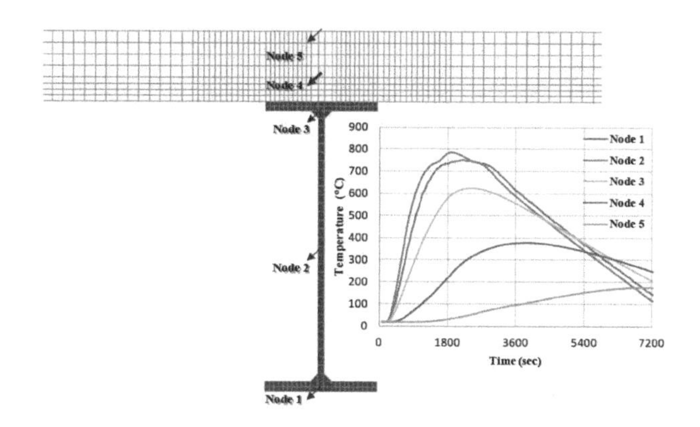

Fig. 6.57: Distribution of temperature inside IPE500 and concrete slab above during heating and then cooling phase.

It is evident that prior to the application of PEF loads to the structure, it is monotonically pushed to arrive at the target displacement which corresponds to the performance level of the structure. The target displacement determined for the structure is 800 mm. The pushover loads are then used to perform the sequential analysis. Figure 6.58 shows the PEF resistance based on scenario 1, when the fire is applied to the first story and then travels vertically with delays of 5 minutes (as a fast spread) and 25 minutes (as a slow spread). To make a comparison, a concurrent fire scenario is also considered. The fire resistance is defined as per the thresholds already introduced.

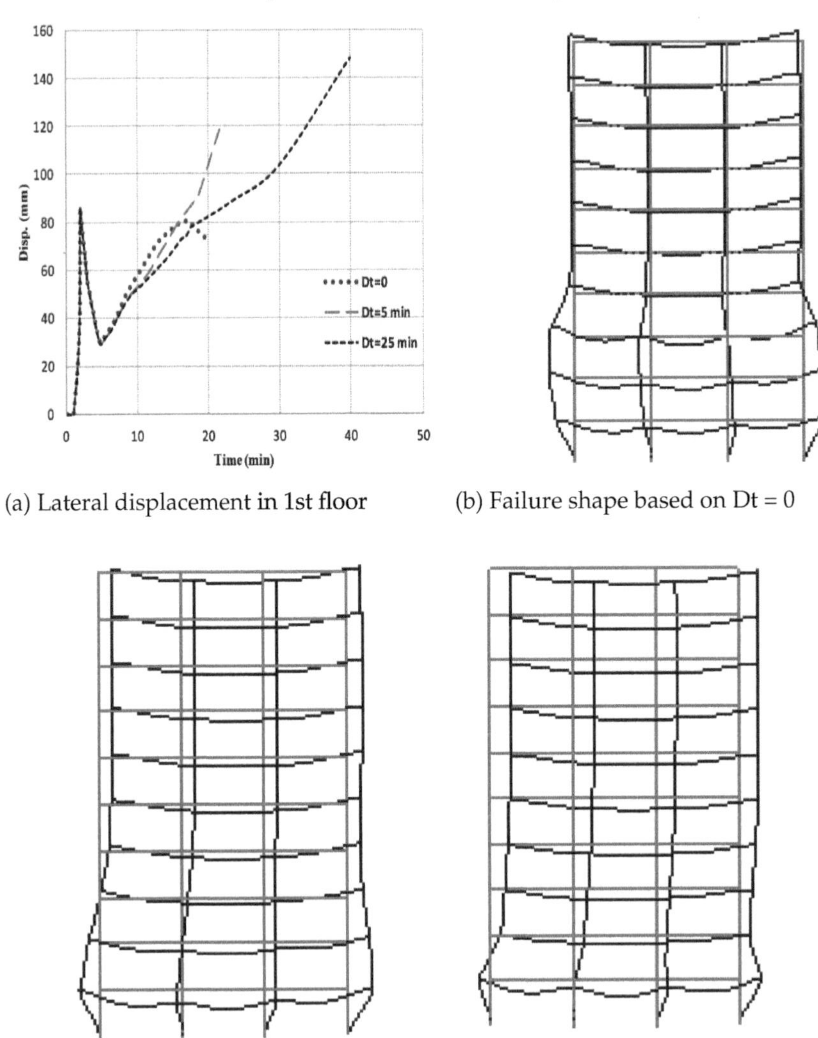

(a) Lateral displacement in 1st floor

(b) Failure shape based on Dt = 0

(c) Failure shape based on Dt = 5 min

(d) Failure shape based on Dt = 25 min

Fig. 6.58: PEF resistance of the frame, based on scenario 1.

As is seen, the PEF resistance of the frame when the fire is concurrently applied is about 20 minutes, and about 23 minutes and 40 minutes when the 5-minute and 25-minute delays are assumed, respectively. In the case of the 25-minute delay and reconsidering the fire curve shown in Fig. 6.56, it can be seen that while the first floor is experiencing cooling, the temperature in the second floor is still rising. However, when there is either no delay or a 5-minute delay in the traveling of the fire, no cooling phase is experienced. Different fashions of applying the fire to the structure lead also to differences in the failure fashions, as shown in Fig. 6.58. It is also mentionable that the sharp increase at the beginning of the figure represents the lateral displacement of the first story after the application of the earthquake loading. This shows a drift value of 2.5%, corresponding to the drift limit at the LS performance level as defined by FEMA356. For the second scenario, Fig. 6.59 shows the lateral displacement versus time plotted for three inter-story fire spread assumptions: no delay, a 5-minute delay, and a 25-minute delay. As can be seen, the fire resistance based on the 25-minute delay is around 45 minutes, which is markedly different from the results of the other scenarios. There is also a difference between the failure shapes, as shown in the figure.

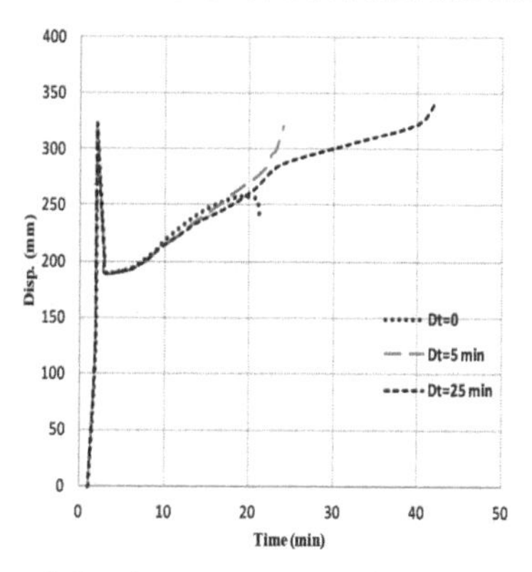

(a) Lateral displacement in 4th floor

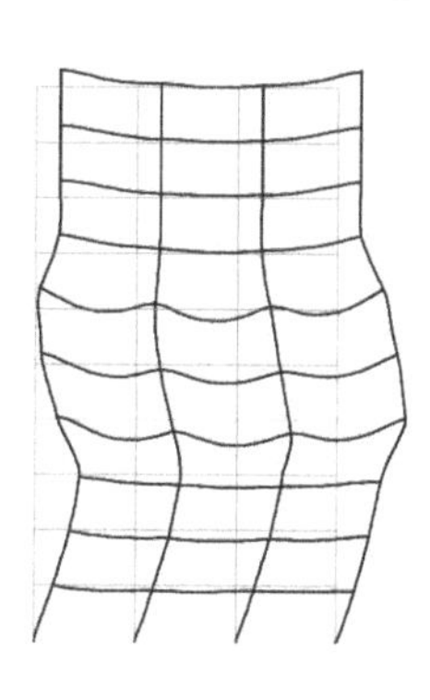

(b) Failure shape based on Dt = 0

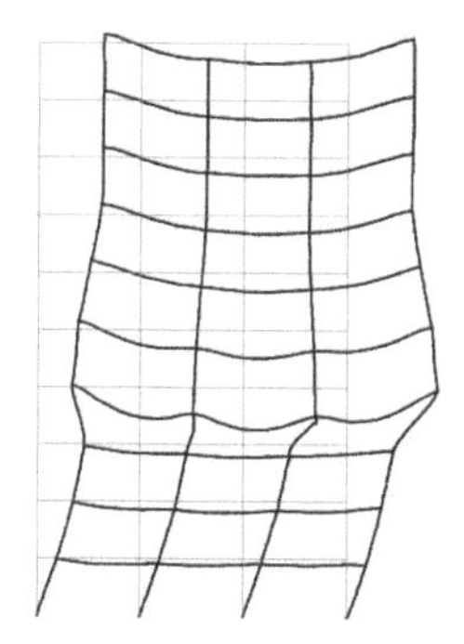

(c) Failure shape based on Dt = 5 min

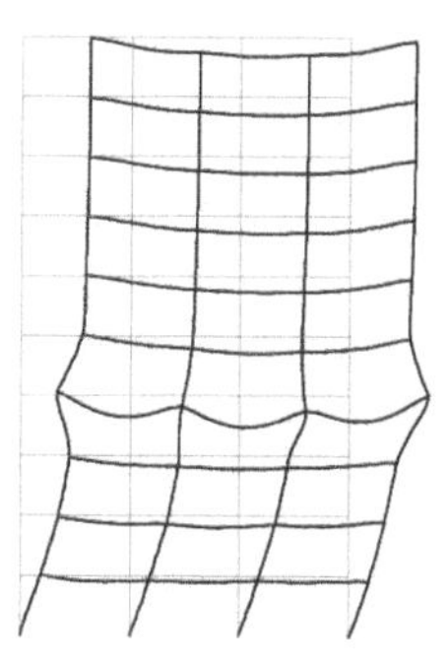

(d) Failure shape based on Dt = 25 min

Fig. 6.59: PEF resistance of the frame, based on scenario 2.

For the third scenario, Fig. 6.60 shows the PEF resistance, along with results for the three mentioned fire spread assumptions. As with the previous scenarios, the fire resistance based on a 25-minutes delay is higher than that in the other scenarios. While it is around 23 minutes when the fire is immediately applied to the stories, it becomes about 28 minutes and 43 minutes in cases with the 5-minute and 25-minute delays, respectively. A notable difference is also seen in terms of the failure shapes for the mentioned scenarios. It can also be seen that, in the case of applying the fire with a 25-minute delay, the seventh floor is experiencing cooling while the eighth floor is heating up. No cooling phase, however, is experienced by the frame when other scenarios are considered.

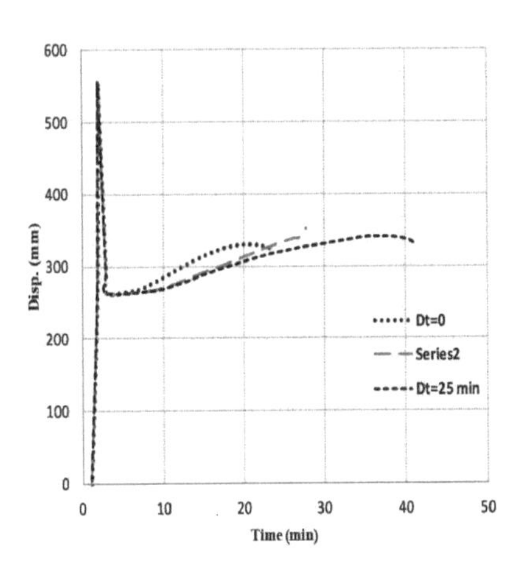

(a) Lateral displacement in 7th floor

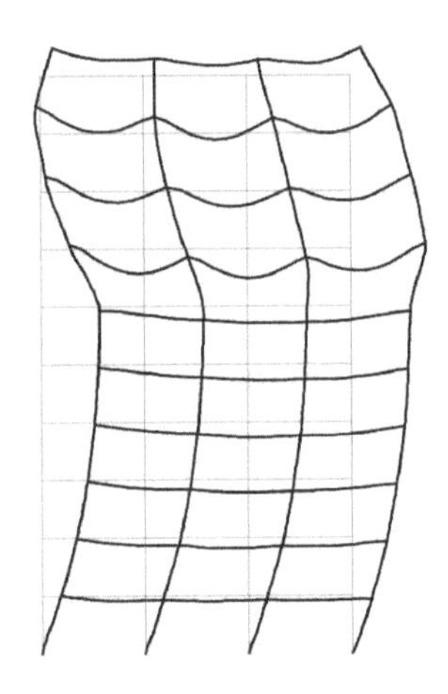

(b) Failure shape based on Dt = 0

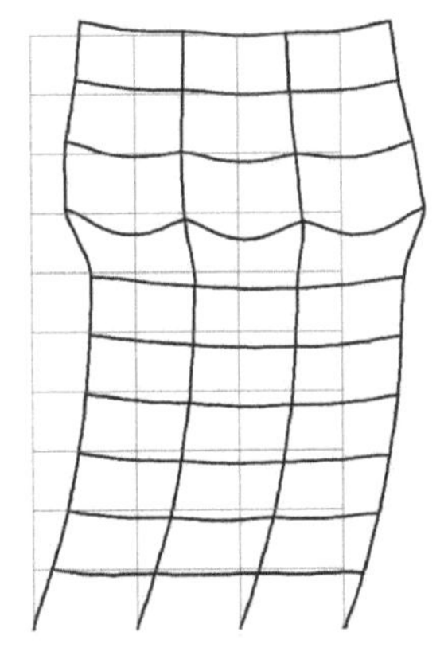

(c) Failure shape based on Dt = 5 min

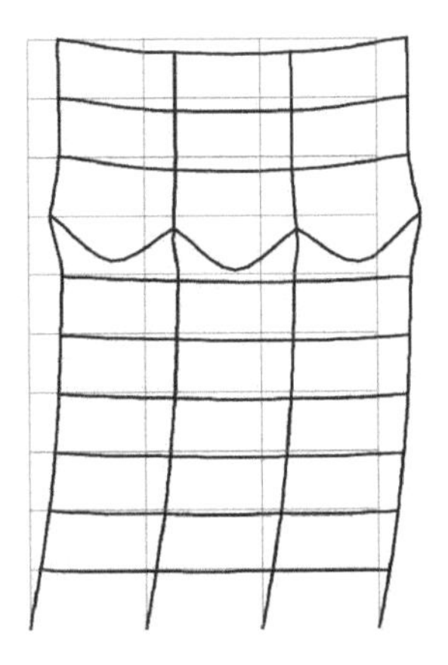

(d) Failure shape based on Dt = 25 min

Fig. 6.60: PEF resistance of the frame, based on scenario 3.

6.10 Behavior of tall steel structures under horizontally traveling fire

This section investigates the response of tall steel structures with large open compartments subject to PEF. As pointed out earlier, the application of parametric fires is limited to small- and medium-sized compartments, in which it is assumed that the temperature is uniform throughout the compartments. In large compartments, however, this assumption is not justifiable.

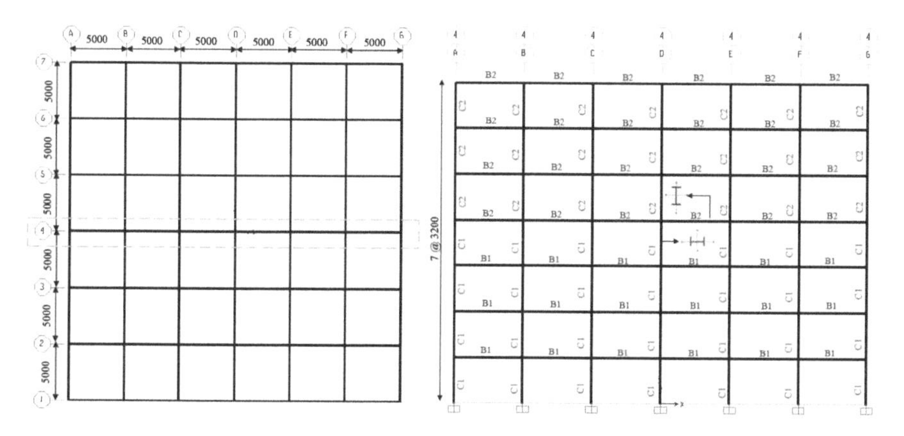

B1: IPE 500, B2: IPE 400, C1: IPE 600, C2: IPE 550

(a) The case study: plan view (b) Cross-sections of beams and columns

Fig. 6.61: The case study: plan view and structural properties [228].

The plan examined in section 6.6 (where a multi-story RC structure was investigated) is used here as a steel skeleton. The structure, however, is designed for seven stories and for a PGA of 0.3g. The structure is dimensioned for load combinations of 7.5 kPa for dead load and 2.0 kPa for live load. A combination of 100% dead load and 20% live load for the designed frame is used to find the required mass for calculating the earthquake load. The plan view and the structural properties are shown in Fig. 6.61. The slab is made of normal-weight concrete with a 100 mm thickness. The thermal properties of the steel and the concrete slab are taken into account using Eurocode 4. Fire sizes of 17%, 50% and 100% (corresponding to the size of the first span, the half of the plan and the entire plan, respectively) are selected. It is assumed that the fire initiates from the left side and spreads to the right, as shown in Fig. 6.62.

The gas phase temperatures for beams and columns are then determined, some of which shown in Fig. 6.63. As the differences between the heating and cooling phases of the columns and beams lead to different thermal load distributions, the effect of the traveling fire on the structure can hence be

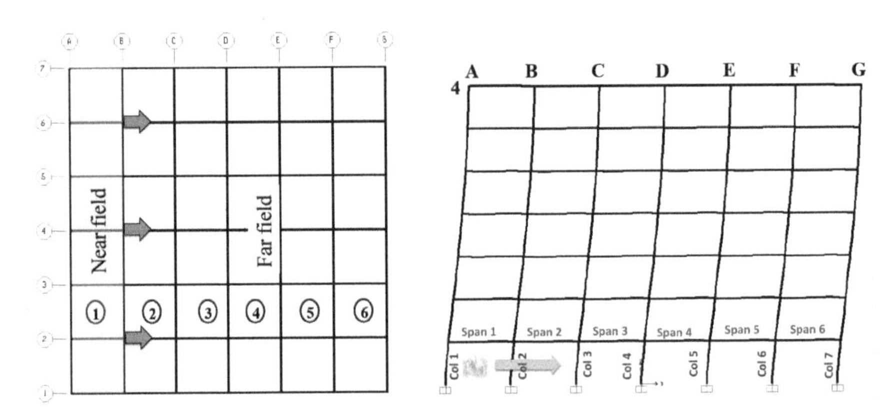

Fig. 6.62: Application of the traveling fire to the case study.

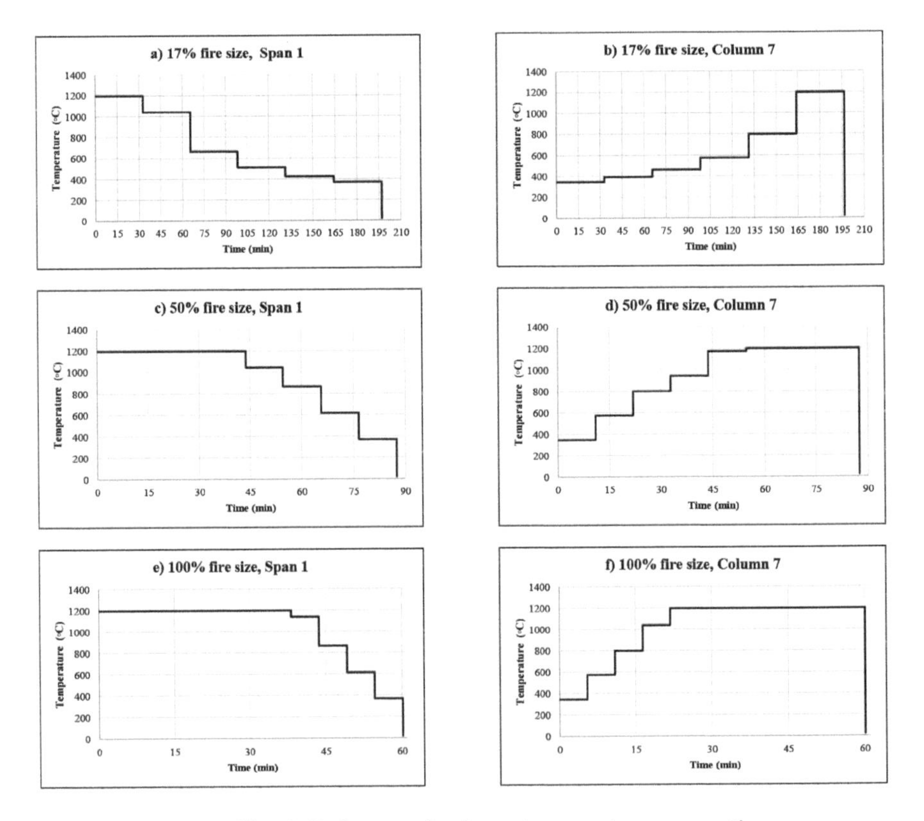

Fig. 6.63: Span and column temperature versus time in 17%, 50.0%, and 100% fire sizes.

investigated. As an example, Fig. 6.64 shows the temperature distribution inside beams B1 and B6 at axes A and G under the 16.7% ($\approx$17%) fire. To draw a comparison, the temperature distribution inside beam B1 under the

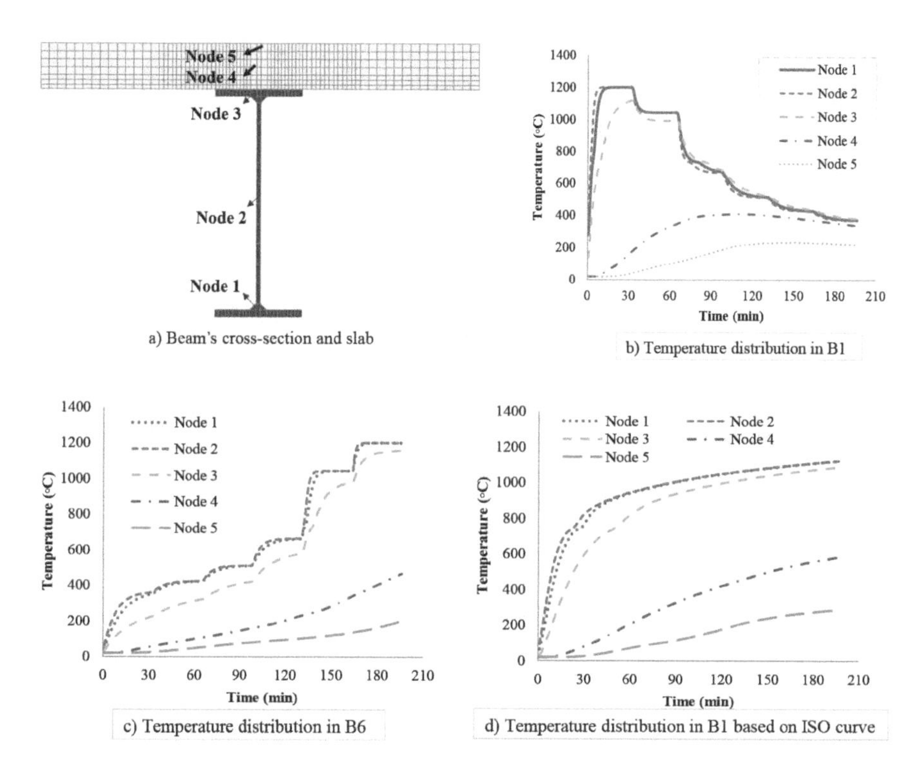

Fig. 6.64: Distribution of temperature inside beams B1 and B6 under the 17% and ISO834 fire curves.

ISO834 fire curve, i.e. a uniform fire, is also illustrated in Fig. 6.64(d). This shows a notable difference between the heat propagation inside the beams under different fire curves.

The determined fire curves are applied to the structure over two scenarios: prior to earthquake, and after earthquake. The seismic analysis is performed using pushover analysis. To achieve this, the structure is pushed to arrive at the target displacement corresponding to the LS level of performance, i.e. 448 mm.

Figure 6.65a, b, and c shows mid-span deflection versus time in B1 based on different fire sizes. As a comparison, the result of the fire analysis under the ISO834 fire curve is also shown (Fig. 6.65d). As is seen in Fig. 6.65a, where the frame is subjected to the 17% fire size, the analysis terminates at 13 minutes. Based on the mid-span deflection and the rate of deflection failure criteria, the beam is considered to have failed if those values exceed 250 mm and 5.56 mm/min, respectively. Based on the mid-span deflection criterion, at 13 minutes the beam cannot be considered to have failed; however, it has failed at 12 minutes if the rate of deflection criterion is considered. The dotted line in the figure shows the results of the fire-alone analysis, indicating that the PEF resistance declines by about 36% at around 19 minutes.

Fig. 6.65b shows the fire resistance of the frame under the 50% fire size. Based on the rate of deflection criterion, which is the predominant criterion here, failure occurs at 9 minutes. This shows a considerably lower fire resistance than for the 16.7% fire size. The dotted line in the figure shows that the fire resistance of the frame under the fire-alone scenario is about 13 minutes. Fig. 6.65c shows the fire resistance of the frame under the 100% traveling fire size. This shows that, based on the rate of deflection criterion, the frame fails at 11 minutes. It is interesting that the fire resistance at the 50% fire size is even lower than at the 100% fire size. A comparison is also made between the fire resistance and the PEF resistance of the frame, showing that the frame fails at around 14 minutes under the fire-alone scenario. The fire resistance of the frame under the uniform fire curve, ISO834, is about 19 minutes, as shown in Fig. 6.65d. However, in terms of the rate of deflection criterion, it fails at 16 minutes. The fire-alone analysis is also performed to provide an understanding about the behavior of the structure, showing that the structure fails at around 22 minutes.

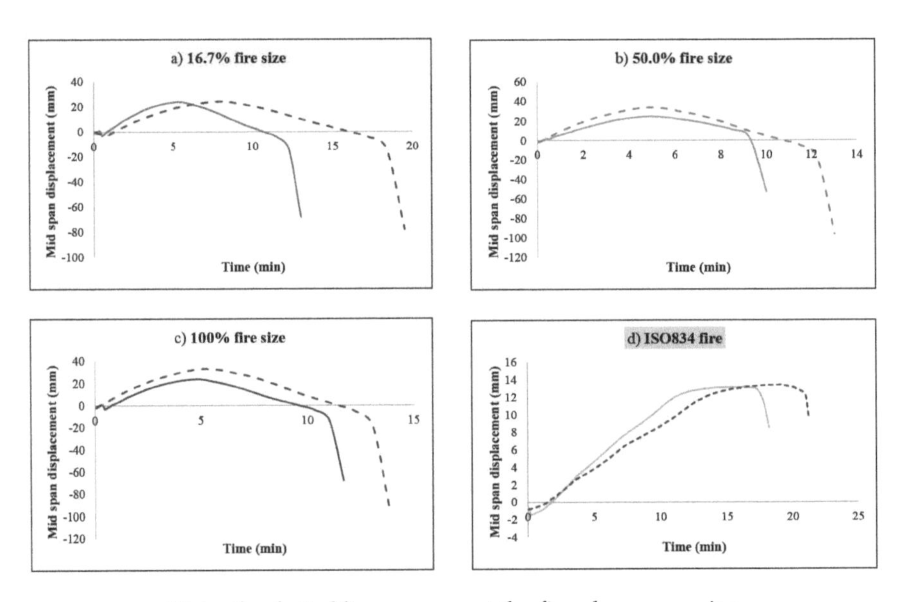

Note: the dotted lines represent the fire-alone scenarios

Fig. 6.65: Fire and PEF resistances of the structure, based on different fire sizes and the ISO834 curve.

While it is evident from the above results that the fire resistance of the frame under a traveling fire, even when the 100% fire size is considered, is much lower than under uniform fire, such as ISO834. This contradicts the traditional belief, which deems that assuming a uniform temperature in a compartment is a conservative assumption and thus can increase the safety margin.

6.11 Response of vertically irregular tall moment-resisting frames under fire

It is accepted that many building structures are irregular, usually as a result of architectural demands. Recorded statistics have confirmed that irregular structures are more vulnerable to earthquakes than regular structures, as explained in section 2.2.3. Previous case studies have shown that earthquake-damaged structures are more vulnerable to fire than intact structures; it could therefore be expected that this vulnerability would even be greater in irregular structures. In this section, a seven-story tall moment-resisting frame, designed for the LS level of performance, is selected. The selected case is loaded and designed based on the AISC code, and the properties of the designed frame are presented in Fig. 6.66. A geometric irregularity is then imposed on the original frame, by changing the horizontal dimension of the lateral resisting systems. As pointed out in Chapter 2, a story is considered vertically irregular if the horizontal dimension of the lateral force resisting systems in a story is more than 130% that of an adjacent story.

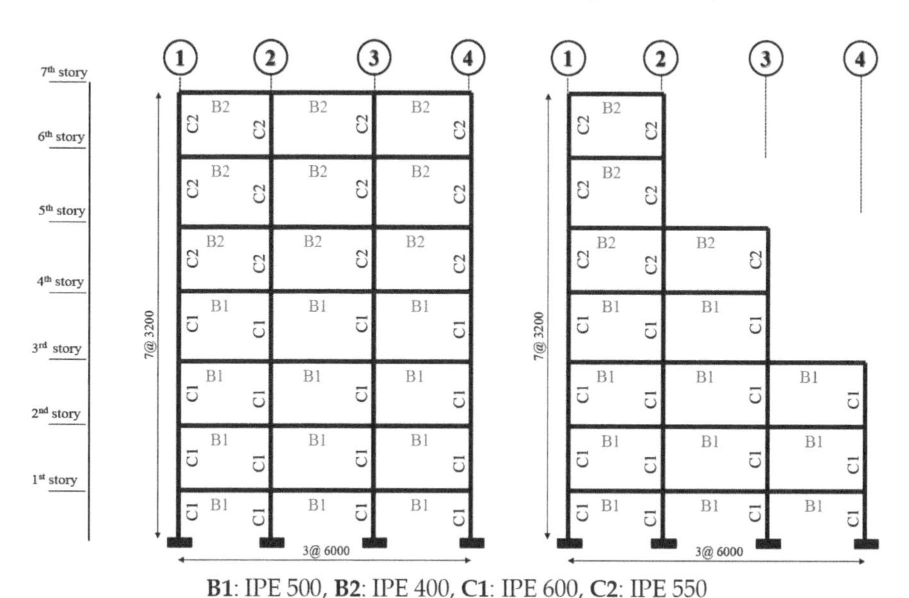

B1: IPE 500, **B2**: IPE 400, **C1**: IPE 600, **C2**: IPE 550

(a) The regular frame (b) The geometric irregular frame

Fig. 6.66: The case study [97].

The original frame is loaded and designed using steel profiles with a yield stress of 240 MPa and a PGA of 0.35g. The frame is dimensioned for the load combinations of 8.0 kPa for dead load and 3.5 kPa for live load. A combination of 100% dead load and 20% live load for the designed frame is used to find the required mass for calculating the earthquake load. For the fire analysis, the thermal expansion coefficient of steel is assumed to be $12\times10^{-6}/°C$. It is assumed that the structural members are not fireproofed when subjected to the fire loads. This assumption closely mimics reality for the PEF analysis, due to the high possibility of damage to fireproofing materials under earthquake loads.

The fire scenarios are shown in Fig. 6.67, whereby the first story in various assumptions is subjected to fire. The fire curve employed here is the generalized exponential curve, as shown in Equation 6.3, where T_0 is assumed to be 20°C, T_{max} is the maximum compartment temperature and α is an optional 'rate of heating'. It is evident that the temperature variation in Equation 6.3 is highly dependent on the adopted values of T_{max} and α, which can directly influence the fire-resistance rating of structures. Nevertheless, this equation is often used for comparative studies and not for determining the fire resistance. If the purpose of an investigation is to account for the fire-resistance rating, more accurate fire curves should be taken into consideration, as explained earlier. As the study here plans only to compare the fire resistances of regular structures and vertically irregular structures, any fire curve can be employed. It is also worth noting that in this equation, the values of T_{max} and α are rooted in the data collected from real compartment fires; thus, they are case-dependent. In some studies, it is mentioned that T_{max} changes in the range of 800 to 1000°C in small and medium compartments [49] and up to 1200°C in large compartments [50], but there are also some studies that have shown that T_{max} in large compartments is lower than in small compartments [51, 52].

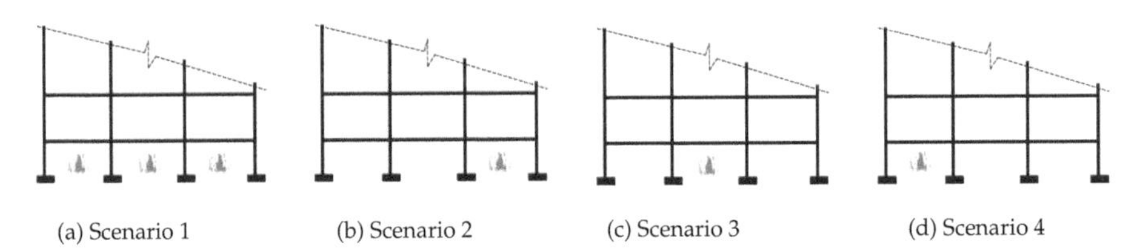

Fig. 6.67: Fire scenarios.

On the other hand, α is proposed based on the fire growth rate, which depends on the fuel type in a compartment, and which varies from very slow to very rapid. It is evident that when a compartment contains highly flammable materials, the fire has a very rapid growth rate. Most urban buildings, such as residential and office structures, are furnished with materials that are not very flammable, and thus the fire growth rate is not considerable. For the case study here, a rate of heating of 0.08 and a maximum temperature of 850°C are assumed proper. Figure 6.68 shows the temperature versus time based on these assumptions for 60 minutes.

To perform the pushover analysis, the structure is subjected to a monotonically increasing lateral load, in order to arrive at a certain level of displacement corresponding to the target displacement. The target displacements at the LS performance level are calculated as 438 mm and 472 mm for the regular and the irregular frames, respectively. Figure 6.69 shows the pushover curves for the mentioned performance level and the response spectrum curve used in the pushover analysis.

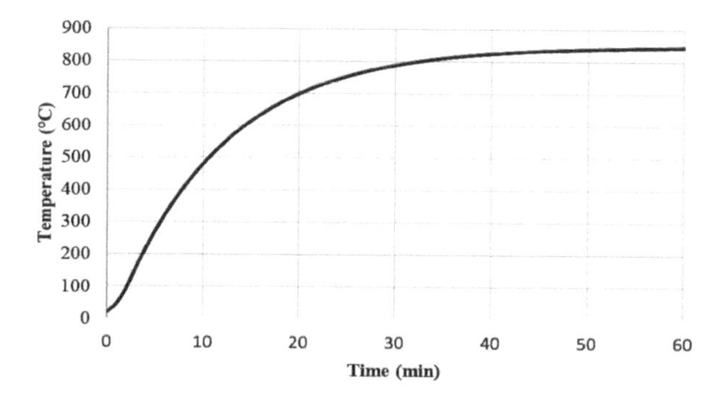

Fig. 6.68: Time-temperature based on generalized exponential curve.

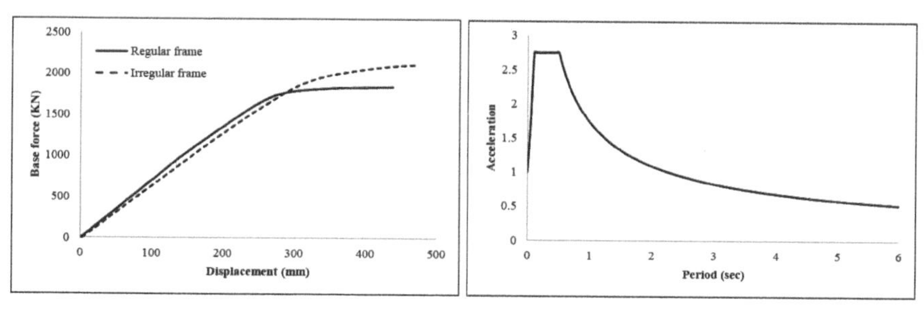

(a) The pushover curve (b) The response spectrum

Fig. 6.69: The pushover curve at the LS level for the regular frame and the geometrically irregular frame.

After conducting the pushover analysis, the structural performance of all structural elements is controlled, based on a number of acceptance criteria. These criteria include deformation-controlled and force-controlled actions, together with inter-story drift. Regarding drift control, Fig. 6.70 shows the drift variations of the regular and irregular frames in the stories. The figure shows the drift values in the stories are lower than the drift allowed for the LS level of performance (that is, 2.5%). However, it is seen from the figure that the drift value of the irregular frame is greater than that of the regular frame. This implicitly shows that the irregular frame has sustained more damage after the earthquake than the regular frame.

Regarding rotation control, the criteria are defined using the force-deformation (or rotation) equations introduced in Chapter 2 by Equations 2.1 to 2.6. Control of the structure after the pushover analysis can confirm whether the acceptance of these criteria complies. At the LS level of performance,

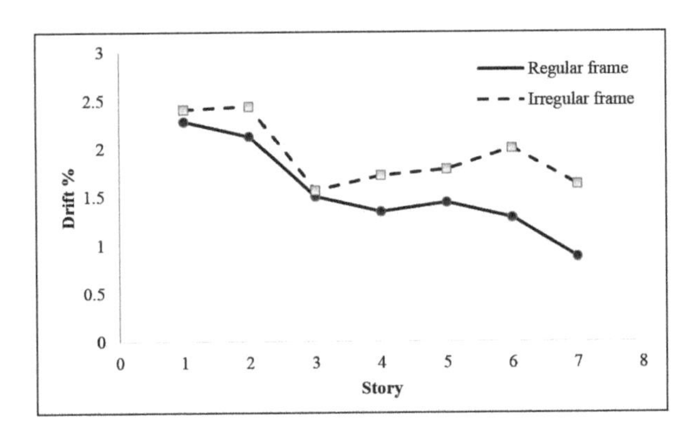

Fig. 6.70: Drift values of the regular and irregular frames [97].

the plastic rotation in seismically-compact beams is between 6θy and 8θy. When the plastic rotation goes beyond 8θy, it corresponds to the CP level of performance. Table 6.8 shows the results of such control, showing that both frames remain under the LS level of performance. From the table, however, it is seen that the damage sustained by the irregular frame is greater than that of the regular frame.

Table 6.8: Control of the acceptance criteria

	Section type	θ_y	*Limit state*	*Acceptance criteria*	*Maximum rotation after the pushover analysis in the **regular** frame*	*Maximum rotation after the pushover analysis in the **irregular** frame*
			IO	0.0144		
Column section	IPE 600	0.0072	LS	0.0504	IO < 0.0437 < LS	IO < 0.0477 < LS
			CP	0.0648		
			IO	0.0158		
	IPE 550	0.0079	LS	0.0553	IO < 0.0483 < LS	IO < 0.0513 < LS
			CP	0.0711		
			IO	0.0172		
Beam section	IPE 500	0.0086	LS	0.0602	IO < 0.0497 < LS	IO < 0.0527 < LS
			CP	0.0774		
			IO	0.0214		
	IPE 400	0.0107	LS	0.0749	0.0204 < IO	IO < 0.0364 < LS
			CP	0.0963		

Figure 6.71 shows the displacement versus time for the regular frame, based on scenario 1, when the fire is applied to all of the bays of the first story. The figure shows that the fire resistance is about 32 minutes under

the fire-alone condition, but that it declines to about 25 minutes under the PEF condition. The sharp increase in the displacements seen in Fig. 6.71b is due to the structures being first laterally pushed to arrive at the target displacement and then being unloaded. Figure 6.71a and b also shows that the frame collapses differently under different fire conditions, in such a way that it fails almost locally under the fire-alone condition, but it fails globally under the PEF condition.

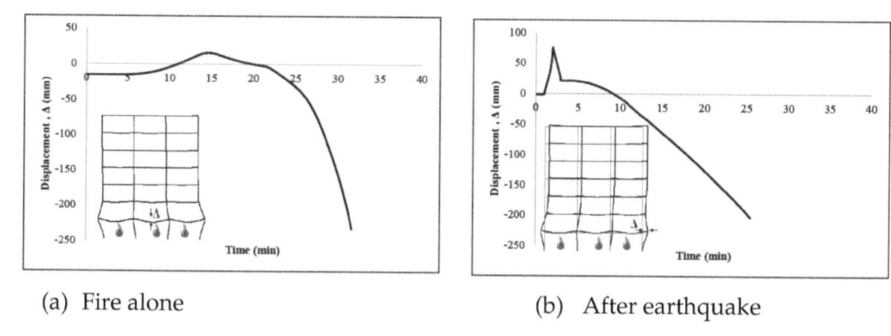

(a) Fire alone (b) After earthquake

Fig. 6.71: Fire resistance of the *regular* frame, based on scenario 1.

Figure 6.72 shows the fire and PEF resistance of the regular frame when scenario 2 is considered. As is seen, the frame fails at 36 minutes under the fire-alone condition but it fails at 34 minutes under the PEF condition. This shows a minor reduction in the resistance of the frame after the PEF. The collapse mode in both situations, i.e. before and after the earthquake, is mainly local. It is evident that, due to symmetry, the fire resistance in scenario 4 is similar to that in scenario 2.

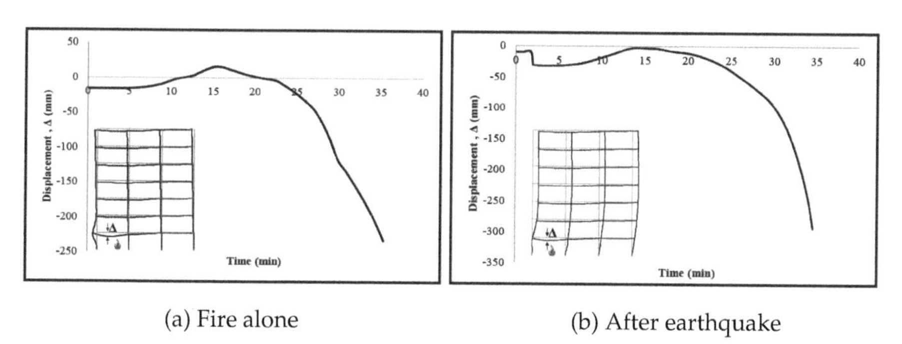

(a) Fire alone (b) After earthquake

Fig. 6.72: Fire resistance of the *regular* frame, based on scenario 2.

In a similar vein, Fig. 6.73 shows the resistance of the regular frame under scenario 3, when only the mid-bay is subjected to fire and PEF. As seen, application of the fire alone to the frame results in a fire resistance of 37 minutes, with local failure being the predominant failure mode. When the frame is exposed to the PEF, the fire resistance declines to about 33 minutes. In that case, the frame again fails locally.

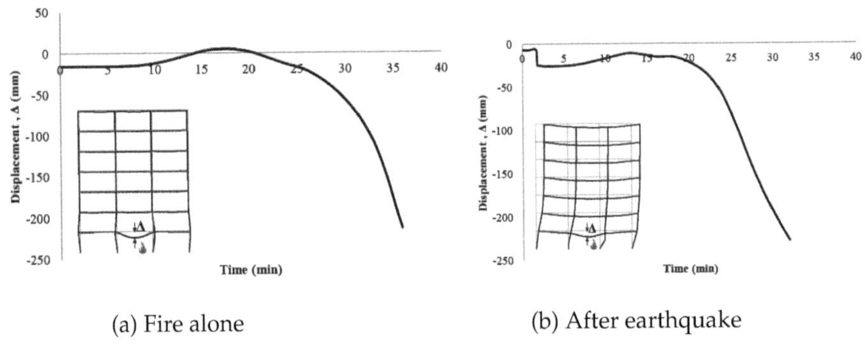

Fig. 6.73: Fire resistance of the *regular* frame, based on scenario 3.

Figure 6.74 shows the fire resistance of the irregular frame under scenario 1, before and after the earthquake. As is seen, under the fire-alone condition the frame fails at 33 minutes, but under the PEF, it fails at 29 minutes. The dominant failure mode is also shown in the figure. Interestingly, the results show that the PEF resistance of the irregular frame is greater than that of the regular frame under the same PEF scenario.

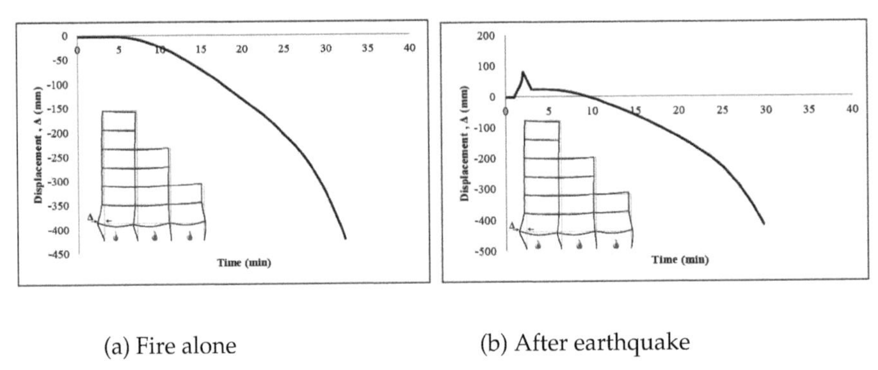

Fig. 6.74: Fire resistance of the *irregular* frame, based on scenario 1.

Figure 6.75 shows the response of the irregular frame under scenario 2, i.e. when the fire is applied to the right bay. As is seen, under the fire-alone condition, the frame fails at 48 minutes, while it fails at 20 minutes under the PEF condition, showing a notable reduction in fire resistance. Regarding the collapse mode, the failure is local when only the fire is considered. The failure mode changes to global when the PEF is applied to the frame.

Figure 6.76 shows the fire resistance of the irregular frame when the fire is applied to the middle bay, prior to and after the earthquake. In that case, the frame fails at 40 minutes and 38 minutes under the fire-alone and the PEF conditions, respectively, showing a minor difference in the fire resistance rating. The frame fails locally in both conditions, before and after the earthquake.

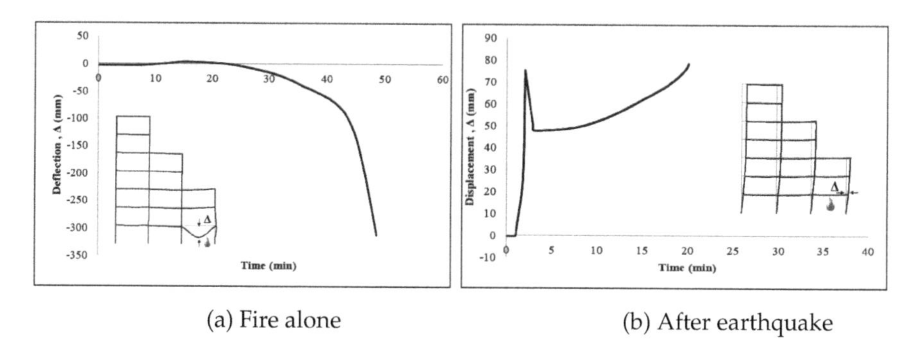

(a) Fire alone (b) After earthquake

Fig. 6.75: Fire resistance of the *irregular* frame, based on scenario 2.

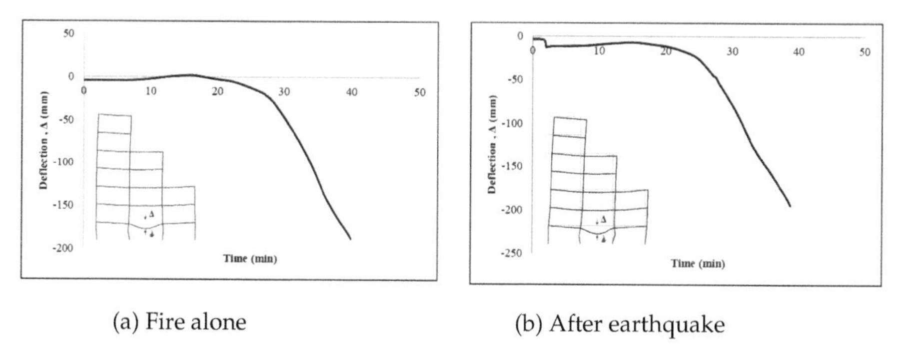

(a) Fire alone (b) After earthquake

Fig. 6.76: Fire resistance of the *irregular* frame, based on scenario 3.

Figure 6.77 shows the fire and PEF resistance of the irregular frame under scenario 4, when only the left bay of the frame is subjected to fire. As is seen, while the fire resistance is about 39 minutes, it declines to about 34 minutes under the PEF loading. For both conditions (before and after the earthquake), the predominant failure mode is local.

The results of the above analyses show that in almost all of the fire scenarios, the irregular frame is more vulnerable to PEF load than the regular frame. This is particularly so when only one bay is exposed to PEF. It is also

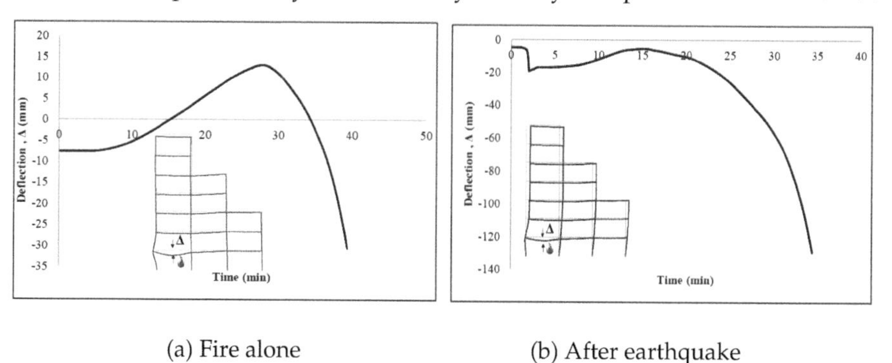

(a) Fire alone (b) After earthquake

Fig. 6.77: Fire resistance of the *irregular* frame, based on scenario 4.

important to consider the failure style. The analyses showed that both the regular and irregular frames failed locally when they were subjected to fire alone. With the PEF, the regular frame failed globally only when all of the bays in a story were subjected to PEF; in the other scenarios, only local collapse was observed. The irregular frame, however, in almost all of the PEF scenarios, failed globally. This confirms the suspicion that irregular frames are more susceptible to global collapse than regular frames in the case of PEF.

Given that the global failure of structures would create a more chaotic situation in urban regions than would local failure, more preventive strategies should thus be put into place when designing irregular structures to withstand extreme loading, such as PEF.

Chapter 7

Risk Management Strategies for Post-earthquake Fire

7.1 Introduction

It was shown in the previous chapter that conventional structures are generally too weak when they encounter PEF loads. This weakness is particularly highlighted in structures designed for the LS and CP levels of performance, as these categories of structures sustain more earthquake damage than other structures designed, for example, for IO level.

To quantify how partially seismic-damaged structures respond under PEF loading, a sequential methodology was introduced and employed, whereby numerous case studies (including steel and RC structures, both low- and high-rise) were investigated. The results of the investigations performed in the previous chapter showed that the PEF resistance of structures can be reduced, even by as much as half, when compared to their resistance to fire alone, i.e. fire prior to earthquake. In bare steel structures, steel materials are often extremely vulnerable to high temperature and can lose their strength and stiffness at a rapid rate. Drift value at different performance levels can also adversely influence the fire resistance, in such a way that the more a structure sustains drift, the greater will be the reduction in fire resistance.

On the other hand, fireproofed steel structures would be expected to show an acceptable response under fire-alone loads. Fireproofing materials are, however, weak under tension and compression. Therefore, when a structural component is subjected to seismic loads, there is a strong possibility of tearing and spalling of the fireproofing materials. Experimental and numerical investigations have shown that when the drift value exceeds about 1%, say when there is some plasticity and residual displacement in structures, the fireproofing materials added to the surface of the structural components will no longer work properly. The PEF resistance of those structures hence declines considerably. Although RC structures generally have a good resistance against fire load, they also might be vulnerable after earthquake if the concrete cover sustains severe damage or spalls off. Thus, almost all RC structures designed for LS or CP level of performance are vulnerable toward PEF, as was shown in the previous chapter.

7.2 Risk management strategies

The above explanation confirms the vital need for adopting mitigation strategies in order to respond properly to PEFs, particularly in urban regions. It is well-accepted that a risk management process, in general, has three steps: risk identification, risk assessment, and risk management. Risk is basically defined as a virtual threat that either restrains the ability of a community from accomplishing a defined operation or reduces the level of safety [352]. Although most risks are unpredictable, this does not necessarily imply that they are unknown; rather, almost all of them can be listed. The assessment of risks can then be performed through qualitative-based, quantitative-based, or semi-quantitative-based methods. The results of a risk assessment can then lead to a decision being made as to whether the risks should be treated. If that decision is affirmative, risk management begins by setting up a series of priorities, to decide which risk is more crucial, before deciding an approach for dealing with the situation. In other words, risk management is a systematic approach to considering the possibilities of unpredictable but realistic disasters before they occur, so that these threats can be avoided, or their impact minimized.

Risk management involves several steps, such as setting up scopes and boundaries, defining the project's timeline, and establishing the objectives and the resources required. The strategies to handle risk are typically classified in order to transfer the risk from one situation to another or to avoid the risk if possible. An alternative is to mitigate the adverse impact while accepting some risk. It must be noted that no exact acceptable risk is set, because whether a risk is acceptable or not is generally dependent on the decision of the policy-makers. For most of the time, whether a risk is acceptable or otherwise depends on the associated level of safety required, which is a case-dependent decision. For example, a higher level of safety is needed for critical infrastructure, such as a high-speed train, than for a residential building, because more lives could be endangered in the case of a disaster. Therefore, every case is designed based on its importance level.

In terms of proposing a strategy for the risk management of PEF, this needs to consider both the effect of an earthquake and the subsequent effect of the fire on structures/inhabitants, to arrive at the decision to either overcome or minimize the risk. Figure 7.1 schematically describes how seismic-damaged buildings are more vulnerable in a PEF situation.

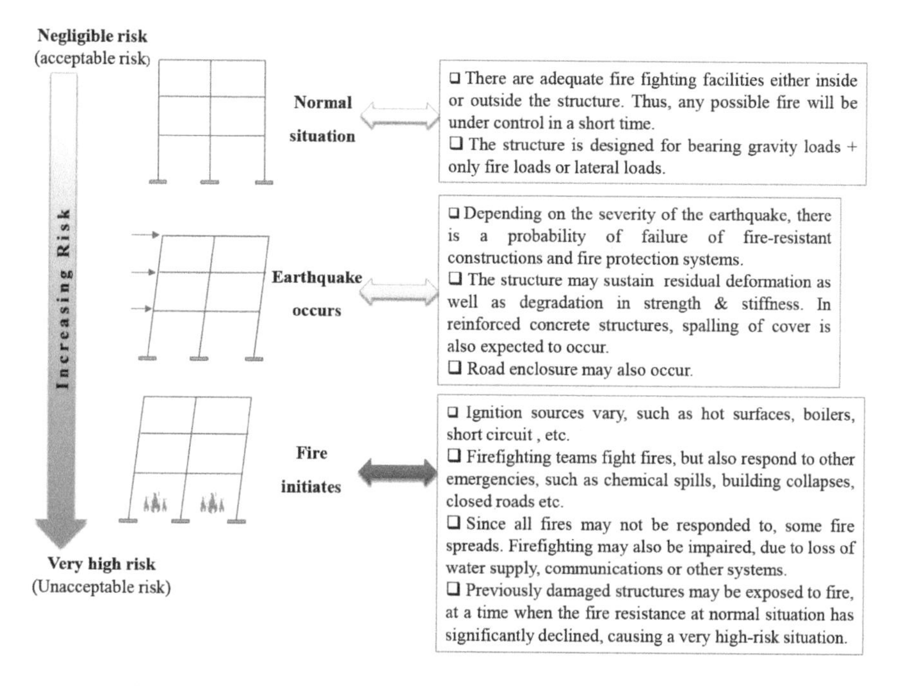

Fig. 7.1: Increasing vulnerability of buildings under PEF [353].

From a different point of view, mitigation strategies can be preventive or responsive. The former addresses mainly steps via which the risk of a PEF event in a specific region is reduced, while the latter addresses the ways via which the level of damage, both casualties and financial losses, is reduced. Given that PEF is a disaster, mitigation strategies can be proposed using a general disaster management plan, as shown in Fig. 7.2. Although the figure shows that the pre- and post-disaster strategies are discrete, they are indeed connected to each other, in a kind of cycle. Hence, while the prevention and preparedness strategies are being put in place, the response and recovery planning might also be under consideration. In combination, these can provide ultimate possible readiness against PEF.

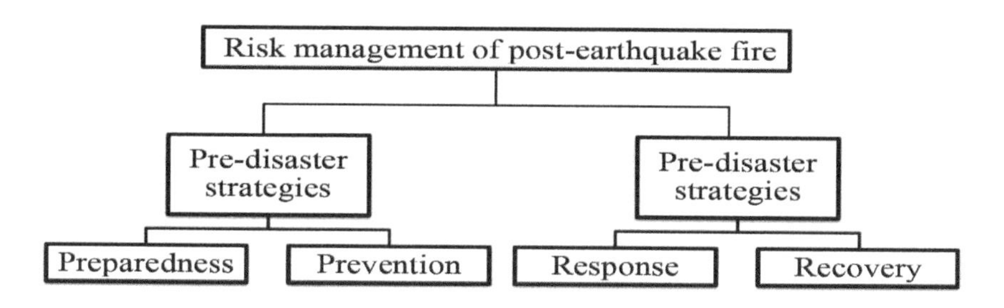

Fig. 7.2: General disaster management plan.

In general, preparedness and prevention strategies comprise those activities that are conducted prior to a disaster and which are adopted to enhance the efficacy of the operational capabilities [354]. To achieve this, the planning must cover: the hazard identification, the countermeasures for aid and rescue; and public awareness programs and training. On the other hand, the response activities are implemented to address the consequences of the disaster once it has occurred. These activities are grouped into three: 1) to identify the potentiality of a disaster in such a way that both people's lives and their property are saved (it is evident that achieving this requires as much safety enhancement as possible); 2) if the disaster occurs, the priority is to save lives and minimize monetary loss as far as possible (it is also evident that if saving property takes priority, much more attention must be paid to enhancing structural safety, e.g. all buildings will be designed as important structures – achievable but costly); and 3) when the response activities are swung into action immediately after the disaster, there is always the possibility of a further disaster leading to even worse outcomes, e.g. PEF [355] (in that case, all emergency activities will beconcentrated on the damaged regions in order to reduce the number of fatalities).

On the other hand, while the response activities are being conducted, the residents of the affected regions should be gradually gaining some capacity to resume their usual activities through adopting some recovery strategies. There is, however, no distinct point at which these recovery activities start. It can be said that adopting recovery strategies can improve the preparedness of a region for future disasters, thus reducing vulnerability. Ideally, there should be a smooth transition from recovery to ongoing development. Recovery activities continue until the affected region is at least back to what it was before the disaster, or maybe until it has become even better [356]. To achieve this, both short- and long-term activities must be adopted. These include: returning vital life-support systems to minimum operating standards (such as medical centers, water supply, etc.); providing temporary housing, public information, and health and safety education; commencing reconstruction; establishing counseling programs; and conducting economic impact studies. Essential information resources and services include data collection relating to rebuilding, and documentation of lessons learned.

7.3 Risk mitigation of PEFs: Macro-scale strategies

In a seismic region, a possible PEF can be managed in advance in two ways, by adopting both macro-scale strategies, where the urban area is directly addressed, and micro-scale strategies, where a building's structure is addressed. A mesoscopic scale can also be adopted, whereby the interaction between macro and micro strategies is addressed.

When employing macro-scale strategies, an endeavor is made to maximize the understanding of achieved results from simulations of various assumed PEF scenarios in the area. To achieve that, all information in the area is carefully collected, including building characteristics (such as types, heights, density, etc.) and the conditions of infrastructure (such as water supply systems). As pointed out in Chapter 1, weather conditions can additionally intensify the development and spread of PEF. Therefore, PEF simulations are enhanced by considering weather conditions as well [357, 358]. The results of such simulations can show how the assumed area would behave under PEF scenarios. Hence, some treatments can be proposed, either to reduce the risk of PEF development and spread, or to evacuate inhabitants from the area to improvetheir safety. These treatments would focus on enhancing the components of the urban area, such as by widening the alleys and streets to avoid the spread of fire through radiation, or by improving the infrastructure, e.g. fire stations and rescue centers.

Having a well-trained population can also considerably reduce the loss from PEF, and a training program should be implemented as part of preparedness activities [359]. This training usually follows a program known as "Community Emergency Response Teams (CERT)". Potential CERT members are first identified among the citizens of a community, and then gradually trained to become prepared to respond in an emergency situation in the community. As documented, the Japanese became pioneers in establishing CERTs, following the PEF of Kantō in 1923. In 1985, a group of Los Angeles City officials were sent to learn about the Japan-wide earthquake preparedness plans. After this, CERTs were rapidly introduced worldwide, to the extent that the author has discovered that more than 98 countries have established CERTs to date. The effectiveness of training has been proven in reducing the risks of numerous natural and human disasters [360, 361].

It is understood that a macro-scale PEF strategy will complete risk identification and risk assessment processes before deciding which mitigation strategies to adopt. Since PEFs, like other disasters, can give rise to very different kinds of damage, a question then arises as to which aspects of the PEF consequences are to be addressed. These consequences can be economic, social and environmental. Potential PEF-mitigation strategies must then be designed to meet the considered aspects. For example, it is assumed here that the social and environmental aspects of PEF in an urban region are worthy of assessment. For this, the first step is to identify and list the potential social and environmental PEF risks, some of the most relevant of which are shown in Fig. 7.3.

The influence of these risks on each other is then understood through a risk assessment process. The risk assessment can be conducted through three methodologies: system analysis, network analysis, and hierarchy analysis; each of these can be quantitative-, qualitative- or semi-quantitative-based. These methodologies are shown schematically in Fig. 7.4. Using system analysis, the effect of each component on another over a chain of events is investigated. Network analysis is similar to system analysis, with the difference that here the events are considered over a period of time. In hierarchy analysis, all components are analyzed over a hierarchical process

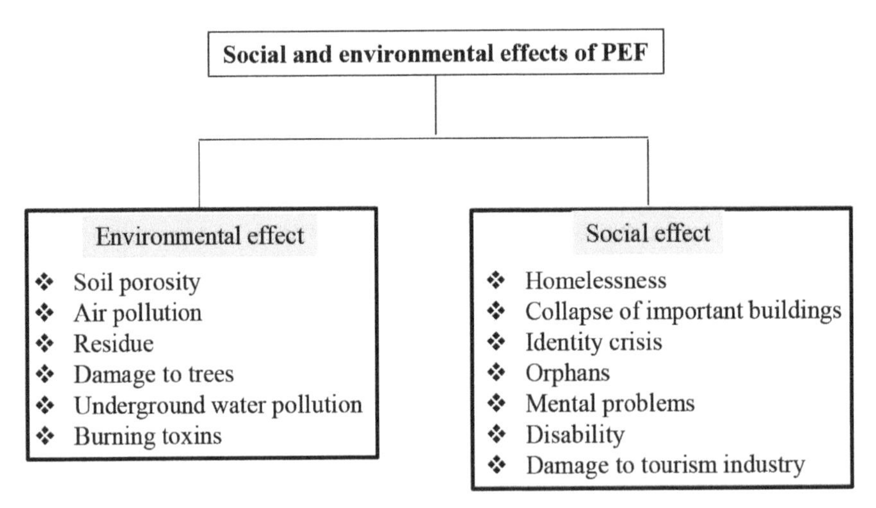

Fig. 7.3: Social and environmental effects of PEF.

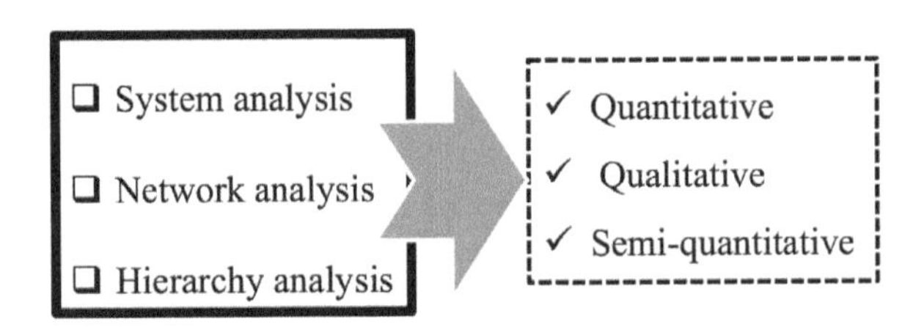

Fig. 7.4: Different analysis methods.

from top to bottom [362]. Referring to the performance-based process (as explained in Chapter 2), it seems that hierarchy analysis follows a performance-based approach to a great extent. In terms of modeling, a PEF event has a great similarity to hierarchy analysis, so here the social and environmental effects of PEF are analyzed using this method.

It is generally accepted that quantitative-based methods are more accurate than qualitative-based methods. Quantitative-based methods, however, are often time-consuming processes. In addition, they need adequate and reliable data, which is not achievable all of the time. As a compromise, semi-quantitative-based methods can be used, such as multi-criteria decision-making (MCDM) [363], which are able to weigh and score a range of criteria. The scores are then prioritized into expertise and related-interest groups. Although using semi-quantitative-based methods can provide some ease, they can also be judged negatively in terms of their accuracy level. In this respect, it is strongly advised to use the results of a qualitative-based method, such as the Delphi method [364], as an input to a semi-quantitative-based method. Figure 7.5 shows frequency of use versus complexity of the abovementioned risk assessment methods.

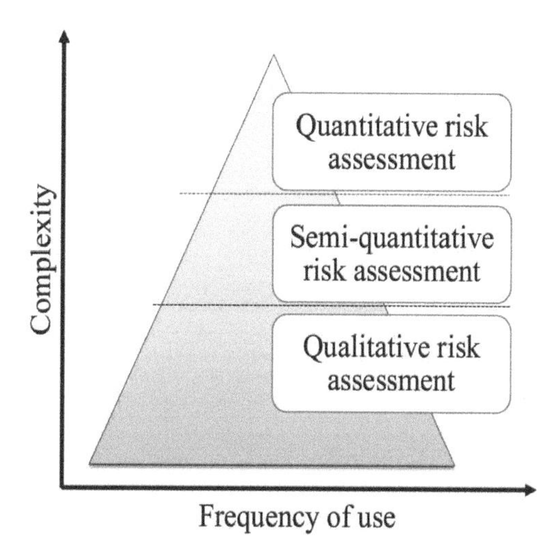

Fig. 7.5: The reliability of different risk assessment methods.

One of the most well-known MCDM techniques is the analytic hierarchy process (AHP) developed by Thomas L. Saaty in the 1970s [365]. The AHP method has been used extensively in solving complex decisions in different fields, such as industry, business and engineering. The AHP method is employed by setting up five stages, as summarized here.

- The problems and objectives are defined.
- The objectives (from a general view to the detailed view) are hierarchically developed from top to bottom, i.e. there might be several layers.
- A simple pair-wise comparison matrix is made for each layer.
- The consistency of the judgements in the previous stage is controlled.
- The relative weight of the components of each level is estimated.

The application of the AHP method to assess the social and environmental effects of PEF is shown in Fig. 7.6. The assessment can be performed through two different approaches: bottom up, or top down. The bottom to top approach is used if the statistical weights of the listed risks are already available. In that case, the assessment is made to find the results of the upper layers, which here are the cost, speed, and necessity of repair, and finally the environmental and social effects. Since in this example, there are no statistics available, the top down approach is followed, as shown.

As seen in Fig. 7.6, there are four layers, which start with the determination of the objective of the assessment, and define the problems, which are the environmental and social effects. As the cost, speed, and necessity of repair are priorities in a PEF management strategy, they are defined in a separate layer. It is worth noting that the necessity of repair can be interpreted as a need for responding to the socioeconomic effects of PEF. The risks defined in the last layer are then compared in pairs, considering their importance to

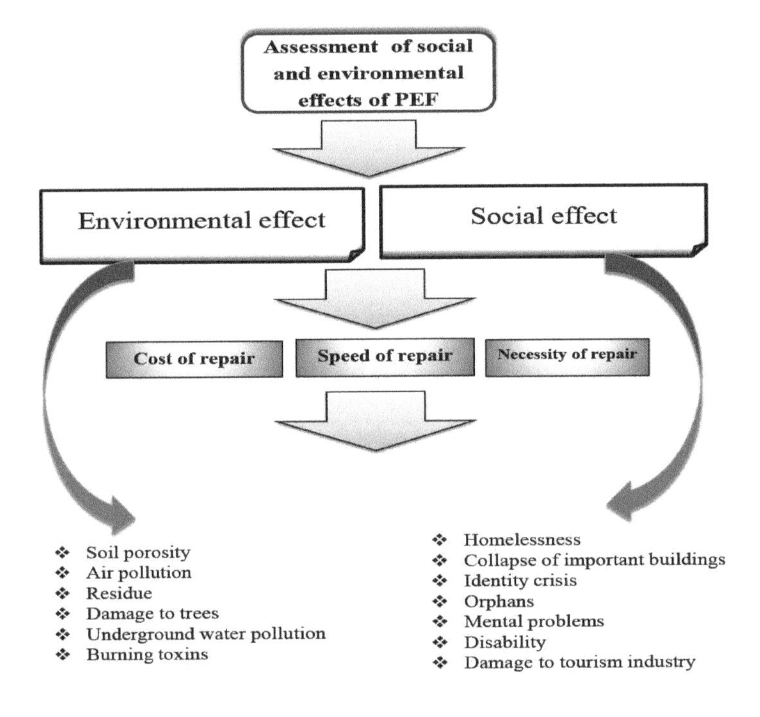

Fig. 7.6: The application of the AHP method to the assessment of social and environmental PEF effects.

a component in the higher layer. The results of the AHP analysis are shown in Fig. 7.7. As seen, the environmental effects make up more than 85% of the whole, meaning that the social effects comprise less than 15%. The relative weight of the environmental effects can be correlated to the sensitivity of the environment to fire; it can also be said that the environmental effects will directly affect a society and its economy.

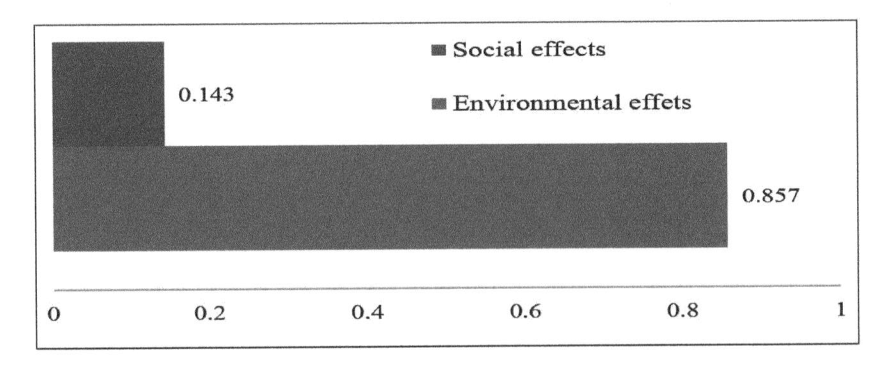

Fig. 7.7: The social and environmental assessment.

The relative weights of cost, speed, and necessity of repair when only the social effects are considered are shown in Fig. 7.8. It is seen that the speed of repair takes priority, followed by the necessity of recovery and finally the cost of recovery.

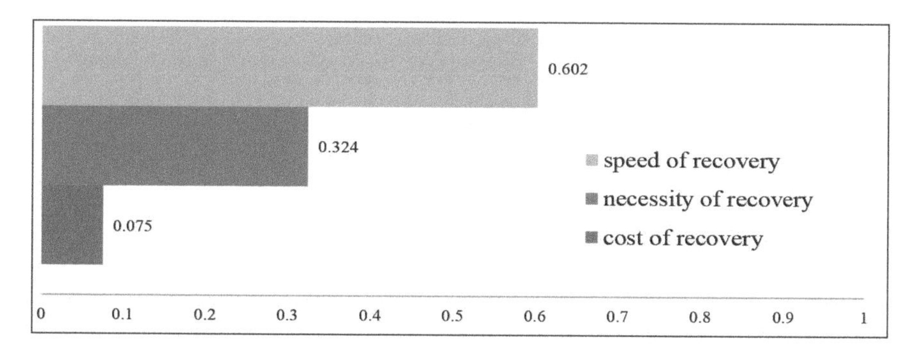

Fig. 7.8: The relative weights of cost, speed, and necessity of repair in social effects.

In a similar vein, the relative weights of cost, speed, and necessity of repair when the environmental effects are considered are determined as shown in Fig. 7.9. As is seen, the necessity of recovery takes priority, which is different to the case for the social effects. The second priority is to accelerate the recovery process and the last priority is the cost of recovery itself.

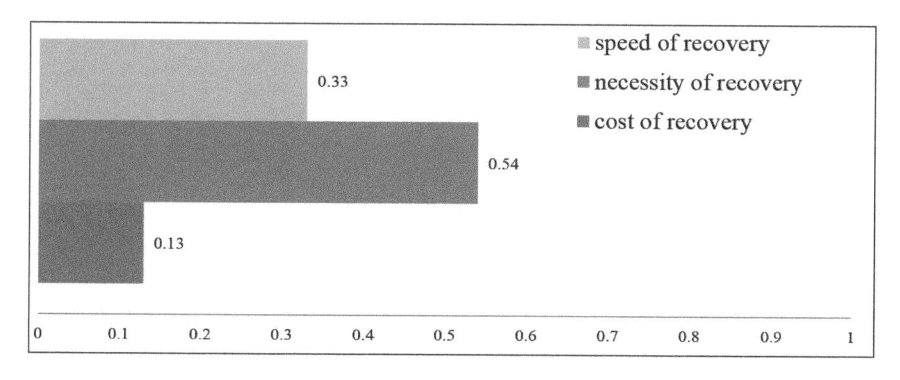

Fig. 7.9: The relative weights of cost, speed, and necessity of repair in environmental effects.

The AHP analysis is now continued, in order to find the relative weights of all parameters shown in Fig. 7.6. The importance value of each parameter is also found in relation to the speed, necessity and cost of recovery. Figure 7.10 shows the results of such an analysis for the social effects and in relation to the cost of recovery. As is seen, tourism is the most important item, while homelessness takes the least priority. The other items are located between these two boundaries.

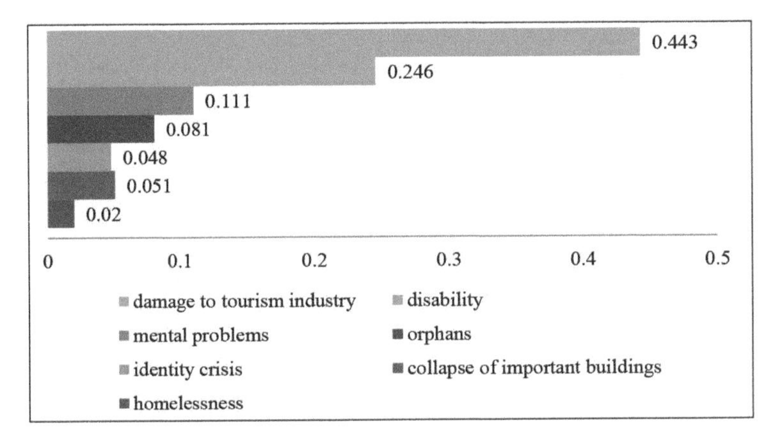

Fig. 7.10: The social effects and the cost of recovery.

From a different point of view, correlations between the necessity of recovery and the social effects are shown in Fig. 7.11. In this respect, disability has first priority, followed by mental health problems. Tourism, however, has the lowest priority, which differs from what is seen in Fig. 7.10.

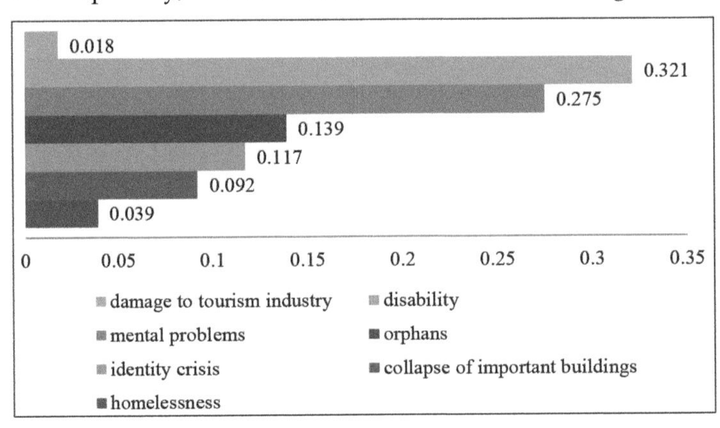

Fig. 7.11: The social effects and the necessity of recovery.

The correlations between the social effects and the speed of recovery are shown in Fig. 7.12. As is seen, in respect to the speed of recovery, mental health problems take priority, followed by tourism. The other items are located in other ranks, with disability coming last in the rankings.

Similar calculations can also be performed on the environmental effects and their correlation to the cost, necessity and speed of recovery. The correlations between the environmental effects and the cost of recovery are shown in Fig. 7.13. As shown, burning toxins take a considerably higher priority compared with other items – more than 50% of the whole. Similarly, Figs. 7.14 and 7.15 show the correlations between the environmental effects

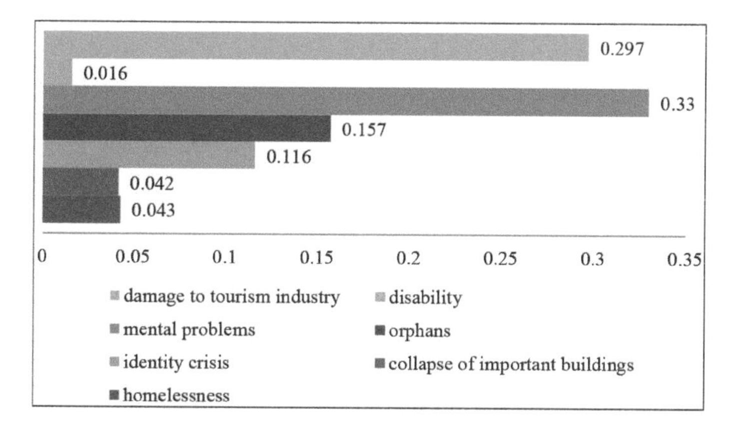

Fig. 7.12: The social effects and the speed of recovery.

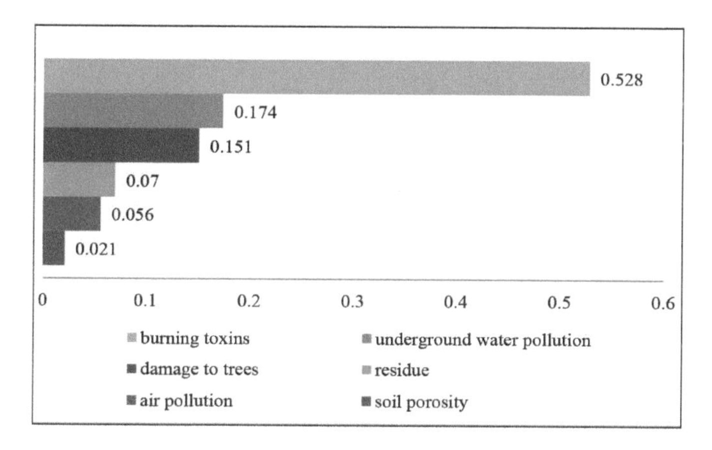

Fig. 7.13: The environmental effects and the cost of recovery.

and the speed and necessity of recovery, respectively. As is seen in Fig. 7.14, burning toxins and underground water pollution are relatively identical, meaning that both take priority in terms of the speed of recovery. As shown in Fig. 7.15, burning toxins also have the maximum relative weight for the necessity of recovery. It can be seen from both figures that soil porosity is not a high priority.

Finally, if all of the items relating to the social and environmental effects based on their relative weights are ordered, the results can be presented in Fig. 7.16. It is evident that burning toxins have the maximum relative weight (about 30%), with underground water pollution in second rank (about 23%). This means that only two of the items among the environmental effects make up more than 50% of the total relative weight. It can thus be concluded that the environmental effects of PEF are, in general, of higher priority than the social effects. These results can provide a tool for making decisions regarding the priorities of different aspects of PEF. While proposing specific solutions for the abovementioned items is of importance, it is not further discussed here, as it is outside of the scope defined for this book.

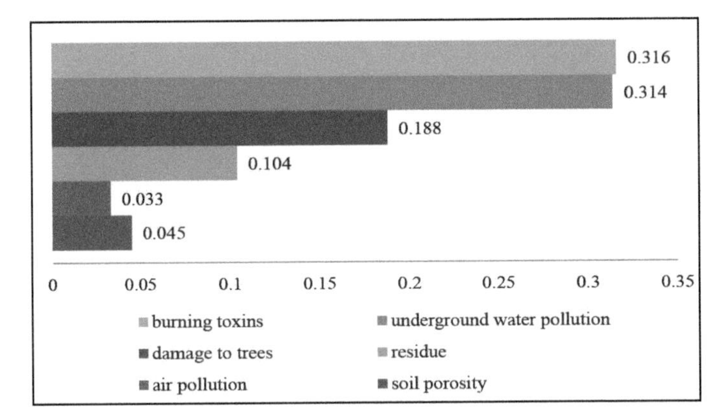

Fig. 7.14: The environmental effects and the speed of recovery.

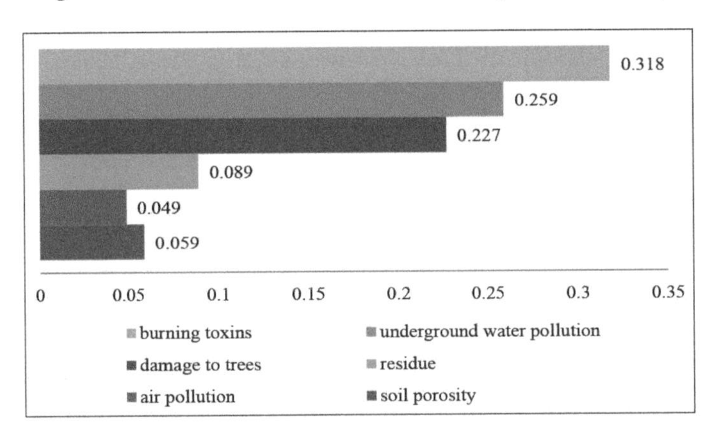

Fig. 7.15: The environmental effects and the necessity of recovery.

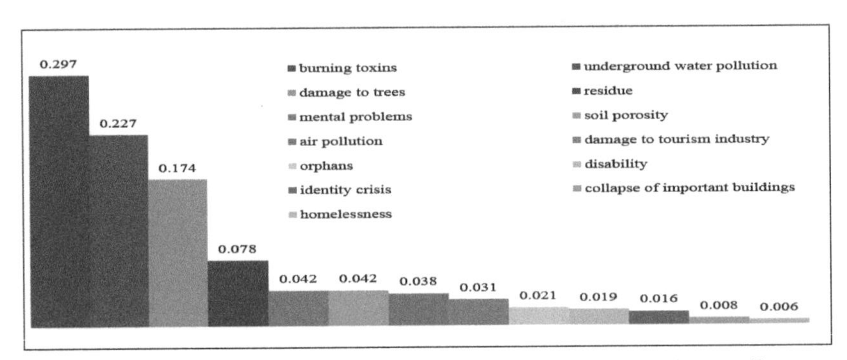

Fig. 7.16: The relative weights of all social and environmental PEF effects.

7.4 Risk mitigation of PEFs: Mesoscopic strategies

The above analyses were performed as a macro-scale risk assessment of PEF, when some general aspects (social and environmental) were considered. Another risk assessment can also be performed at a lower level, when the role of active and passive fire fighting facilities is taken into account. They can be prioritized in order to propose mitigation strategies.

It is understood that a fire risk can be influenced by a number of factors, grouped as follows: those factors that can reduce the risk of developing an ignition (Group 1); those factors that can hamper the spread of a developed fire inside the building (Group 2); and those factors that can reduce the risk of a fire spreading outside the building, in order to prevent potential conflagration (Group 3). These factors are shown in Fig. 7.17.

Group 1 includes fire-extinguishment facilities, such as sprinklers and vertical pipes, which can reduce the risk of fire developing after an earthquake,

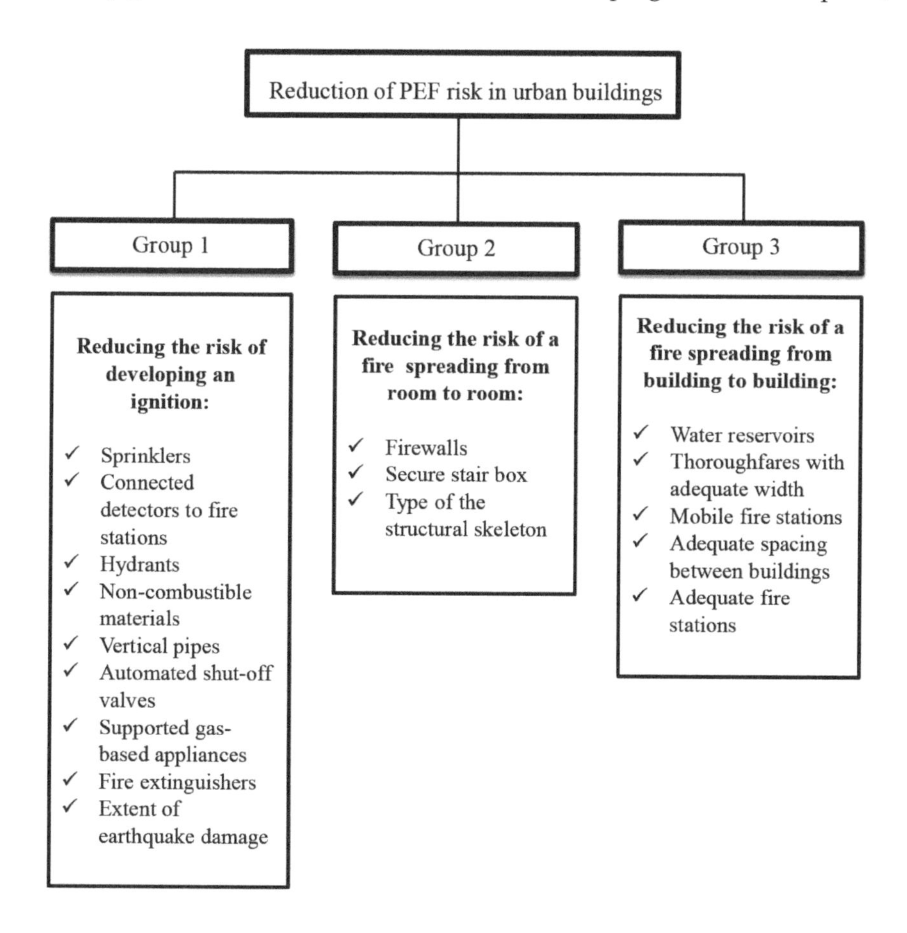

Fig. 7.17: Some factors that can reduce the PEF risk in urban regions.

provided that they remain workable. Reducing the amount of combustible materials (often this material is the furniture in use in the building) can also reduce the risk of fire, since it is evident that the fire's intensity depends on the fire load density. If the fire develops, then it might spread from one room to the next. In that case, the role of firewalls and secure stair boxes can be of importance in providing more time for inhabitants to be evacuated safely. The materials which might be used to create the skeleton can also be important. These factors make up Group 2. Finally, if the fire cannot be extinguished in a building, then there is a possibility that it will spread to other buildings. To respond to such a situation, the factors mentioned in Group 3 come into play. For example, adequate spacing between two adjacent buildings can act as a hindrance, standby and mobile fire stations can suppress a fire before it becomes uncontrollable, and providing water reservoirs is useful when the water supply system has sustained earthquake damage and hence is not functioning.

The factors listed in the above groups can be prioritized as per their relative importance, using the AHP method. The results of such an analysis can be used as a pre-disaster management plan for addressing a possible PEF event. The plan can be proposed at both building scale and regional scale. The results of the AHP analysis performed for the mentioned groups are shown in Figs. 7.18 to 7.21.

As is seen in Fig. 7.18, using automated shut-off valve systems can markedly reduce the risk of an ignition developing. Shut-off valve systems can be established in dispatch centers at regional and/or building level. These systems are particularly useful in high-rise buildings where the

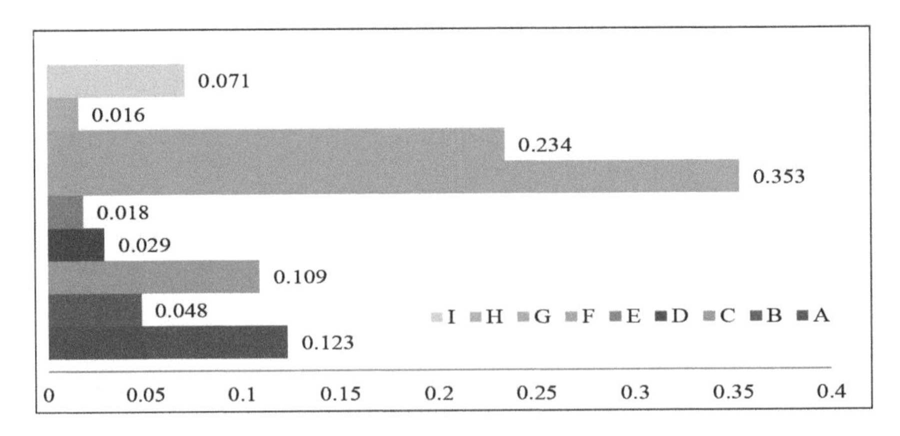

A: Sprinklers, **B**: Detectors connected to fire stations, **C**: Hydrants, **D**: Non-combustible materials, **E**: Vertical pipes, **F**: Automated shut-off valves, **G**: Supported gas-fueled appliances, **H**: Fire extinguishers, **I**: Extent of earthquake damage

Fig. 7.18: The relative importance of factors that can reduce the risk of an ignition developing.

number of residents is high. Supporting any gas-fueled appliances in a way that they cannot move during the shock can also considerably reduce the risk of an ignition developing. Adopting this strategy would not impose substantial extra cost to inhabitants, so it is reasonable to see this as an achievable strategy in the short term. Sprinkler systems can also be useful in reducing fire risk, although most would sustain significant damage after a shock. Nevertheless, it would be beneficial if earthquake-flexible sprinkler systems could be established. There are some solutions for supporting the pipes of sprinkler systems in such a way that they can remain workable after earthquake (as shown in Fig. 7.19 as an example).

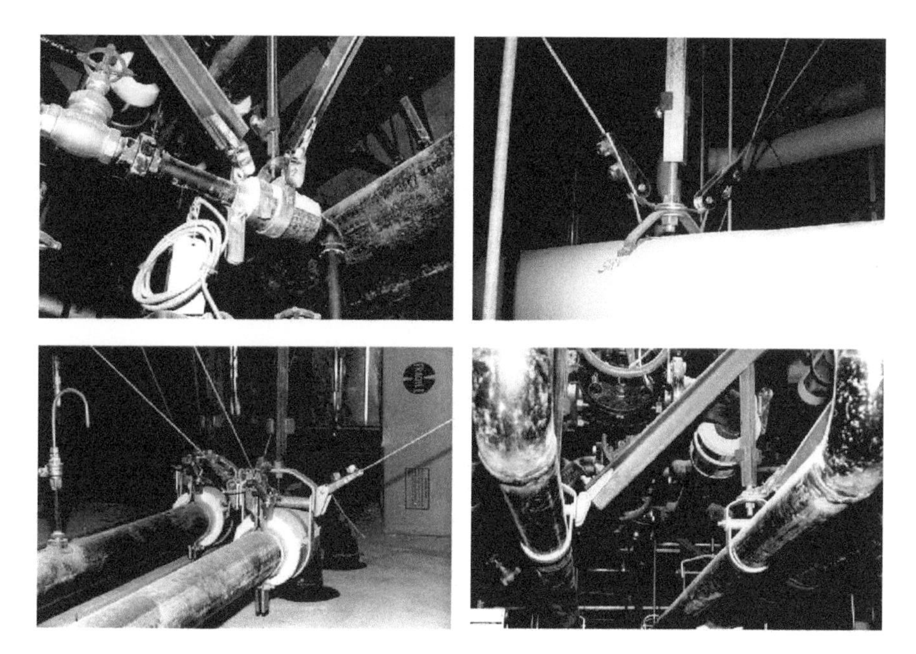

Fig. 7.19: Examples of pipes in buildings, supported to resist earthquake loads (photo courtesy of H. Vosough).

It is evident that taking into account the three factors discussed above could notably reduce the risk of any ignition developing. The other factors listed in Fig 7.18 are shown to be not as effective. For example, fire extinguishers are generally available in buildings, but they are not very effective in relative terms.

Once the fire has started, other factors become important. Figure 7.20 shows the relative importance of some factors in reducing the risk of fire spreading from room to room. It is seen that the collapse of internal partitions can markedly increase the spread of fire, such that it alone accounts for more than 55% of the total weight. The type of structural skeleton is also important in reducing fire spread – if there are no fireproofing materials (either insulation materials spread on the surface of steel structures or concrete cover of RC structures), the fire might destroy the structure sooner and thus spread to other parts. Firewalls can also hamper the spread of fire from one point to the next, although their relative weights are not considerable compared with the first two factors. Having a secure stair box has the least weight. However, this does not reflect the important role of stair boxes in protecting the residents' lives during a fire event.

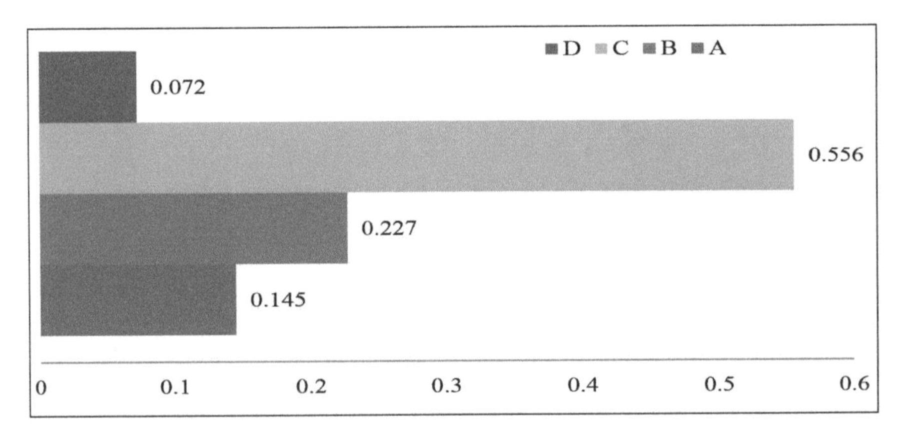

A: Firewalls, **B**: Type of structural skeleton, **C**: Collapse of partitions, **D**: Secure stair box

Fig. 7.20: The relative importance of factors that can reduce the risk of fire spreading from room to room.

The spread of fire to other buildings can be influenced by urban facilities, as shown in Fig. 7.21. As seen, water reservoirs have the most important role in reducing the risk of fire spread, reducing that risk by more than 55% before other factors are taken into account. As pointed out in Chapter 1, it has been observed from most large earthquakes in the past that urban water supply systems sustain considerable damage and thus rapidly go out of service. In that case, water reservoirs can be effective substitutes. Providing adequately wide streets and other thoroughfares can also reduce the risk of fire spreading by about 21%. In other words, dense urban regions are more susceptible to fire spread than non-dense regions (obviously, changing the nature of dense urban regions is not an easy task and cannot be achieved cheaply). For example, it is believed that the disastrous dimensions of the PEF in Kobe, Japan in 1995 could have been reduced if there had been a 20 m street between the buildings. The risk of fire spread can also worsen in the presence of favorable weather conditions, such as high wind. Figure 7.21 also shows that mobile fire stations can provide some help in reducing the risk of a fire spreading. The personnel of mobile stations can be more rapidly dispatched to fire scenes, and thus the fires can be brought under control much sooner. This strategy has been adopted in densely-populated regions of Karaj city, in Iran (Fig. 7.22), with very useful results. Considering that it is difficult for firefighters to pass through densely-settled regions even in a normal situation, the worth of mobile fire stations is further highlighted in the situation of a PEF. Mobile stations can be distributed throughout urban regions in such a way that, while the firefighters are performing their daily duties, they can also rapidly attend the scene in an urgent situation like a PEF. Providing adequate distance between buildings and increasing the number of fire stations can also reduce the risk of a fire spreading, as shown in Fig. 7.21.

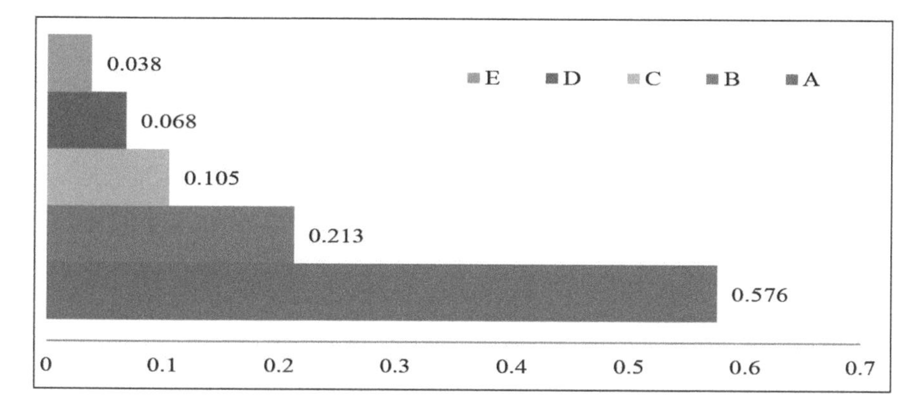

A: Water reservoirs, **B**: Thoroughfares with adequate width, **C**: Mobile fire stations, **D**: Adequate spacing between buildings, **E**: Adequate number of fire stations

Fig. 7.21: The relative importance of some factors that can reduce the risk of fire spreading from building to building.

The results shown in Figs. 7.18, 7.20, and 7.21 taken together confirm that there are three factors, inside and outside of buildings, which can significantly reduce the risk of PEF. These factors are: providing automated shut-off gas valve systems; preventing the collapse of internal partitions; and making water available for responding to a possible PEF in the absence of functioning water supply systems. Nevertheless, it is understood that attaining some of these goals is not an easy task. For example, according to most seismic codes, it is allowed that internal partitions of ordinary urban structures can fail when subjected to the design earthquake. Therefore, if there is a need to keep these partitions intact, a major revision has to be made to seismic codes. In addition, although buildings' skeleton types can affect the PEF risk, convincing the population or the urban policy-makers to confine their choices is not easily achievable. These all point to the difficulties in adopting regional-scale strategies for PEF, as taking a multiplicity of factors into account simultaneously results in very sophisticated situations.

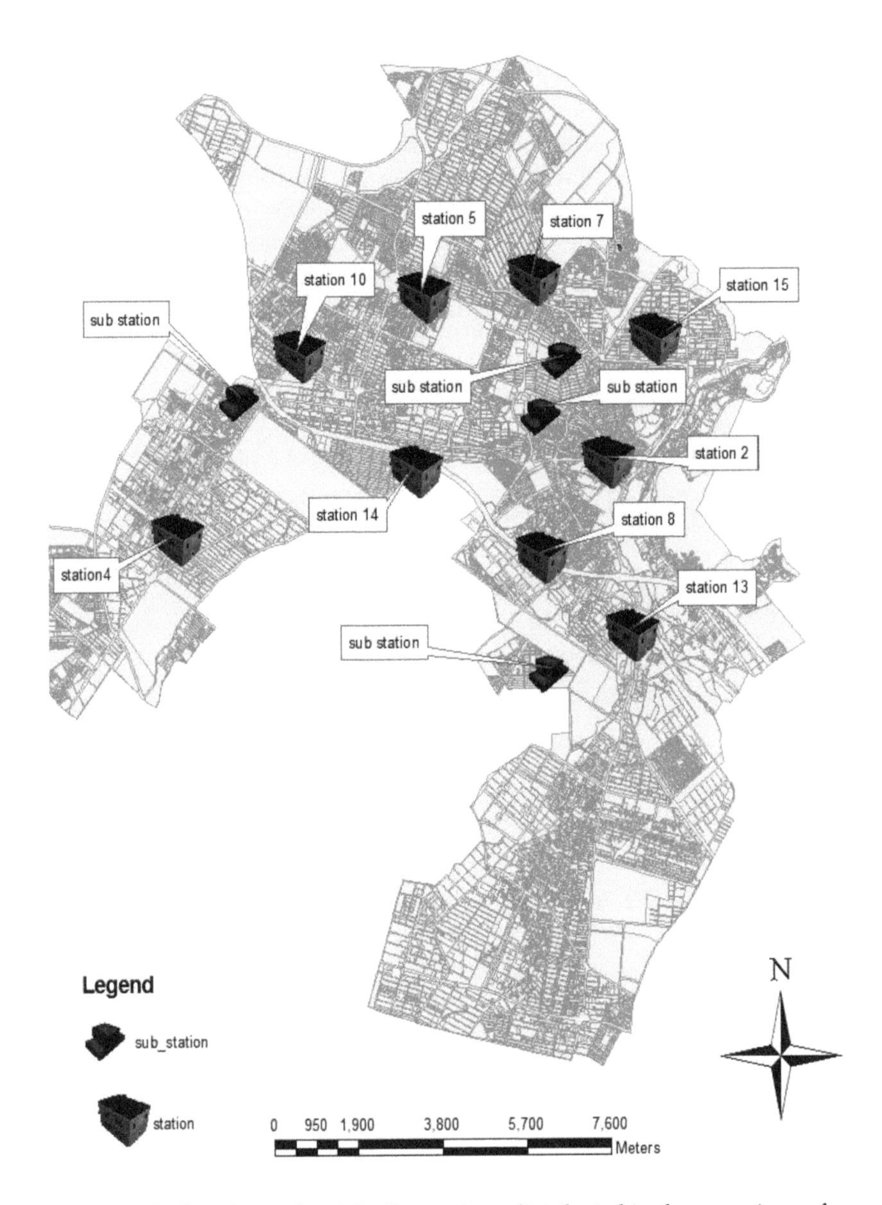

Fig. 7.22: The locations of mobile fire stations distributed in dense regions of Karaj city (Map courtesy of Karaj Fire Department).

7.5 Risk mitigation of PEF: Building-scale strategies

PEF risk management strategies can also be implemented over a micro scale when the stability of building structures is directly addressed. To do that, some practical structural engineering solutions are proposed in order to increase the fire resistance of partially seismic-damaged structures. The mitigation strategies cover both existing structures and those yet to be designed, making separate engineering solutions necessary. In order to achieve this, a criterion must be set, and the PEF resistance of a structure must be increased until it can meet thatset criterion. This criterion can be defined as the Safety Guaranteed Time Target (SGTT) where the structural stability over a defined time is guaranteed. When the PEF resistance of a building is lower than the defined SGTT, mitigation strategies must be adopted to increase that resistance. The SGTT is dependent on various factors, such as the availability of fire facilities, road closure conditions, the number of fire events that occur after the shock, etc. The SGTT is thus absolutely case-dependent – it is defined for a specific region where all possibility of PEF development and spread, along with the weather conditions, are involved. Here, assuming that the SGTT has already been defined for an assumed region, a systematic application can be proposed for both existing structures and those that will be designed in the future, as shown in Fig. 7.23.

The flowchart begins by defining the SGTT as the criterion, which is then followed by assigning the seismic performance level to the building. A nonlinear seismic analysis is then performed to see whether the performance criteria are met. If so, the PEF analysis is performed using sequential analysis, as explained in a previous chapter. The PEF resistance of the

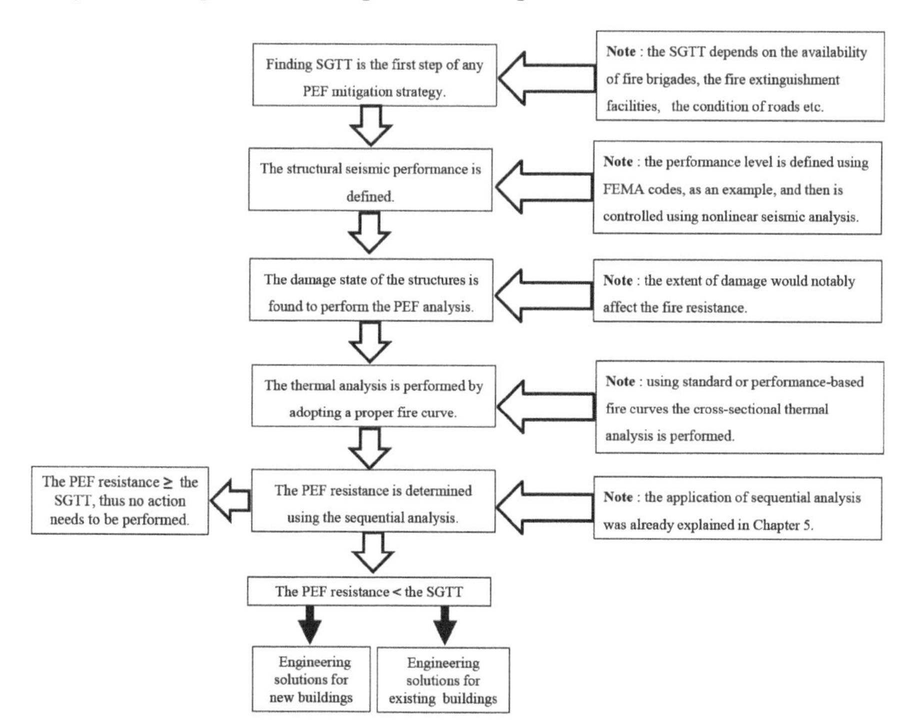

Fig. 7.23: The systematic application of PEF mitigation at building scale [353].

analysed building is now compared with the defined SGTT. If the SGTT is lower than the PEF resistance of the building, no further action is required because the structural stability is adequate for a specific duration of a fire exposure. However, when the SGTT is greater than the PEF resistance, the structural stability of the building needs to be improved. The theory behind this necessity can be found in almost all seismic codes, which mention that all buildings classified as "Ordinary Buildings" (such as residential, low-density commercial and office buildings) are designed to meet the Life Safety level for the design earthquake. This condition remains, even when a PEF situation is encountered, until the situation becomes under control or until inhabitants of the building have been safely evacuated. In the following sections, different solutions are proposed for structures that exist now and for those that are yet to be designed.

7.5.1 A PEF factor to improve the fire resistance of buildings yet to be designed

For new structures that do not meet the Life Safety requirement under a PEF loading and a defined SGTT, one can propose the use of a PEF factor to improve the structural resistance at the time of designing for earthquake. This factor will be applied to the base shear, increasing it so that the structure is detailed for a higher load [366]. In a normal design, using the "Equivalent Static" method, the base shear is calculated using Equation 7.1.

$$V = C.W \tag{7.1}$$

where C is called the earthquake coefficient, which depends on the structural characteristics, and W is the sum of permanent loads and a portion of imposed loads. The base shear (V) is then distributed along the height of the building depending on the mass of every story. A PEF coefficient (C_{PEF}) is therefore proposed which modifies V as dependent on SGTT, as shown in Equation 7.2.

$$V = C_{PEF}.C.W \tag{7.2}$$

Figure 7.24 shows a systematic approach to evaluating the PEF resistance based on C_{PEF}. It is evident that, prior to an earthquake and its following fire, buildings should have adequate resistance against gravity loads. When subjected to design earthquake loads, the defined performance level should be controlled.The expectation is that the structure, designed based on the codes using appropriate detailing, will remain within the boundaries of its intended performance levels if subjected to the design earthquake. If this building is then exposed to fire, which is to be brought under control or extinguished within the SGTT, it is expected that no failure will occur before the SGTT. Thus, if the defined SGTT level is met, it is not necessary to increase the base-shear force. Otherwise, an increase needs to be considered, using a CPEF > 1.0. This operation is terminated whenever the SGTT level is met.

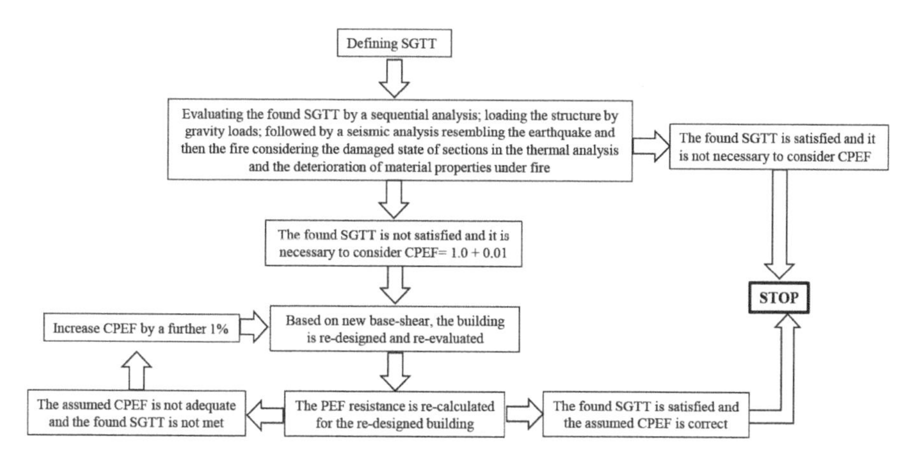

Fig. 7.24: Using C_{PEF} for improving the PEF resistance of new buildings [367].

An example of using C_{PEF} to improve the PEF resistance of RC buildings is shown below. Frame B of the case study that was explained in section 6.5 is again selected. As previously explained, the frame was designed for a PGA of 0.3g and for LS level of performance as a residential occupancy. The PEF resistance of the frame was determined as per the natural fire curve, while the presence or absence of fire fighting facilities was also involved in the thermal analysis. The frame and its PEF resistance are shown in Fig. 7.25. As is seen, the frame fails at 42 minutes. Assuming that the determined SGTT is lower than the PEF resistance of the frame, it is therefore not necessary to take any further action, as it can be supposed that the fire will be extinguished prior to it causing the frame to fail. If, however, the SGTT is assumed to be 90 minutes (for instance), the data presented in Fig. 7.25b confirms that the

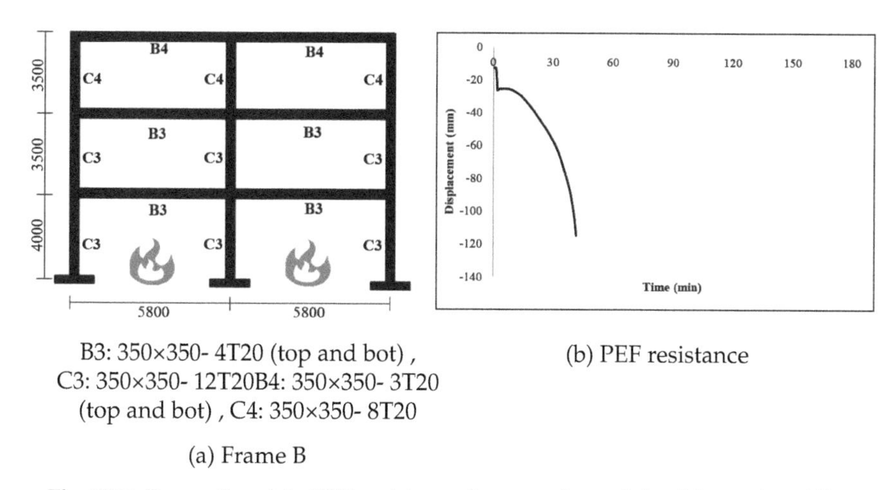

B3: 350×350- 4T20 (top and bot) , C3: 350×350- 12T20B4: 350×350- 3T20 (top and bot) , C4: 350×350- 8T20

(a) Frame B

(b) PEF resistance

Fig. 7.25: Frame B and its PEF resistance (case study explained in section 6.5).

frame does not satisfy the defined SGTT; thus the frame shall be redesigned using $C_{PEF}(t) > 1.0$. Using trial and error, according to the flowchart shown in Fig. 7.24, a series of analyses are performed again, in order to obtain a value for C_{PEF} that will meet the defined SGTT. Therefore, the frame is redesigned using $C_{PEF} > 1.0$ and then re-evaluated. After several iterations of trial and error, a C_{PEF} of 1.15 is found for the frame, which puts it within the defined SGTT. Figure 7.26 shows the redesigned frame and the PEF resistance of the frame with applied C_{PEF}.

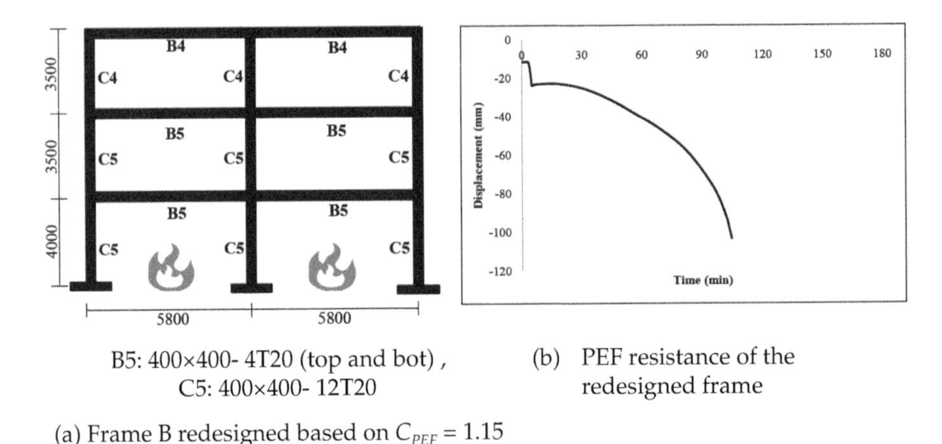

B5: 400×400- 4T20 (top and bot), C5: 400×400- 12T20

(a) Frame B redesigned based on $C_{PEF} = 1.15$

(b) PEF resistance of the redesigned frame

Fig. 7.26: The redesigned Frame B and its PEF resistance after the application of C_{PEF}.

7.5.2 Retrofitting existing buildings

Structures that are in service can be strengthened to improve their PEF resistance, via modifications ensuring that their structural components do not enter the plastic range and, as such, do not suffer from earthquake damage. In structures suffering only minor damage, the extent of heat transfer inside the cross-sections is reduced, resulting in a lower reduction in strength. Retrofitting RC structures against earthquake loads can be implemented, for instance, by sizing the cross-sections up. This was suggested in the previous section, regarding the design of new structures. While simple to propose in theory, this would be an expensive and ineffective method in practice when applied to existing structures. In reality, options for retrofitting are often limited. The main question arising concerns which structural element should be retrofitted – should both beams and columns be considered, or should only one of these be taken into consideration? To answer these questions, it is important to perform a sensitivity analysis, investigating whether the beams or the columns are more sensitive when exposed to fire loads. This sensitivity analysis was performed for the case study previously explained in section 6.4 of Chapter 6, where two RC portal frames were examined under PEF loading. As shown there, beams are more sensitive to fire than columns.

This result pertains to the recommendation in most seismic codes that beams be deliberately designed to be weaker than columns; thus beam elements sustain more earthquake damage than columns. As earthquake damage and PEF resistance are correlated to each other, it was reasonable to expect that beams would fail first and columns after. To ensure that this conclusion is also applicable to multi-story structures and does not pertain only to one-bay portal frames, a two-bay two-story RC frame is selected for investigation, with properties as shown in Fig. 7.27.

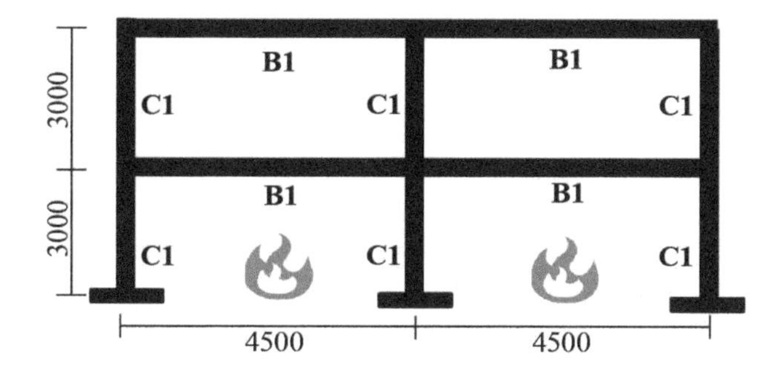

B1: 300×300- 3T20 (top and bot) , C1: 300×300- 8T20

Fig. 7.27: Geometric properties of frame and fire scenario (dimensions are in mm).

The mechanical and thermal properties of the frame are similar to those for the other case studies examined. The standard fire curve, ISO834, is used for the thermal analysis. The seismic analysis is performed using a conventional pushover analysis, and the PEF analysis is performed using sequential analysis. The results are shown in Fig. 7.28, where the mid-span deflection versus time for the left beam of the first story is plotted.

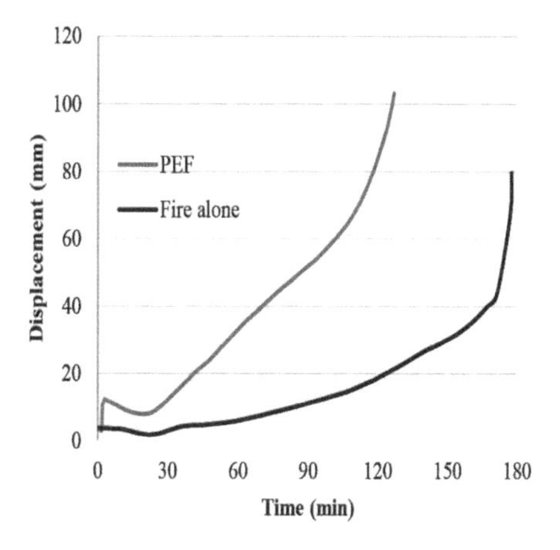

Fig. 7.28: Fire and PEF resistance of the frame.

To perform the sensitivity analysis, the PEF load is applied to the beams and columns, separately. The results of these analyses are shown in Fig. 7.29. As is seen, the PEF resistance of the frame when only the beams are exposed to fire is notably lower than when only the columns are exposed to fire, showing that the beams are more sensitive to the PEF than the columns. Analogously, viewing the results of Fig. 7.25 together with those presented in Fig. 7.28 shows that the PEF resistance of the frame depends largely on the resistance of the beams. Therefore, if any retrofitting strategy is to be adopted, the beams elements would take priority.

A practical retrofitting method that has been widely used in recent decades is to employ fiber-reinforced polymers (FRPs), either glass-based (GFRP) or carbon-based (CFRP). Here, the application of CFRP to the beams' extremities, in order to relocate plastic hinges away from the column face toward the beams, is investigated. The concept of this relocation is shown schematically in Fig. 7.30. It is evident that an earthquake causes the maximum moment and shear forces to be created near the joints, all of which then come under the PEF loads. Therefore, if it is possible to distribute the total loads to two points instead of one, there should be some improvement to the PEF resistance. It should be mentioned that the hinges will necessarily be relocated over the length of the beam and not the column. Firstly, as

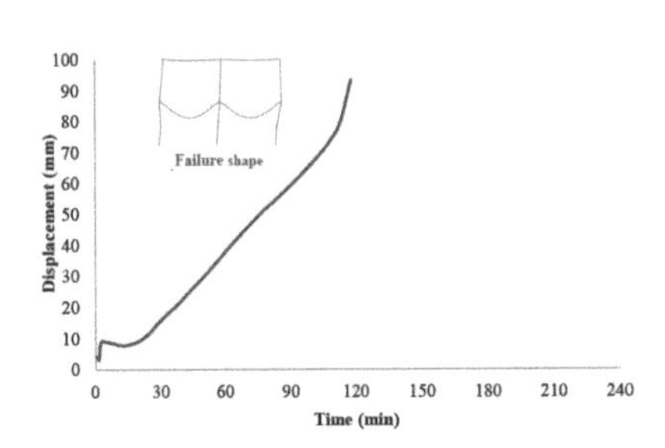

(a) Only the beams are exposed to fire

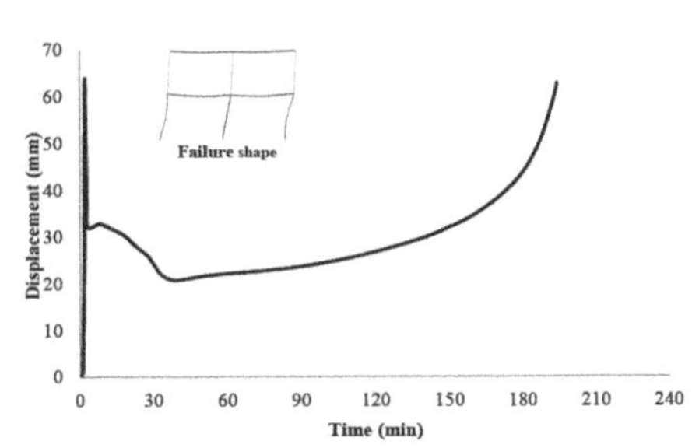

(b) Only the columns are exposed to fire

Fig. 7.29: PEF resistance of the frame when the beams and the columns are exposed separately.

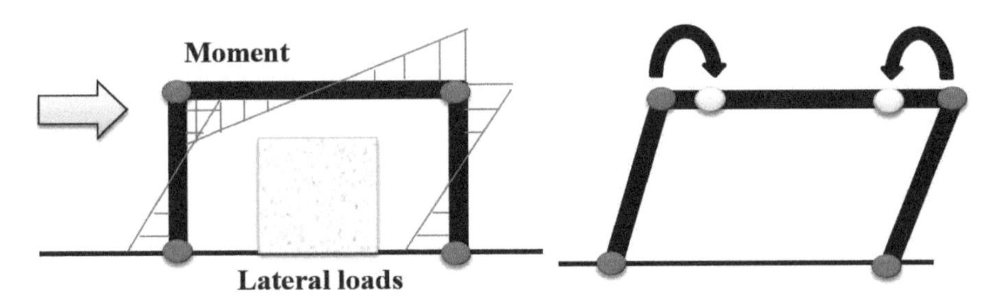

Fig. 7.30: The concept of plastic hinge relocation.

seen from the sensitivity analysis above, the beams are more vulnerable to failure than the columns. Thus, the application of hinge relocation theory to the columns would not be useful. Secondly, relocating the hinges toward the column length is extremely harmful for the structural stability, and can therefore increase the risk of shear failure.

It is supposed that the application of hinge relocation theory can also increase the lateral load-carrying capacity of the structure, decrease ductility and energy dissipation, and therefore reduce the damage caused by earthquake [368]. In practice, an increase in stiffness can lead to a reduction in the lateral displacement. An investigation conducted on 54 RC specimens by Bousselham [369] showed that external retrofitting of RC joints by FRPs can significantly improve the strength and ductility, and can decrease the energy dissipation. Mahini and Ronagh [370, 371] showed that appropriate strengthening of the joints with FRP can move plastic hinges away from the joints towards the beam. They also showed that movement of the hinges could help increase the load-carrying capacity of the structures. Failure in the beam leads to a localized failure, and this is less devastating than a global failure that ensues from failure of the joints/column. Le-Trung *et al.*[372], in an experimental study on exterior beam-column joints, showed that strengthening using CFRP materials could effectively improve the seismic performance of the joints, particularly in terms of lateral strength and ductility. Different configurations of CFRP sheets, such as T-shape, L-shape, and X-shape, however, showed different benefits in terms of ductility and strength. Another experimental study carried out by Attari *et al.* [373] showed that external strengthening of beam-column joints using different composite materials, such as GFRP and CFRP, could change the ductility of the joints.

However, FRPs often do not behave favorably under fire conditions [374, 375]. FRPs are made of polymers and as well understood, polymers generally soften and melt below 200°C. In addition, the temperature rise causes the resins to decompose, leading to a rapid degradation of the strength and stiffness of FRP and leading to subsequent de-bonding [376-378]. Being fireproof, however, is not important here, since it is assumed that when the structure is exposed to PEF, the FRPs will rapidly be damaged. Hence, the FRP has no role in the load-carrying capacity after PEF. The intention is rather to strengthen the structure satisfactorily, in such a way that the seismic damage can be limited or transferred and, as a result, the damage resulting from PEF will be delayed.

A test conducted on an RC specimen is performed to investigate whether, in practice, the application of CFRP to the joint of an RC structure would achieve the desired aim. The specimen was previously described in section 2.2.1 in Fig. 2.13, where an RC joint was subjected to cyclic loading in order to monitor the extent of damage under various performance levels, such as IO, LS, and CP. The specimen used here has an identical configuration except that it is laminated with CFRPs in the vicinity of the joint. The retrofitted specimen is shown in Fig. 7.31. The length of the CFRP is equal to the beam depth, i.e. 175 mm. The same level of cyclic loads as used for the original specimen is applied to the laminated specimen, so that the extent and the location of the damage can be observed. The cyclic load curve and the envelope curve are shown in Fig. 7.32.

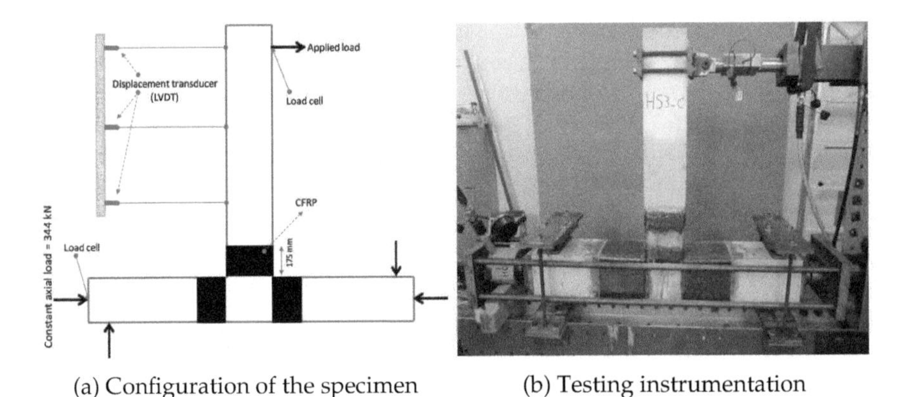

(a) Configuration of the specimen (b) Testing instrumentation

Fig. 7.31: CFRP laminated specimen under cyclic loading.

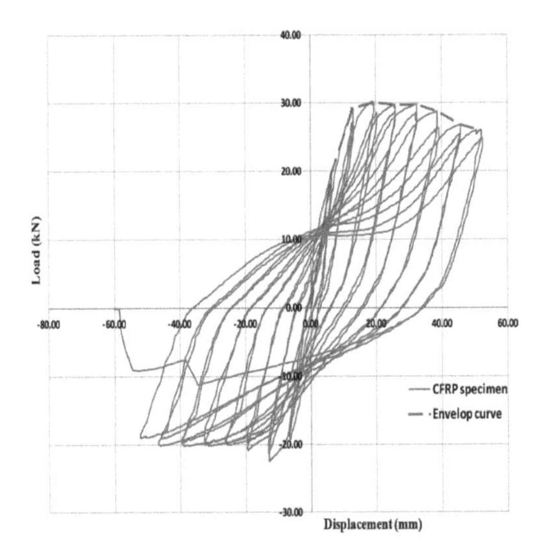

Fig. 7.32: Lateral load–displacement curve of the CFRP specimen.

Figure 7.33 shows the CFRP specimen's damage state at different levels of performance. No damage was observed in the laminated region, which shows that the plastic hinge relocated away from the joint. However, the same level of damage was seen to occur in locations just after the laminated region. The extent of damage in each case was approximately similar to that observed in the original specimen.

Considering the results for the CFRP specimen, and using the fire curve employed in evaluating the fire resistance of the original specimen, the fire resistance of the retrofitted specimen is now determined. A comparison can thus be made between the resulting PEF resistances of the retrofitted specimen and the original specimen (see Fig. 6.15). Using the methodology explained in section 6.3, and the systematic application of PEF analysis presented in Fig. 6.14, the PEF analysis of the retrofitted specimen is performed. Figure 7.34 shows the fire resistance of the retrofitted specimen at LS and CP performance levels. As is seen, the fire resistance of the retrofitted specimen is considerably higher than that of the original specimen. While the fire resistance of the original specimen at the LS level is about 32 minutes, the laminated specimen shows a fire resistance of around 43 minutes, an improvement of about 34%. Similarly, while the fire resistance of the original specimen at the CP level is around 15 minutes, it increases to around 23 minutes in the laminated specimen, an improvement of about 53%. Therefore, the laminated specimen is characterized by a higher fire resistance, even after sustaining significant damage.

(a) IO level of performance (b) LS level of performance (c) CP level of performance

Fig. 7.33: Damage of retrofitted specimens at different performance levels.

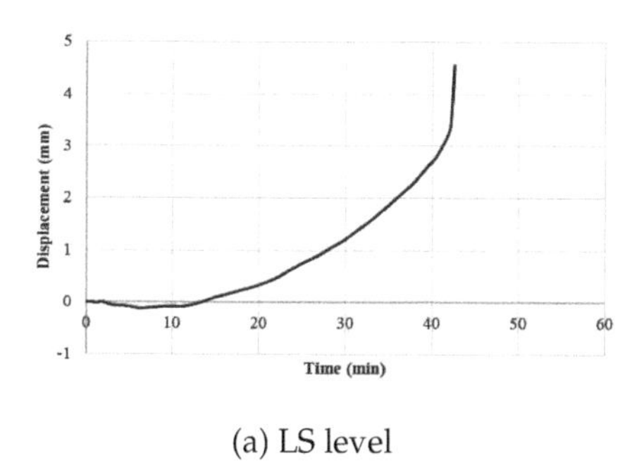

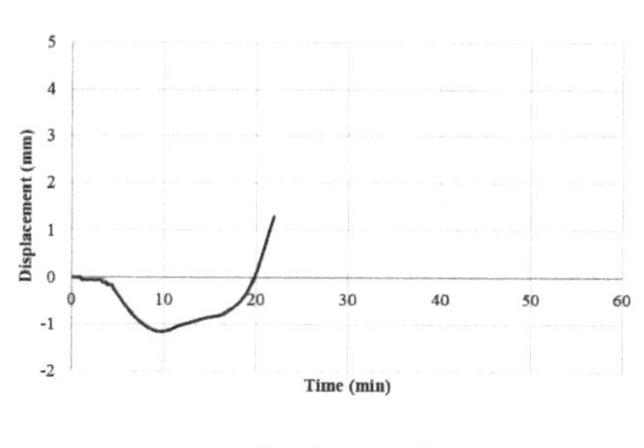

(a) LS level (b) CP level

Fig. 7.34: Fire resistance of the CFRP retrofitted specimen at LS and CP performance levels.

A detailed finite element modeling of the CFRP retrofitted joints is explained hereafter, to provide more understanding about the theory of plastic hinge relocation. The external joint of the first story of the frame shown in Fig. 7.27 is extracted and is recreated in ANSYS, with the aim of displacing the plastic hinge away from the column-beam interface into the beam span. To this aim, the joints at the mid-level, i.e. the level that is exposed to the fire, are retrofitted by attaching equal layers of CFRP layers to both sides of the web of the beam for the full depth (h) of the beam and extending the same h distance into the beam from the beam-column interface. There are two joints, a T-joint (located at the external side of the frame), and a F-joint (located at the intermediate of the frame) for that mid-level. The details of the T-joint are shown in Fig. 7.35, while the mechanical properties of the concrete, steel reinforcement and CFRP are provided in Table 7.1.

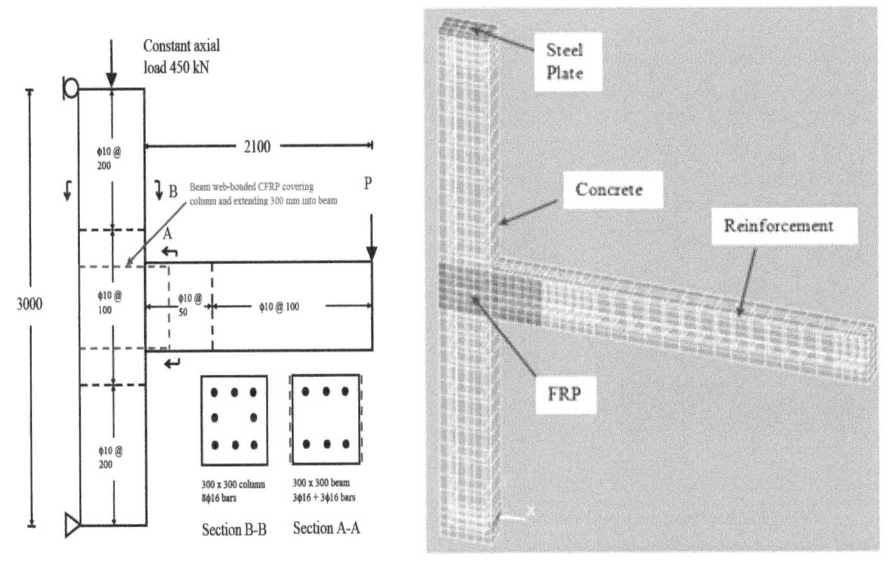

(a) Geometry of joint, reinforcement details and CFRP location (all dimensions in mm)

(b) Finite element model of retrofitted joint

Fig. 7.35: The details of the T-joint.

Finite element models are then prepared for the T-joint and the F-joint with increasing numbers of CFRP layers. Details of the finite element models for the T-joint are shown in Fig. 7.35b. Due to the vertical symmetry of the joints, only half of each joint was modeled, in order to reduce computational time. The joints are loaded by displacing the top of the beam (representing the location of the beam mid-span in the actual frame) vertically, to beyond the yielding point of the beam rebars in the joint, as shown by a clear slope change in the force versus displacement graph of the joint. Figure 7.36a shows the strain distribution of the top middle rebar of the original unretrofitted T-joint versus the strain distribution in the joint retrofitted with two layers of

Table 7.1: Mechanical properties of concrete, reinforcement steel and CFRP fibers

Concrete

Compressive Strength, f'_c (MPa)	Tensile Strength, f_r (MPa)	Young's modulus, E (MPa)
25	3	23,750

Reinforcement Steel

Yield Strength, f_y (MPa)	Ultimate Strength, f_u (MPa)	Young's modulus, E (MPa)
500	550	200,000

CFRP

Tensile Strength, f_{fr} (MPa)	Young's modulus E, (MPa)	Thickness per layer, t_f (mm)	CFRP length into beam span, l_f (mm)
3900	240,000	0.165	300

CFRP on each face of the web. The original T-joint had a strain exceeding the yield strain of reinforcement (0.002 mm/mm) near the face of the column, showing that reinforcement yielding and plastic hinge formation had occurred there. Two layers of CFRP are just adequate to displace the plastic hinge to beyond 300 mm from the face of the column, i.e. beyond the end of the CFRP layers. Similarly, for the F-joint, Fig. 7.36b shows that three layers of CFRP on each face of the web are required to displace the plastic hinge to beyond 300 mm from the face of the column.

It is accepted that applying CFRPs to the joints increases the stiffness of the structure and decreases the lateral displacement. Therefore, while the PEF resistance of the retrofitted structure is determined, the additional stiffness

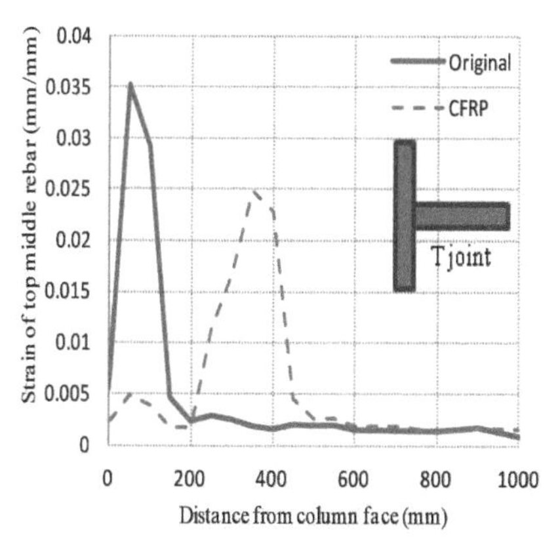

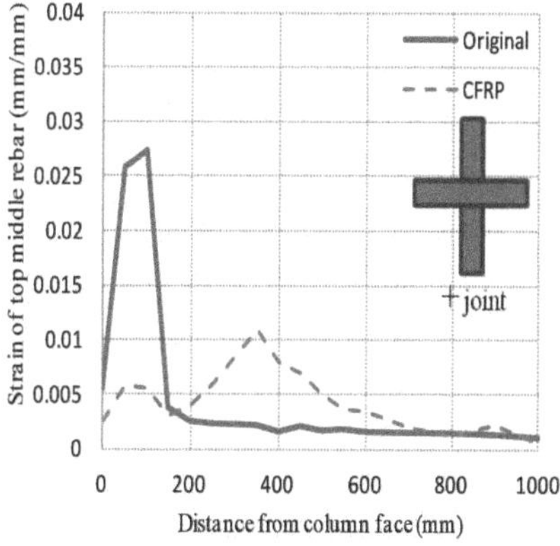

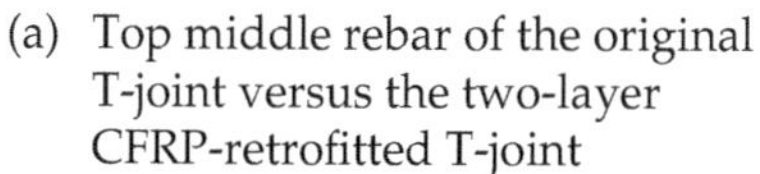
(a) Top middle rebar of the original T-joint versus the two-layer CFRP-retrofitted T-joint

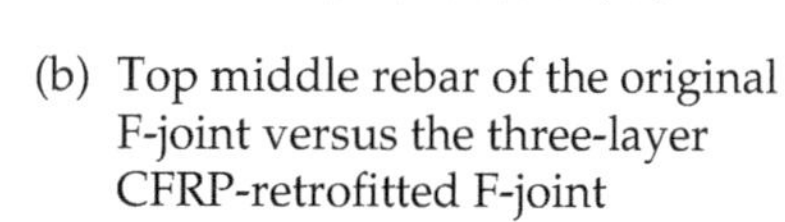
(b) Top middle rebar of the original F-joint versus the three-layer CFRP-retrofitted F-joint

Fig. 7.36: Strain distribution in the original and CFRP-retrofitted joints [379].

must be considered as well. To do that, the equivalent additional stiffness of the CFRPs applied on the beams' webs is added to the original stiffness of the structure, while at the same time, the plastic hinges are defined to occur immediately at the end of the CFRP within the beam region. The validity of this assumption has been proven by work conducted by Mahini and Ronagh [370, 371]. Performing a nonlinear static analysis, the additional stiffness is calculated by drawing a comparison between the moment rotation curves of the structure, with and without the CFRPs. It is also worth mentioning that before the retrofitting strategy is adopted, the original structure is analyzed according to the target displacement previously determined. When the laminated structure is analyzed, however, the total base shear has already been calculated and thus, the analyses should be performed based on the calculated base shear, instead of on the target displacement. In addition, after retrofitting the structure, there might be some improvement in the plastic hinge conditions, e.g. some hinges shift from the LS to IO level of performance. This, in turn, can shift the applied fire frontiers to the cross-sections. For example, Fig. 7.37a shows the frame that was introduced in Fig. 7.27, but which is now retrofitted with CFRP as a means of improving its PEF resistance. As shown, the CFRPs are applied to the joints of the first story, where it is assumed the fire occurs. The PEF of the retrofitted frame is shown in Fig. 7.37b. As shown, the retrofitted frame has a much higher PEF resistance than the original frame.

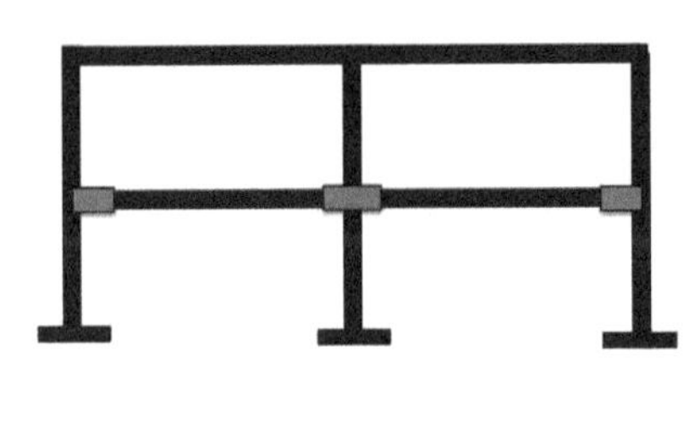

(a) The CFRP configuration of the frame

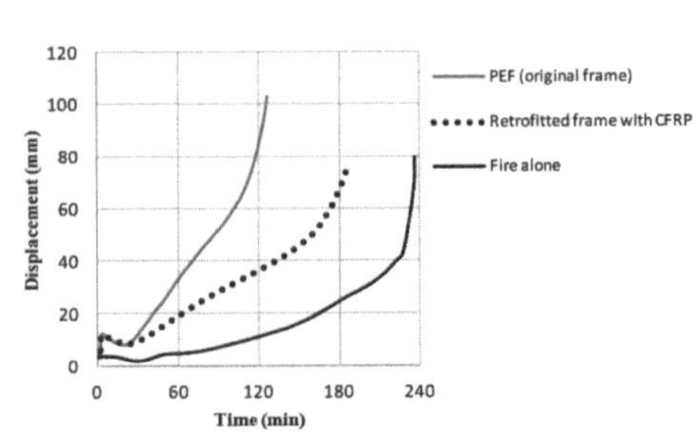

(b) Fire and PEF resistance of the original original and CFRP-retrofitted frames

Fig. 7.37: The retrofitted frame and its PEF resistance.

The method explained above can be employed as a systematic approach to improve the PEF resistance of existing RC structures. A flowchart can be proposed to show the method for improving the PEF of existing structures (Fig. 7.38).

As shown, the flowchart starts with determining the SGTT and then continues with evaluating the PEF resistance of the structure. If the time taken for the structure to collapse is greater than the SGTT, no further action is required. However, for those structures that cannot meet the SGTT, the retrofitting strategy should be adopted in such a way that their PEF resistance becomes greater than the SGTT. It is evident that in a multi-story structure, the number of joints that are retrofitted determines the level of improvement in the PEF resistance – if more joints are strengthened, the strength of the structure will be higher. The number of joints to be retrofitted with CFRP is dictated by the defined SGTT. For example, in the five-story RC structure shown in Fig. 7.39, the PEF resistance has been accounted for when the first two stories are on fire, i.e. about 2 hours. Assuming different SGTTs, different CFRP configurations (as shown in Fig. 7.40a and b) can be considered in order

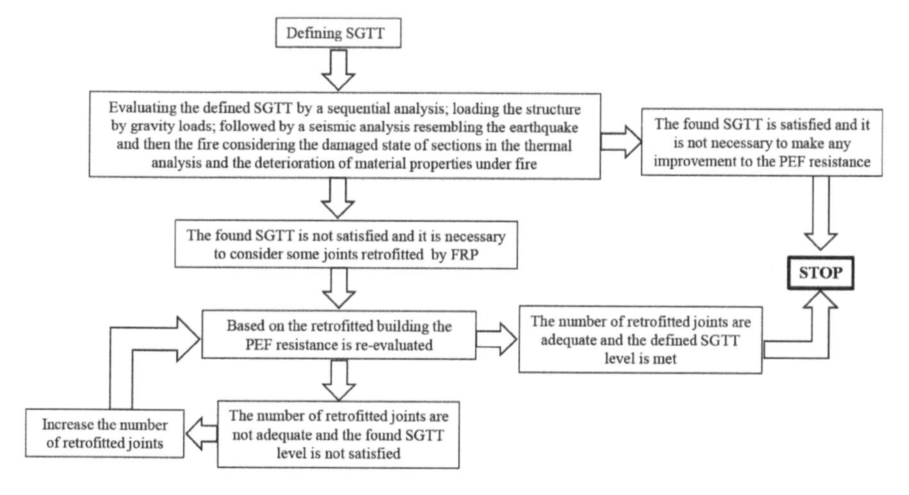

Fig. 7.38: Using CFRP for improving the PEF resistance of existing buildings [380].

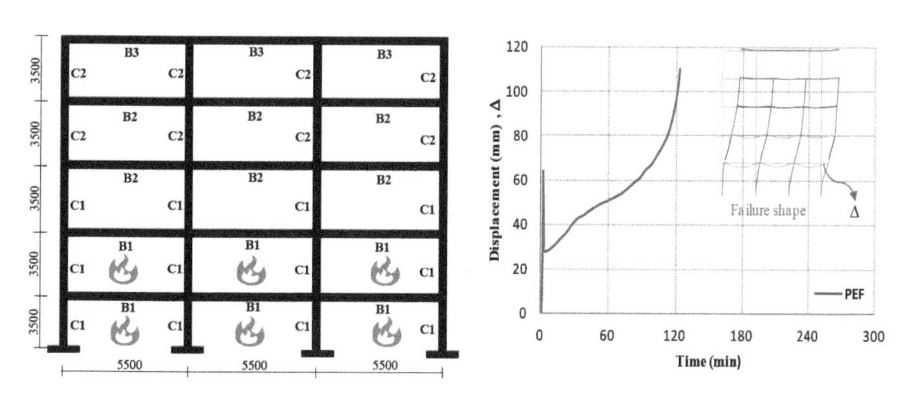

B1: 450×450- 6T20 (top) and 4T20 (bot) ,
B2: 400×400- 5T20 (top) and 3T20 (bot) ,
B3: 350×350- 4T20 (top) and 3T20 (bot) ,
C1: 450×450- 16T20 , C2: 400×400- 12T20

(a) The frame's properties

(b) PEF resistance under the fire scenario assumed

Fig. 7.39: The geometric properties and the PEF resistance.

to meet the SGTT determined. Figure 7.40c shows the PEF resistance of the frame, determined based on different CFRP configurations. As an example, if an SGTT of 150 minutes is required, Case 1 would be appropriate, while for an SGTT of 180 minutes, Case 2 should be used.

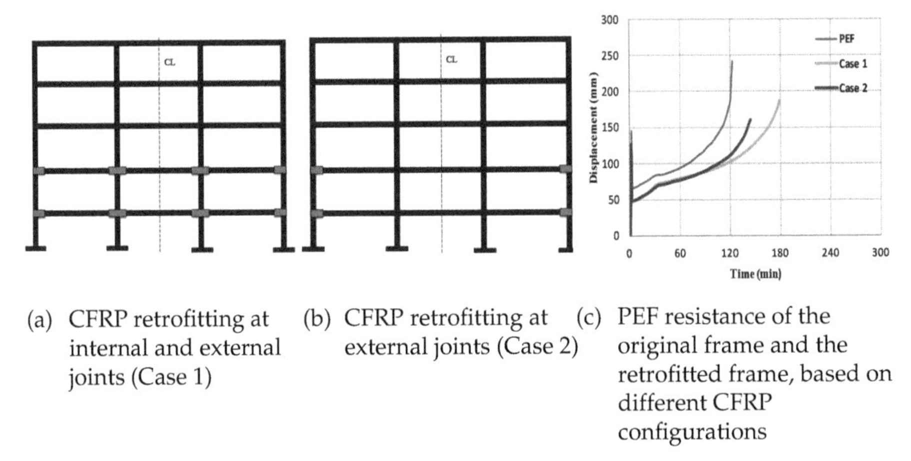

(a) CFRP retrofitting at internal and external joints (Case 1)
(b) CFRP retrofitting at external joints (Case 2)
(c) PEF resistance of the original frame and the retrofitted frame, based on different CFRP configurations

Fig. 7.40: Different configurations of CFRP and the PEF resistance.

7.5.3 Using firewalls as a means to improve the PEF resistance of RC structures

In the previous chapters, it was pointed out that urban structures sustain both structural and non-structural damage after an earthquake event. The extent of damage, either non-structural or structural, varies from one performance level to another. The structural damage at various performance levels was explained in Chapter 2. In Chapter 5, it was then shown that the structural damage incurred could affect the PEF resistance of structures, varying in degree from slightly to extensively. Based on the extent of damage, some remedies were proposed for new and existing structures in the previous sections.

An issue as yet unconsidered in the PEF analysis is how the presence or absence of exterior and interior walls can affect the PEF resistance of structures. It is strongly recommended in most seismic codes that external walls shall be permanently connected to the structural components in a way that they withstand the effects of earthquake movements. In terms of post-disaster planning, this recommendation can, for example, help to keep the roads open and thus enable the rescue teams to access the areas of damage more easily, and therefore to provide the required help more quickly [381]. Although post-earthquake rescue operations are beyond the scope of this book, it is useful to refer to Table 7.2, where the chance of survival is related to the rescue time.

Table 7.2: The correlation between rescue time and chance of being alive for people trapped under rubble [367]

Rescue time	*Chance of being alive*
30 minutes	99.3%
1 day	81.0%
2 days	53.7%
3 days	36.7%
4 days	19.0%
5 days	7.4%

The condition of external walls can also be important from a fire engineering point of view, since the destruction of external walls can change the rate of burning of the combustible materials inside the buildings, meaning that the fire would change from being fuel-controlled to ventilation-controlled.

In terms of the condition of internal walls after earthquake, in buildings designed for IO level of performance, it is expected that the partitions remain stable after earthquake, since any failure would adversely affect the serviceability of the buildings (in contradiction with the design aim). In buildings designed for LS level of performance, however, the collapse of partitions is often allowed. Assuming that the fire loads are uniformly distributed within a story, loss of partitions would lead to the creation of an open area, which would allow the fire not only to consume the newly-available combustible materials, but also to spread through the story at a rapid rate.

To respond to this issue, one solution could be to consider firewalls as non-structural components that could improve the structural stability against fire loads. The assumption when using firewalls is that they will have a defined fire resistance for a specific fire duration, e.g. one hour. If firewalls are used as the perimeter walls of a building, they should remain stable after earthquake, as discussed above. If firewalls are used inside buildings, they might be considered as partitions and thus, they might be designed to allow collapse after earthquake. Hence, as a simple but effective PEF mitigation strategy, firewalls could be designed to be connected to the structural components, meaning that they will remain stable after earthquake. In most seismic codes, it is advised that the presence of non-structural components, such as external and internal walls, must not cause the natural lateral movement of vertical components to be limited, i.e. the non-structural components should be connected to horizontal elements, such as beams. If the design aim is that the non-structural components (in this case, firewalls) will remain after the shock, they would not affect the structural response after the earthquake, while they can also play their assigned role as fire protection.

An example of using a firewall to improve the PEF resistance of RC structures follows. Figure 7.41 shows a one-story RC structure and the geometric properties of the frames. The structure is divided into two parts by a firewall. The thermal characteristics of the materials considered are shown in Table 7.3. The characteristic fire load density of 1100 MJ/m² is assumed, to account for the thermal actions. The structure is designed using the ACI 318-08 code for a PGA of 0.30 g, while a load combination of 5.0 kPa and 1.5 kPa for dead load and live load are considered, respectively. The compressive strength of concrete is assumed to be 24 MPa and the yield stress of longitudinal and transverse reinforcing bars is assumed to be 400 MPa.

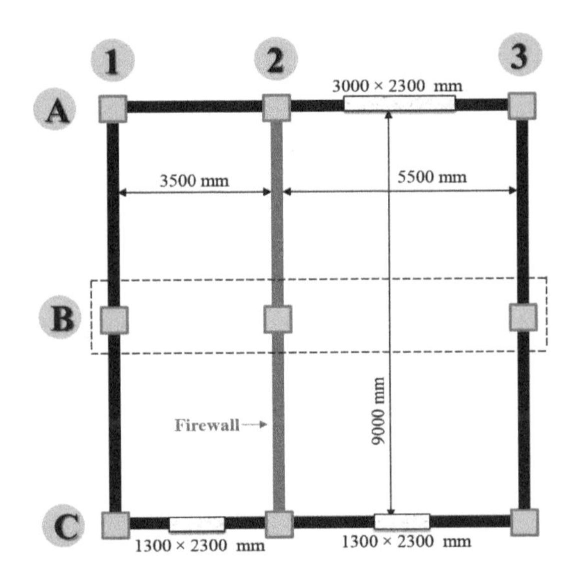

columns: 300×300- 8T20; beams: 300×300- 4T20 (top and bot)

Fig. 7.41: The one-story RC structure divided into two parts by a firewall [382].

Table 7.3: Thermal characteristics of the materials considered

Description	*Specific heat (J/kg K)*	*Material's density (kg/m³)*	*Thermal conductivity (W/mK)*
Floor and Roof	1000	2300	1.6
External walls	1000	2000	1.2
Firewalls	1000	1500	1.15

For the fire analysis, three scenarios are considered: 1) the left side exposed to fire; 2) the right side exposed to fire; and 3) the entire frame exposed to fire, while no firewall is considered. The fire curves are determined using the Eurocode 1 Annex A fire curve. As well, the effects of active measures (on-site or off-site facilities) are ignored. The analysis is performed for two situations – before earthquake, and after earthquake. When only a part of the frame is subjected to fire, it is the middle column that is exposed to fire (from one side only) where applicable. Table 7.4 shows summarized calculations of the fire curves for the three fire scenarios. Figure 7.42 shows the time-temperature curves, based the abovementioned calculations.

Table 7.4: Summarized calculations of fire curves

Scenario 1 (Left side)	*Scenario 2 (Right side)*	*Scenario 3 (Entire frame)*
$A_{floor} = 63.0$ m^2	$A_{floor} = 99.0$ m^2	$A_{floor} = 162$ m^2
$A_{external\ walls} = 51.2$ m^2	$A_{external\ walls} = 54.1$ m^2	$A_{external\ walls} = 115.2$ m^2
$A_{fire\ wall} = 28.8$ m^2	$A_{fire\ wall} = 28.8$ m^2	There is no firewall
$A_t = 143$ m^2	$A_t = 191.8$ m^2	$A_t = 277.2$ m^2
$A_v = 3$ m^2	$A_v = 9.9$ m^2	$A_v = 12.9$ m^2
$h_w = 2.3$ m	$h_w = 2.3$ m	$h_w = 2.3$ m
$b_{avg} = 1692$	$b_{avg} = 1732$	$b_{avg} = 1775$
Opening factor = 0.031 m$^{1/2}$	Opening factor = 0.078 m$^{1/2}$	Opening factor = 0.070 m$^{1/2}$
Γ factor = 0.297	Γ factor = 1.71	Γ factor = 1.33
$q_{t,d} = 242$	$q_{t,d} = 284$	$q_{t,d} = 321$
$\theta_{max} = 829$°C	$\theta_{max} = 897$°C	$\theta_{max} = 931$°C

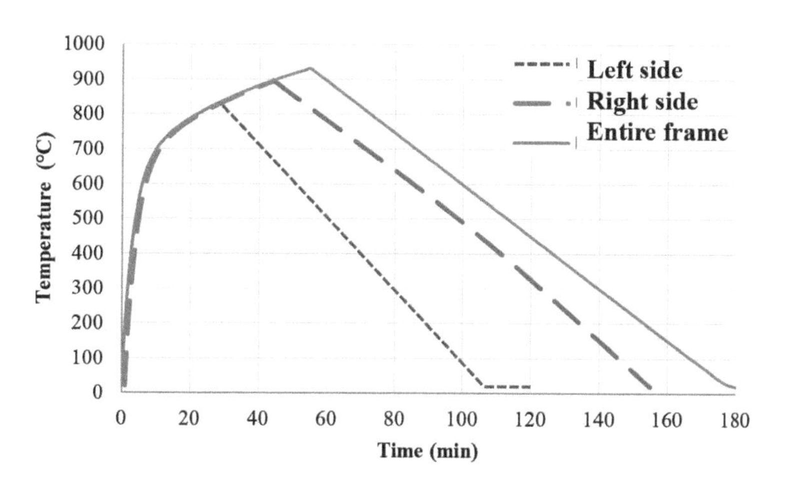

Fig. 7.42: Time-temperature curves, based on natural fire curves.

Over the fire analyses, no failure was observed in any of the three scenarios, which explicitly confirms the adequate fire resistance of the frame before the earthquake. The PEF analyses are performed using sequential analysis, the results of which are shown in Fig. 7.43. In these figures, the first sharp displacement is due to the pushover analysis performed. Figure 7.43a shows that, under the first PEF scenario, the frame does not collapse, as none of the failure criteria are reached. The failure criteria were explained and used in the previous chapter. However, under the second PEF scenario, and based on the rate of deflection criterion, the frame fails at around 50 minutes, as shown in Fig. 7.43b. Also, in the case where the entire frame is exposed to fire, (i.e. when the firewall has been destroyed after earthquake), and based on the rate of deflection criterion, the PEF resistance substantially reduces to about 25 minutes, as shown in Fig. 7.43c.

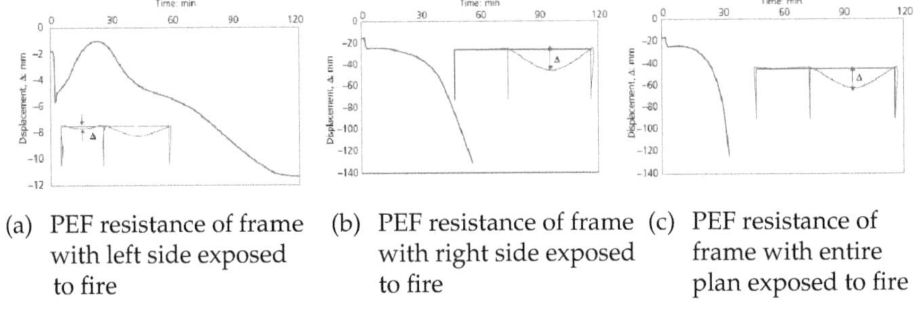

(a) PEF resistance of frame with left side exposed to fire (b) PEF resistance of frame with right side exposed to fire (c) PEF resistance of frame with entire plan exposed to fire

Fig. 7.43: PEF resistance of the structure under various fire scenarios.

The results from the above scenarios confirm that there is a strong case for using firewalls as a means to enhance the PEF resistance of structures and, as a result, reducing the failure risk. However, the role of non-structural componentsas a tool to reduce the risk of post-earthquake events is ignored in most current seismic codes. It is also evident that the suitable location of firewalls to provide the required PEF resistance can be determined using comparative analyses.

7.6 Postscript and final remarks

The book, to this point, has presented information regarding PEF events in urban regions from different aspects.

There was an examination of the weaknesses of fire fighting facilities, established both inside and outside of urban structures, in their ability to battle against possible PEFs. These weaknesses are largely due to most urban facilities having been mainly designed to resist only one extreme condition (such as fire *or* earthquake). Two sequential disasters (such as a PEF) can therefore pose a major threat to the safety of urban residents, in part since the workability of facilities to extinguish fires is not guaranteed after an earthquake. This claim is in the line with ample historical records of PEF events, some of which were addressed in the preceding chapters. Reducing the risk of ignition after earthquake is one important way to prevent a PEF disaster, and some strategies for this are discussed. However, these measures are not a guarantee, and post-earthquake fire events will still occur.

Structural robustness is hence the last line of defence. However, most urban structures sustain considerable damage from earthquakes. This damage is not necessarily due to shortcomings in the seismic codes, but rather, is due to design objectives which allow that most urban structures can sustain a significant level of earthquake damage, as long as the life safety of inhabitants is guaranteed. This guarantee is limited to the earthquake event itself, and takes no account of possible post-earthquake events. It is clear, then, that consideration should be given to the difference between structural responses under fire and under PEF loads (the investigations in this book show that a structure's PEF resistance is significantly lower than its fire resistance).

It is therefore of utmost importance to provide risk mitigation strategies to alleviate the adverse effects of PEF. These strategies can be aimed at regional scale (where reducing the spread of fire, both within and between buildings can be addressed), and at building scale (where enhancing the PEF resistance of structures can be addressed).

The following points summarize the strategies proposed in this chapter.

- **Strategies at macro scale**

It is believed that most large-scale disasters have long-term (also called tertiary) effects on urbanism. These effects have particular influence on the environment and society, both of which can take a very long time to recover. A number of comparative analyses were performed, in order to understand which aspects of these environmental or social effects should take priority in having attention paid to them.

- **Mesoscopic strategies**

The capacity of a PEF to develop and then to spread can be influenced by the presence or absence of various fire fighting facilities, both those established within buildings and those that exist in the wider urban area. From a number of comparative analyses, it was possible to understand which factors have more influence on a PEF's development, and which have more influence on its spread between compartments and then between buildings. Based on the results of the analyses, priorities were given for reducing the risk of a PEF's development and spread.

- **Building-scale strategies**

Alongside the abovementioned risk mitigation strategies, different strategies can also be adopted at building scale. The strategies are different for existing buildings and for those yet to be designed. For existing buildings, the use of FRPs (based on the plastic hinge relocation theory) brought about some improvements in the PEF resistance of structures. For new buildings, a PEF factor (larger than 1.0) was proposed, such that applying it to the base shear of structures led to an upgrading in the safety margins in the case of a PEF. Bearing in mind that non-structural components can hamper the spread of fire from one compartment to the next, another building-scale strategy was proposed. The role of firewalls as non-structural elements that are useful in this regard was particularly underlined, but only if they are designed to remain stable after an earthquake. The analyses performed showed that firewalls remaining intact after an earthquake can provide further fire resistance to structures.

As the last point, it needs to be reiterated that the investigations performed and the remedies proposed in this book aim to familiarize various groups of researchers, such as structural engineers, fire engineers, and urban managers, with the phenomenological behavior of urban structures under PEF loadings when multiple scenarios and structural systems are considered. The information provided here, however, cannot entirely address all of the possible scenarios and aspects of PEFs. It is hoped that the contents of this book have helped to elucidate the importance and complexity of this post-earthquake disaster.

Appendix A

An RC Portal Frame Subjected to Post-earthquake Fire

An RC portal frame; L=1.5H; L= span length and H = 3.0m
Input file for SAFIR2011
Frame analyzed in a STATIC mode
Note: the Flat0.fct is supposed to represent the effect of the earthquake load and no M_NODE is associated to it.

```
NNODE       61
NDIM         2
NDDLMAX      3
FROM         1   TO    61   STEP    2    NDDL     3
FROM         2   TO    60   STEP    2    NDDL     1
END_NDDL
STATIC     PURE_NR
NLOAD        2
OBLIQUE      0
COMEBACK        0.0001
NMAT         2
ELEMENTS
BEAM        30          4
NG           2
    NFIBER      900
END_ELEM
      NODES
         NODE            1      0.00000    0.00000
         GNODE          21      0.00000    3.00000    1
         GNODE          41      4.50000    3.00000    1
         GNODE          61      4.50000    0.00000    1
FIXATIONS
            BLOCK        1         F0        F0         F0
            BLOCK       61         F0        F0         F0
```

```
END_FIX
    NODOFBEAM
C25-3L.tem                          : SECTION 1
  TRANSLATE          1     1
  TRANSLATE          2     2
  END_TRANS
C25-3R.tem                          : SECTION 2
  TRANSLATE          1     1
  TRANSLATE          2     2
  END_TRANS
B25-3S.tem                          : SECTION 3
  TRANSLATE          1     1
  TRANSLATE          2     2
END_TRANS
B25-3S.tem                          : SECTION 4
  TRANSLATE          1     1
  TRANSLATE          2     2
END_TRANS
ELEM                 1     1     2     3     2
ELEM                 2     3     4     5     1
GELEM                9    17    18    19     1     2
ELEM                10    19    20    21     2
ELEM                11    21    22    23     4
ELEM                12    23    24    25     3
GELEM               19    37    38    39     3     2
ELEM                20    39    40    41     4
ELEM                21    41    42    43     2
ELEM                22    43    44    45     1
GELEM               29    57    58    59     1     2
ELEM                30    59    60    61     2
PRECISION              1.E-3
LOADS
FUNCTION            F1
DISTRBEAM                 11                0.     -22000
GDISTRBEAM                21                0.     -22000        1
END_LOAD
FUNCTION  Flat0.fct
NODELOAD   21  1.00 0.00  0.00
END_LOAD
MATERIALS
CALCONC_EN
                0.2       24.0E6        0.0E6      0.
STEELEC3EN
  210.E9        0.3       300.0E6       1200.      0.
```

```
TIME
                    1.          1.
                    5           180
                    300         14400
    END_TIME
LARGEDISPL
                EPSTH
IMPRESSION
    TIMEPRINT
                  1.14400
END_TIMEPR
    PRINTMN
PRINTREACT
PRINTDEPL
PRINTTMPRT
PRNSIGMABM          10      2
PRINTET             10      2
PRNEPSMBM           10      2
PRNEIBEAM
```

THERMAL calculation in B25

NOTE : THREE SIDES ARE EXPOSED TO FIRE

```
        NNODE       96      1
        NDIM         2
     NDOFMAX         1
EVERY_NODE           1
  END_NDOF
  TEMPERAT
     TETA          0.9
  TINITIAL        20.0
  MAKE.TEM

     NMAT            2

  ELEMENTS
     SOLID         900
        NG           2
     NVOID           0
END_ELEM

  NODES
     NODE            1     0.0000     0.0000
  GNODE             31     0.0000     0.2500     1
   REPEAT           31     0.00833    0.0000     30
   NODELINE                0.125      0.
    YC_ZC                  0.125      0.
```

```
FIXATIONS
   END_FIX

   NODOFSOLID
     ELEM            1     1     2    33    32    1    0.
   GELEM            30   900   901   932   931    1    0.    31
    REPEAT          30     1    29
    NEW_MAT        125     2
    NEW_MAT        155     2
    NEW_MAT        126     2
    NEW_MAT        156     2
    NEW_MAT        725     2
    NEW_MAT        755     2
    NEW_MAT        726     2
    NEW_MAT        756     2
    NEW_MAT        425     2
    NEW_MAT        455     2
    NEW_MAT        426     2
    NEW_MAT        456     2
    NEW_MAT        145     2
    NEW_MAT        175     2
    NEW_MAT        146     2
    NEW_MAT        176     2
    NEW_MAT        445     2
    NEW_MAT        475     2
    NEW_MAT        446     2
    NEW_MAT        476     2
    NEW_MAT        745     2
    NEW_MAT        775     2
    NEW_MAT        746     2
    NEW_MAT        776     2

FRONTIER
 F     1     FIRE.fct   NO         NO    NO
 GF    871   FIRE.fct   NO         NO    NO            30
 F     1     NO         NO         NO    FIRE.fct
 GF    30    NO         NO         NO    FIRE.fct      1
 F     30    NO         NO         F20   NO
 GF    900   NO         NO         F20   NO            30
 F     871   NO         FIRE.fct   NO    NO
 GF    900   NO         FIRE.fct   NO    NO     1
     END_FRONT

SYMMETRY
 END_SYM
```

```
PRECISION  0.002

MATERIALS
    CALCONC_EN
                                2400.       60.       25.        4.       0.8       0.
    STEELEC3EN                   25.         4.       0.7

       TIME
                                 12.    14400.
 END_TIME

OUTPUT
    TIMEPRINT
                        60.    14400.
    END_TIMEPR
```

THERMAL calculation in C25x25-8T16

NOTE : THREE SIDES ARE EXPOSED TO FIRE

```
NNODE              961
NDIM                 2

   NDOFMAX           1
EVERY_NODE           1
END_NDOF
   TEMPERAT
     TETA          0.9
   TINITIAL       20.0
   MAKE.TEM

    NMAT             2

   ELEMENTS
     SOLID         900
     NG              2
    NVOID            0
   END_ELEM

   NODES
     NODE            1      0.0000       0.0000
     GNODE          31      0.0000       0.2500          1
     REPEAT         31      0.00833      0.0000         30
   NODELINE      0.125          0.
     YC_ZC       0.125          0.
```

```
FIXATIONS
   END_FIX

NODOFSOLID
      ELEM            1     1     2    33    32    1    0.
   GELEM             30   900   901   932   931    1    0.    31
   REPEAT            30     1    29
   NEW_MAT          125     2
   NEW_MAT          155     2
   NEW_MAT          126     2
   NEW_MAT          156     2
   NEW_MAT          725     2
   NEW_MAT          755     2
   NEW_MAT          726     2
   NEW_MAT          756     2
   NEW_MAT          425     2
   NEW_MAT          455     2
   NEW_MAT          426     2
   NEW_MAT          456     2
   NEW_MAT          145     2
   NEW_MAT          175     2
   NEW_MAT          146     2
   NEW_MAT          176     2
   NEW_MAT          445     2
   NEW_MAT          475     2
   NEW_MAT          446     2
   NEW_MAT          476     2
   NEW_MAT          745     2
   NEW_MAT          775     2
   NEW_MAT          746     2
   NEW_MAT          776     2
   NEW_MAT          135     2
   NEW_MAT          165     2
   NEW_MAT          136     2
   NEW_MAT          166     2
   NEW_MAT          735     2
   NEW_MAT          765     2
   NEW_MAT          736     2
   NEW_MAT          766     2

FRONTIER
  F     1    FIRE.fct    NO   NO   NO
  GF  871    FIRE.fct    NO   NO   NO      30
  F     1    NO          NO   NO   FIRE.fct
  GF   30    NO          NO   NO   FIRE.fct 1
```

```
 F      30    NO   NO          F20   NO
 GF     900   NO   NO          F20   NO   30
 F      871   NO   FIRE.fct    NO    NO
 GF     900   NO   FIRE.fct    NO    NO   1
  END_FRONT

SYMMETRY
    END_SYM

PRECISION              0.002

   MATERIALS
CALCONC_EN
               2400.      60.      25.      4.      0.8      0.
STEELEC3EN
                 25.       4.      0.7
     TIME
                 12.   14400.
   END_TIME

OUTPUT
    TIMEPRINT
              60.   14400.
    END_TIMEPR
```

Appendix B

A 5-story Moment Resisting Steel Frame Subjected to Post-earthquake Fire

A five FLOORS steel moment resisting frame, 3 BAYS ANALYZED BY SAFIR PROGRAM. THE FRAME IS SUBJECTED TO GRAVITY LOADS IN ALL THE STORIES AND DIFFERENT POSITIONS OF FIRE. THE FIRE PATTERN IS ISO834 WITHOUT DECAY.

Frame analyzed in a STATIC mode

```
    NNODE        746
      NDIM       2
NDDLMAX          3
    FROM    1     TO   261   STEP   2   NDDL   3
    FROM    2     TO   260   STEP   2   NDDL   1
    FROM  262     TO   360   STEP   2   NDDL   3
    FROM  263     TO   361   STEP   2   NDDL   1
    FROM  362     TO   460   STEP   2   NDDL   3
    FROM  363     TO   461   STEP   2   NDDL   1
    FROM  462     TO   480   STEP   2   NDDL   1
    FROM  463     TO   479   STEP   2   NDDL   3
    FROM  481     TO   499   STEP   2   NDDL   1
    FROM 482      TO   498   STEP   2   NDDL   3
    FROM 500      TO   518   STEP   2   NDDL   1
    FROM 501      TO   517   STEP   2   NDDL   3
    FROM 519      TO   537   STEP   2   NDDL   1
    FROM  520     TO   536   STEP   2   NDDL   3
    FROM 538      TO   556   STEP   2   NDDL   1
    FROM 539      TO   555   STEP   2   NDDL   3
    FROM  557     TO   575   STEP   2   NDDL   1
    FROM 558      TO   574   STEP   2   NDDL   3
    FROM 576      TO   594   STEP   2   NDDL   1
```

```
  FROM      577       TO    593    STEP    2    NDDL    3
  FROM      595       TO    613    STEP    2    NDDL    1
  FROM      596       TO    612    STEP    2    NDDL    3
  FROM      614       TO    632    STEP    2    NDDL    1
  FROM 615            TO    631    STEP    2    NDDL    3
  FROM      633       TO    651    STEP    2    NDDL    1
  FROM 634            TO    650    STEP    2    NDDL    3
  FROM 652            TO    670    STEP    2    NDDL    1
  FROM 653            TO    669    STEP    2    NDDL    3
  FROM      671       TO    689    STEP    2    NDDL    1
  FROM      672       TO    688    STEP    2    NDDL    3
  FROM      690       TO    708    STEP    2    NDDL    1
  FROM      691       TO    707    STEP    2    NDDL    3
  FROM      709       TO    727    STEP    2    NDDL    1
  FROM      710       TO    726    STEP    2    NDDL    3
  FROM      728       TO    746    STEP    2    NDDL    1
  FROM      729       TO    745    STEP    2    NDDL    3
  FROM      102       TO    120    STEP    1    NDDL    0
  FROM 122            TO    140    STEP    1    NDDL    0
  FROM 142            TO    160    STEP    1    NDDL    0
     END_NDDL

 STATIC    PURE_NR
  NLOAD      6
 OBLIQUE     0
COMEBACK0.0001
  NMAT       1
ELEMENTS
  BEAM       350   8
   NG        2
  NFIBER     12750
END_ELEM

    NODES
  NODE        1     0.00000    0.00000
  GNODE      21     0.00000    4.00000     1
  GNODE      41     0.00000    7.50000     1
  GNODE      61     0.00000    11.0000     1
  GNODE      81     0.00000    14.5000     1
  GNODE     101     0.00000    18.0000     1
  GNODE     161     15.0000    18.0000     1
  GNODE     181     15.0000    14.5000     1
  GNODE     201     15.0000    11.0000     1
  GNODE     221     15.0000    7.50000     1
  GNODE     241     15.0000    4.00000     1
```

GNODE	261	15.0000	0.00000	1
NODE	262	5.00000	0.00000	
GNODE	282	5.00000	4.00000	1
GNODE	302	5.00000	7.50000	1
GNODE	322	5.00000	11.0000	1
GNODE	342	5.00000	14.5000	1
GNODE	361	5.00000	17.8500	1
NODE	362	10.00000	0.00000	
GNODE	382	10.00000	4.00000	1
GNODE	402	10.00000	7.50000	1
GNODE	422	10.00000	11.0000	1
GNODE	442	10.00000	14.5000	1
GNODE	461	10.00000	17.8500	1
NODE	462	0.25000	4.00000	
GNODE	480	4.75000	4.00000	1
NODE	481	5.25000	4.00000	
GNODE	499	9.75000	4.00000	1
NODE	500	10.25000	4.00000	
GNODE	518	14.75000	4.00000	1
NODE	519	0.25000	7.50000	
GNODE	537	4.75000	7.50000	1
NODE	538	5.25000	7.50000	
GNODE	556	9.75000	7.50000	1
NODE	557	10.25000	7.50000	
GNODE	575	14.75000	7.50000	1
NODE	576	0.25000	11.0000	
GNODE	594	4.75000	11.0000	1
NODE	595	5.25000	11.0000	
GNODE	613	9.75000	11.0000	1
NODE	614	10.25000	11.0000	
GNODE	632	14.75000	11.0000	1
NODE	633	0.25000	14.5000	
GNODE	651	4.75000	14.5000	1
NODE	652	5.25000	14.5000	
GNODE	670	9.75000	14.5000	1
NODE	671	10.25000	14.5000	
GNODE	689	14.75000	14.5000	1
NODE	690	0.25000	18.0000	
GNODE	708	4.75000	18.0000	1
NODE	709	5.25000	18.0000	
GNODE	727	9.75000	18.0000	1
NODE	728	10.25000	18.0000	
GNODE	746	14.75000	18.0000	1

```
FIXATIONS
   BLOCK      1     F0     F0     F0
   BLOCK      261   F0     F0     F0
   BLOCK      262   F0     F0     F0
   BLOCK      362   F0     F0     F0
END_FIX
   NODOFBEAM
C40x30-3R.tem                    ! SECTION 1
TRANSLATE         1      1
END_TRANS
C40x30-4.tem                     ! SECTION 2
TRANSLATE         1      1
END_TRANS
C40x30-3R.tem                    ! SECTION 3
TRANSLATE         1      1
END_TRANS
B30x30-3S.tem                    ! SECTION 4
TRANSLATE         1      1
END_TRANS
C40x30-0.tem                     ! SECTION 5
TRANSLATE         1      1
END_TRANS
B30x30-0.tem                     ! SECTION 6
TRANSLATE         1      1
END_TRANS
C35x30-0.tem                     ! SECTION 7
TRANSLATE         1      1
END_TRANS
B25x25-0.tem                      ! SECTION 8
TRANSLATE         1      1
END_TRANS
ELEM     1     1     2     3     1
ELEM     2     3     4     5     1
GELEM    9     17    18    19    1     2
ELEM     10    19    20    21    1           !!!!
ELEM     11    21    22    23    5           !!!!
ELEM     12    23    24    25    5
GELEM    19    37    38    39    5     2
ELEM     20    39    40    41    5           !!!!
ELEM     21    41    42    43    5           !!!!
ELEM     22    43    44    45    5
GELEM    29    57    58    59    5     2
ELEM     30    59    60    61    5     2     !!!!
ELEM     31    61    62    63    7           !!!!
ELEM     32    63    64    65    7
```

GELEM	39	77	78	79	7	2	
ELEM	40	79	80	81	7		!!!!
ELEM	41	81	82	83	7		!!!!
ELEM	42	83	84	85	7		
GELEM	49	97	98	99	7	2	
ELEM	50	99	100	101	7	2	!!!!
ELEM	51	101	690	691	8		!!!!
ELEM	52	691	692	693	8		
GELEM	59	706	706	707	8	2	
ELEM	60	707	708	121	8		!!!!
ELEM	61	121	709	710	8		!!!!
ELEM	62	710	711	712	8		
GELEM	69	724	725	726	8	2	
ELEM	70	726	727	141	8		!!!!
ELEM	71	141	728	729	8		!!!!
ELEM	72	729	730	731	8		
GELEM	79	743	744	745	8	2	
ELEM	80	745	746	161	8		!!!!
ELEM	81	161	162	163	7		!!!!
ELEM	82	163	164	165	7		
GELEM	89	177	178	179	7	2	
ELEM	90	179	180	181	7		!!!!
ELEM	91	181	182	183	7		!!!!
ELEM	92	183	184	185	7		
GELEM	99	197	198	199	7	2	
ELEM	100	199	200	201	7		!!!!
ELEM	101	201	202	203	5		!!!!
ELEM	102	203	204	205	5		
GELEM	109	217	218	219	5	2	
ELEM	110	219	220	221	5		!!!!
ELEM	111	221	222	223	5		!!!!
ELEM	112	223	224	225	5		
GELEM	119	237	238	239	5	2	
ELEM	120	239	240	241	5		!!!!
ELEM	121	241	242	243	3		!!!!
ELEM	122	243	244	245	3		
GELEM	129	257	258	259	3	2	
ELEM	130	259	260	261	3		!!!!
ELEM	131	21	462	463	4		!!!!
ELEM	132	463	464	465	4		
GELEM	139	477	478	479	4	2	
ELEM	140	479	480	282	4		!!!!
ELEM	141	282	481	482	4		!!!!
ELEM	142	482	483	484	4		
GELEM	149	496	497	498	4	2	

ELEM	150	498	499	382	4		!!!!
ELEM	151	382	500	501	4		!!!!
ELEM	152	501	502	503	4		
GELEM	159	515	516	517	4	2	
ELEM	160	517	518	241	4		!!!!
ELEM	161	41	519	520	6		!!!!
ELEM	162	520	521	522	6		
GELEM	169	534	535	536	6	2	
ELEM	170	536	537	302	6		!!!!
ELEM	171	302	538	539	6		!!!!
ELEM	172	539	540	541	6		
GELEM	179	553	554	555	6	2	
ELEM	180	555	556	402	6		!!!!
ELEM	181	402	557	558	6		!!!!
ELEM	182	558	559	560	6		
GELEM	189	572	573	574	6	2	
ELEM	190	574	575	221	6		!!!!
ELEM	191	61	576	577	6		!!!!
ELEM	192	577	578	579	6		
GELEM	199	591	592	593	6	2	
ELEM	200	593	594	322	6		!!!!
ELEM	201	322	595	596	6		!!!!
ELEM	202	596	597	598	6		
GELEM	209	610	611	612	6	2	
ELEM	210	612	613	422	6		!!!!
ELEM	211	422	614	615	6		!!!!
ELEM	212	615	616	617	6		
GELEM	219	629	630	631	6	2	
ELEM	220	631	632	201	6		!!!!
ELEM	221	81	633	634	8		!!!!
ELEM	222	634	635	636	8		
GELEM	229	648	649	650	8	2	
ELEM	230	650	651	342	8		!!!!
ELEM	231	342	652	653	8		!!!!
ELEM	232	653	654	655	8		
GELEM	239	667	668	669	8	2	
ELEM	240	669	670	442	8		!!!!
ELEM	241	442	671	672	8		!!!!
ELEM	242	672	673	674	8		
GELEM	249	686	687	688	8	2	
ELEM	250	688	689	181	8		!!!!
ELEM	251	262	263	264	2		!!!!
ELEM	252	264	265	266	2		
GELEM	259	278	279	280	2	2	
ELEM	260	280	281	282	2		!!!!

```
ELEM        261   282   283   284   5         !!!!
ELEM        262   284   285   286   5
GELEM       269   298   299   300   5    2
ELEM        270   300   301   302   5         !!!!
ELEM        271   302   303   304   5         !!!!
ELEM        272   304   305   306   5
GELEM       279   318   319   320   5    2
ELEM        280   320   321   322   5         !!!!
ELEM        281   322   323   324   7         !!!!
ELEM        282   324   325   326   7
GELEM       289   338   339   340   7    2
ELEM        290   340   341   342   7         !!!!
ELEM        291   342   343   344   7         !!!!
ELEM        292   344   345   346   7
GELEM       299   358   359   360   7    2
ELEM        300   360   361   121   7         !!!!
ELEM        301   362   363   364   2         !!!!
ELEM        302   364   365   366   2
GELEM       309   378   379   380   2    2
ELEM        310   380   381   382   2         !!!!
ELEM        311   382   383   384   5         !!!!
ELEM        312   384   385   386   5
GELEM       319   398   399   400   5    2
ELEM        320   400   401   402   5         !!!!
ELEM        321   402   403   404   5         !!!!
ELEM        322   404   405   406   5
GELEM       329   418   419   420   5    2
ELEM        330   420   421   422   5         !!!!
ELEM        331   422   423   424   7         !!!!
ELEM        332   424   425   426   7
GELEM       339   438   439   440   7    2
ELEM        340   440   441   442   7         !!!!
ELEM        341   442   443   444   7         !!!!
ELEM        342   444   445   446   7
GELEM       349   458   459   460   7    2
ELEM        350   460   461   141   7         !!!!
PRECISION 1.E-3
LOADS
FUNCTION          F1
DISTRBEAM         131   0.    -50000
GDISTRBEAM              160   0.     -50000   1
DISTRBEAM         161   0.    -50000
GDISTRBEAM              190   0.     -50000   1
DISTRBEAM         191   0.    -50000
GDISTRBEAM              220   0.     -50000   1
```

```
DISTRBEAM       221    0.     -50000
GDISTRBEAM      250    0.     -50000              1
DISTRBEAM       51     0.     -50000
GDISTRBEAM      80     0.     -50000              1
END_LOAD
FUNCTION        Flat1.fct
NODELOAD        21     1.00   0.00     0.00
NODELOAD        282    1.00   0.00     0.00
NODELOAD        382    1.00   0.00     0.00
NODELOAD        241    1.00   0.00     0.00
END_LOAD
FUNCTION        Flat2.fct
NODELOAD        41     1.00   0.00     0.00
NODELOAD        302    1.00   0.00     0.00
NODELOAD        402    1.00   0.00     0.00
NODELOAD        221    1.00   0.00     0.00
END_LOAD
FUNCTION        Flat3.fct
NODELOAD        61     1.00   0.00     0.00
NODELOAD        322    1.00   0.00     0.00
NODELOAD        422    1.00   0.00     0.00
NODELOAD        201    1.00   0.00     0.00
END_LOAD
FUNCTION        Flat4.fct
NODELOAD        81     1.00   0.00     0.00
NODELOAD        342    1.00   0.00     0.00
NODELOAD        442    1.00   0.00     0.00
NODELOAD        181    1.00   0.00     0.00
END_LOAD
FUNCTION        Flat5.fct
NODELOAD        101    1.00   0.00     0.00
NODELOAD        121    1.00   0.00     0.00
NODELOAD        141    1.00   0.00     0.00
NODELOAD        161    1.00   0.00     0.00
END_LOAD
MATERIALS
STEELEC3EN
          210.E9    0.3    240.0E6           1200.    0.
TIME
                    5.     180.
             100    3600
  END_TIME
LARGEDISPL
          EPSTH
IMPRESSION
```

```
        TIMEPRINT
                               1.    3600
END_TIMEPR
        PRINTMN
PRINTREACT
PRINTDEPL
PRINTTMPRT
PRNSIGMABM                     10   2
PRINTET                        10   2
PRNEPSMBM                      10   2
PRNEIBEAM
```

Profile: Type: B25x25-0
Fire curve: FISO - Exposed Faces:0 0 0 0
Protected with: (None) 0 mm 0 mm 2 mm;

```
  NNODE 183
    NDIM  2
  NDOFMAX  1
    FROM      1    TO        183 STEP      1 NDOF       1
  END_NDOF
  TEMPERAT
    TETA            0.9
  TINITIAL         20.0
  MAKE.TEM
    NORENUM
B25x25-0.tem
    NMAT            1
  ELEMENTS
    SOLID         124
        NG          2
    NVOID           0
END_ELEM
  NODES
      NODE               1        0.1250          -0.1250
      NODE               2        0.1250          -0.1140
      NODE               3        0.1250          -0.1030
      NODE               4        0.1250          -0.0919
      NODE               5        0.1250          -0.0809
      NODE               6        0.1250          -0.0699
      NODE               7        0.1250          -0.0589
      NODE               8        0.1250          -0.0479
      NODE               9        0.1250          -0.0369
      NODE              10        0.1250          -0.0258
      NODE              11        0.1250          -0.0148
```

NODE	12	0.1250	-0.0050
NODE	13	0.1250	0.0000
NODE	14	0.1250	0.0050
NODE	15	0.1250	0.0148
NODE	16	0.1250	0.0258
NODE	17	0.1250	0.0369
NODE	18	0.1250	0.0479
NODE	19	0.1250	0.0589
NODE	20	0.1250	0.0699
NODE	21	0.1250	0.0809
NODE	22	0.1250	0.0919
NODE	23	0.1250	0.1030
NODE	24	0.1250	0.1140
NODE	25	0.1250	0.1250
NODE	26	0.1175	-0.1250
NODE	27	0.1175	-0.1140
NODE	28	0.1175	-0.1030
NODE	29	0.1175	-0.0919
NODE	30	0.1175	-0.0809
NODE	31	0.1175	-0.0699
NODE	32	0.1175	-0.0589
NODE	33	0.1175	-0.0479
NODE	34	0.1175	-0.0369
NODE	35	0.1175	-0.0258
NODE	36	0.1175	-0.0148
NODE	37	0.1175	-0.0050
NODE	38	0.1175	0.0000
NODE	39	0.1175	0.0050
NODE	40	0.1175	0.0148
NODE	41	0.1175	0.0258
NODE	42	0.1175	0.0369
NODE	43	0.1175	0.0479
NODE	44	0.1175	0.0589
NODE	45	0.1175	0.0699
NODE	46	0.1175	0.0809
NODE	47	0.1175	0.0919
NODE	48	0.1175	0.1030
NODE	49	0.1175	0.1140
NODE	50	0.1175	0.1250
NODE	51	0.1100	-0.1250
NODE	52	0.1100	-0.1140
NODE	53	0.1100	-0.1030
NODE	54	0.1100	-0.0919
NODE	55	0.1100	-0.0809
NODE	56	0.1100	-0.0699

```
NODE    57     0.1100    -0.0589
NODE    58     0.1100    -0.0479
NODE    59     0.1100    -0.0369
NODE    60     0.1100    -0.0258
NODE    61     0.1100    -0.0148
NODE    62     0.1100    -0.0050
NODE    63     0.1100     0.0000
NODE    64     0.1100     0.0050
NODE    65     0.1100     0.0148
NODE    66     0.1100     0.0258
NODE    67     0.1100     0.0369
NODE    68     0.1100     0.0479
NODE    69     0.1100     0.0589
NODE    70     0.1100     0.0699
NODE    71     0.1100     0.0809
NODE    72     0.1100     0.0919
NODE    73     0.1100     0.1030
NODE    74     0.1100     0.1140
NODE    75     0.1100     0.1250
NODE    76     0.1002    -0.0050
NODE    77     0.1002     0.0000
NODE    78     0.1002     0.0050
NODE    79     0.0801    -0.0050
NODE    80     0.0801     0.0000
NODE    81     0.0801     0.0050
NODE    82     0.0601    -0.0050
NODE    83     0.0601     0.0000
NODE    84     0.0601     0.0050
NODE    85     0.0401    -0.0050
NODE    86     0.0401     0.0000
NODE    87     0.0401     0.0050
NODE    88     0.0200    -0.0050
NODE    89     0.0200     0.0000
NODE    90     0.0200     0.0050
NODE    91     0.0000    -0.0050
NODE    92     0.0000     0.0000
NODE    93     0.0000     0.0050
NODE    94    -0.0200    -0.0050
NODE    95    -0.0200     0.0000
NODE    96    -0.0200     0.0050
NODE    97    -0.0401    -0.0050
NODE    98    -0.0401     0.0000
NODE    99    -0.0401     0.0050
NODE   100    -0.0601    -0.0050
NODE   101    -0.0601     0.0000
```

NODE	102	-0.0601	0.0050
NODE	103	-0.0801	-0.0050
NODE	104	-0.0801	0.0000
NODE	105	-0.0801	0.0050
NODE	106	-0.1002	-0.0050
NODE	107	-0.1002	0.0000
NODE	108	-0.1002	0.0050
NODE	109	-0.1100	-0.1250
NODE	110	-0.1100	-0.1140
NODE	111	-0.1100	-0.1030
NODE	112	-0.1100	-0.0919
NODE	113	-0.1100	-0.0809
NODE	114	-0.1100	-0.0699
NODE	115	-0.1100	-0.0589
NODE	116	-0.1100	-0.0479
NODE	117	-0.1100	-0.0369
NODE	118	-0.1100	-0.0258
NODE	119	-0.1100	-0.0148
NODE	120	-0.1100	-0.0050
NODE	121	-0.1100	0.0000
NODE	122	-0.1100	0.0050
NODE	123	-0.1100	0.0148
NODE	124	-0.1100	0.0258
NODE	125	-0.1100	0.0369
NODE	126	-0.1100	0.0479
NODE	127	-0.1100	0.0589
NODE	128	-0.1100	0.0699
NODE	129	-0.1100	0.0809
NODE	130	-0.1100	0.0919
NODE	131	-0.1100	0.1030
NODE	132	-0.1100	0.1140
NODE	133	-0.1100	0.1250
NODE	134	-0.1175	-0.1250
NODE	135	-0.1175	-0.1140
NODE	136	-0.1175	-0.1030
NODE	137	-0.1175	-0.0919
NODE	138	-0.1175	-0.0809
NODE	139	-0.1175	-0.0699
NODE	140	-0.1175	-0.0589
NODE	141	-0.1175	-0.0479
NODE	142	-0.1175	-0.0369
NODE	143	-0.1175	-0.0258
NODE	144	-0.1175	-0.0148
NODE	145	-0.1175	-0.0050
NODE	146	-0.1175	0.0000

```
NODE        147    -0.1175    0.0050
NODE        148    -0.1175    0.0148
NODE        149    -0.1175    0.0258
NODE        150    -0.1175    0.0369
NODE        151    -0.1175    0.0479
NODE        152    -0.1175    0.0589
NODE        153    -0.1175    0.0699
NODE        154    -0.1175    0.0809
NODE        155    -0.1175    0.0919
NODE        156    -0.1175    0.1030
NODE        157    -0.1175    0.1140
NODE        158    -0.1175    0.1250
NODE        159    -0.1250   -0.1250
NODE        160    -0.1250   -0.1140
NODE        161    -0.1250   -0.1030
NODE        162    -0.1250   -0.0919
NODE        163    -0.1250   -0.0809
NODE        164    -0.1250   -0.0699
NODE        165    -0.1250   -0.0589
NODE        166    -0.1250   -0.0479
NODE        167    -0.1250   -0.0369
NODE        168    -0.1250   -0.0258
NODE        169    -0.1250   -0.0148
NODE        170    -0.1250   -0.0050
NODE        171    -0.1250    0.0000
NODE        172    -0.1250    0.0050
NODE        173    -0.1250    0.0148
NODE        174    -0.1250    0.0258
NODE        175    -0.1250    0.0369
NODE        176    -0.1250    0.0479
NODE        177    -0.1250    0.0589
NODE        178    -0.1250    0.0699
NODE        179    -0.1250    0.0809
NODE        180    -0.1250    0.0919
NODE        181    -0.1250    0.1030
NODE        182    -0.1250    0.1140
NODE        183    -0.1250    0.1250
NODELINE           0          0
YC_ZC              0          0
FIXATIONS
END_FIX
NODOFSOLID
ELEM     1     1     26     27     2     1     0.
ELEM     2     2     27     28     3     1     0.
ELEM     3     3     28     29     4     1     0.
```

ELEM	4	4	29	30	5	1	0.
ELEM	5	5	30	31	6	1	0.
ELEM	6	6	31	32	7	1	0.
ELEM	7	7	32	33	8	1	0.
ELEM	8	8	33	34	9	1	0.
ELEM	9	9	34	35	10	1	0.
ELEM	10	10	35	36	11	1	0.
ELEM	11	11	36	37	12	1	0.
ELEM	12	12	37	38	13	1	0.
ELEM	13	13	38	39	14	1	0.
ELEM	14	14	39	40	15	1	0.
ELEM	15	15	40	41	16	1	0.
ELEM	16	16	41	42	17	1	0.
ELEM	17	17	42	43	18	1	0.
ELEM	18	18	43	44	19	1	0.
ELEM	19	19	44	45	20	1	0.
ELEM	20	20	45	46	21	1	0.
ELEM	21	21	46	47	22	1	0.
ELEM	22	22	47	48	23	1	0.
ELEM	23	23	48	49	24	1	0.
ELEM	24	24	49	50	25	1	0.
ELEM	25	26	51	52	27	1	0.
ELEM	26	27	52	53	28	1	0.
ELEM	27	28	53	54	29	1	0.
ELEM	28	29	54	55	30	1	0.
ELEM	29	30	55	56	31	1	0.
ELEM	30	31	56	57	32	1	0.
ELEM	31	32	57	58	33	1	0.
ELEM	32	33	58	59	34	1	0.
ELEM	33	34	59	60	35	1	0.
ELEM	34	35	60	61	36	1	0.
ELEM	35	36	61	62	37	1	0.
ELEM	36	37	62	63	38	1	0.
ELEM	37	38	63	64	39	1	0.
ELEM	38	39	64	65	40	1	0.
ELEM	39	40	65	66	41	1	0.
ELEM	40	41	66	67	42	1	0.
ELEM	41	42	67	68	43	1	0.
ELEM	42	43	68	69	44	1	0.
ELEM	43	44	69	70	45	1	0.
ELEM	44	45	70	71	46	1	0.
ELEM	45	46	71	72	47	1	0.
ELEM	46	47	72	73	48	1	0.
ELEM	47	48	73	74	49	1	0.
ELEM	48	49	74	75	50	1	0.

ELEM	49	61	76	62	0	1	0.
ELEM	50	62	76	77	63	1	0.
ELEM	51	63	77	78	64	1	0.
ELEM	52	64	78	65	0	1	0.
ELEM	53	76	79	80	77	1	0.
ELEM	54	77	80	81	78	1	0.
ELEM	55	79	82	83	80	1	0.
ELEM	56	80	83	84	81	1	0.
ELEM	57	82	85	86	83	1	0.
ELEM	58	83	86	87	84	1	0.
ELEM	59	85	88	89	86	1	0.
ELEM	60	86	89	90	87	1	0.
ELEM	61	88	91	92	89	1	0.
ELEM	62	89	92	93	90	1	0.
ELEM	63	91	94	95	92	1	0.
ELEM	64	92	95	96	93	1	0.
ELEM	65	94	97	98	95	1	0.
ELEM	66	95	98	99	96	1	0.
ELEM	67	97	100	101	98	1	0.
ELEM	68	98	101	102	99	1	0.
ELEM	69	100	103	104	101	1	0.
ELEM	70	101	104	105	102	1	0.
ELEM	71	103	106	107	104	1	0.
ELEM	72	104	107	108	105	1	0.
ELEM	73	106	119	120	0	1	0.
ELEM	74	106	120	121	107	1	0.
ELEM	75	107	121	122	108	1	0.
ELEM	76	108	122	123	0	1	0.
ELEM	77	109	134	135	110	1	0.
ELEM	78	110	135	136	111	1	0.
ELEM	79	111	136	137	112	1	0.
ELEM	80	112	137	138	113	1	0.
ELEM	81	113	138	139	114	1	0.
ELEM	82	114	139	140	115	1	0.
ELEM	83	115	140	141	116	1	0.
ELEM	84	116	141	142	117	1	0.
ELEM	85	117	142	143	118	1	0.
ELEM	86	118	143	144	119	1	0.
ELEM	87	119	144	145	120	1	0.
ELEM	88	120	145	146	121	1	0.
ELEM	89	121	146	147	122	1	0.
ELEM	90	122	147	148	123	1	0.
ELEM	91	123	148	149	124	1	0.
ELEM	92	124	149	150	125	1	0.
ELEM	93	125	150	151	126	1	0.

```
  ELEM     94   126   151   152   127   1   0.
  ELEM     95   127   152   153   128   1   0.
  ELEM     96   128   153   154   129   1   0.
  ELEM     97   129   154   155   130   1   0.
  ELEM     98   130   155   156   131   1   0.
  ELEM     99   131   156   157   132   1   0.
  ELEM    100   132   157   158   133   1   0.
  ELEM    101   134   159   160   135   1   0.
  ELEM    102   135   160   161   136   1   0.
  ELEM    103   136   161   162   137   1   0.
  ELEM    104   137   162   163   138   1   0.
  ELEM    105   138   163   164   139   1   0.
  ELEM    106   139   164   165   140   1   0.
  ELEM    107   140   165   166   141   1   0.
  ELEM    108   141   166   167   142   1   0.
  ELEM    109   142   167   168   143   1   0.
  ELEM    110   143   168   169   144   1   0.
  ELEM    111   144   169   170   145   1   0.
  ELEM    112   145   170   171   146   1   0.
  ELEM    113   146   171   172   147   1   0.
  ELEM    114   147   172   173   148   1   0.
  ELEM    115   148   173   174   149   1   0.
  ELEM    116   149   174   175   150   1   0.
  ELEM    117   150   175   176   151   1   0.
  ELEM    118   151   176   177   152   1   0.
  ELEM    119   152   177   178   153   1   0.
  ELEM    120   153   178   179   154   1   0.
  ELEM    121   154   179   180   155   1   0.
  ELEM    122   155   180   181   156   1   0.
  ELEM    123   156   181   182   157   1   0.
  ELEM    124   157   182   183   158   1   0.
  FRONTIER
END_FRONT
  SYMMETRY
    END_SYM
  PRECISION      1.E-3
  MATERIALS
  STEELEC3
                 25.     9.     0.50
     TIME
                 12.   2400.
  END_TIME
    IMPRESSION
    TIMEPRINT
                 60.   2400.
END_TIMEPR
```

```
Profile: Type: B25x25-3s
Fire curve: FIRE.FCT - Exposed Faces:1 1 1 0
Protected with: (None) 0 mm 0 mm 0 mm;
         NNODE          183
            NDIM          2
     NDOFMAX              1
            FROM          1        TO          183 STEP     1  NDOF     1
     END_NDOF
     TEMPERAT
            TETA        0.9
        TINITIAL       20.0
     MAKE.TEM
        NORENUM
B25x25-3S.tem
            NMAT          1
     ELEMENTS
            SOLID       124
               NG         2
            NVOID         0
 END_ELEM
     NODES
     NODE          1        0.1250        -0.1250
     NODE          2        0.1250        -0.1140
     NODE          3        0.1250        -0.1030
     NODE          4        0.1250        -0.0919
     NODE          5        0.1250        -0.0809
     NODE          6        0.1250        -0.0699
     NODE          7        0.1250        -0.0589
     NODE          8        0.1250        -0.0479
     NODE          9        0.1250        -0.0369
     NODE         10        0.1250        -0.0258
     NODE         11        0.1250        -0.0148
     NODE         12        0.1250        -0.0050
     NODE         13        0.1250         0.0000
     NODE         14        0.1250         0.0050
     NODE         15        0.1250         0.0148
     NODE         16        0.1250         0.0258
     NODE         17        0.1250         0.0369
     NODE         18        0.1250         0.0479
     NODE         19        0.1250         0.0589
     NODE         20        0.1250         0.0699
     NODE         21        0.1250         0.0809
     NODE         22        0.1250         0.0919
     NODE         23        0.1250         0.1030
     NODE         24        0.1250         0.1140
```

NODE	25	0.1250	0.1250
NODE	26	0.1175	-0.1250
NODE	27	0.1175	-0.1140
NODE	28	0.1175	-0.1030
NODE	29	0.1175	-0.0919
NODE	30	0.1175	-0.0809
NODE	31	0.1175	-0.0699
NODE	32	0.1175	-0.058
NODE	33	0.1175	-0.0479
NODE	34	0.1175	-0.0369
NODE	35	0.1175	-0.0258
NODE	36	0.1175	-0.0148
NODE	37	0.1175	-0.0050
NODE	38	0.1175	0.0000
NODE	39	0.1175	0.0050
NODE	40	0.1175	0.0148
NODE	41	0.1175	0.0258
NODE	42	0.1175	0.0369
NODE	43	0.1175	0.0479
NODE	44	0.1175	0.0589
NODE	45	0.1175	0.0699
NODE	46	0.1175	0.0809
NODE	47	0.1175	0.0919
NODE	48	0.1175	0.1030
NODE	49	0.1175	0.1140
NODE	50	0.1175	0.1250
NODE	51	0.1100	-0.1250
NODE	52	0.1100	-0.1140
NODE	53	0.1100	-0.1030
NODE	54	0.1100	-0.0919
NODE	55	0.1100	-0.0809
NODE	56	0.1100	-0.0699
NODE	57	0.1100	-0.0589
NODE	58	0.1100	-0.0479
NODE	59	0.1100	-0.0369
NODE	60	0.1100	-0.0258
NODE	61	0.1100	-0.0148
NODE	62	0.1100	-0.0050
NODE	63	0.1100	0.0000
NODE	64	0.1100	0.0050
NODE	65	0.1100	0.0148
NODE	66	0.1100	0.0258
NODE	67	0.1100	0.0369
NODE	68	0.1100	0.0479
NODE	69	0.1100	0.0589

NODE	70	0.1100	0.0699
NODE	71	0.1100	0.0809
NODE	72	0.1100	0.0919
NODE	73	0.1100	0.1030
NODE	74	0.1100	0.1140
NODE	75	0.1100	0.1250
NODE	76	0.1002	-0.0050
NODE	77	0.1002	0.0000
NODE	78	0.1002	0.0050
NODE	79	0.0801	-0.0050
NODE	80	0.0801	0.0000
NODE	81	0.0801	0.0050
NODE	82	0.0601	-0.0050
NODE	83	0.0601	0.0000
NODE	84	0.0601	0.0050
NODE	85	0.0401	-0.0050
NODE	86	0.0401	0.0000
NODE	87	0.0401	0.0050
NODE	88	0.0200	-0.0050
NODE	89	0.0200	0.0000
NODE	90	0.0200	0.0050
NODE	91	0.0000	-0.0050
NODE	92	0.0000	0.0000
NODE	93	0.0000	0.0050
NODE	94	-0.0200	-0.0050
NODE	95	-0.0200	0.0000
NODE	96	-0.0200	0.0050
NODE	97	-0.0401	-0.0050
NODE	98	-0.0401	0.0000
NODE	99	-0.0401	0.0050
NODE	100	-0.0601	-0.0050
NODE	101	-0.0601	0.0000
NODE	102	-0.0601	0.0050
NODE	103	-0.0801	-0.0050
NODE	104	-0.0801	0.0000
NODE	105	-0.0801	0.0050
NODE	106	-0.1002	-0.0050
NODE	107	-0.1002	0.0000
NODE	108	-0.1002	0.0050
NODE	109	-0.1100	-0.1250
NODE	110	-0.1100	-0.1140
NODE	111	-0.1100	-0.1030
NODE	112	-0.1100	-0.0919
NODE	113	-0.1100	-0.0809
NODE	114	-0.1100	-0.0699

NODE	115	-0.1100	-0.0589
NODE	116	-0.1100	-0.0479
NODE	117	-0.1100	-0.0369
NODE	118	-0.1100	-0.0258
NODE	119	-0.1100	-0.0148
NODE	120	-0.1100	-0.0050
NODE	121	-0.1100	0.0000
NODE	122	-0.1100	0.0050
NODE	123	-0.1100	0.0148
NODE	124	-0.1100	0.0258
NODE	125	-0.1100	0.0369
NODE	126	-0.1100	0.0479
NODE	127	-0.1100	0.0589
NODE	128	-0.1100	0.0699
NODE	129	-0.1100	0.0809
NODE	130	-0.1100	0.0919
NODE	131	-0.1100	0.1030
NODE	132	-0.1100	0.1140
NODE	133	-0.1100	0.1250
NODE	134	-0.1175	-0.1250
NODE	135	-0.1175	-0.1140
NODE	136	-0.1175	-0.1030
NODE	137	-0.1175	-0.0919
NODE	138	-0.1175	-0.0809
NODE	139	-0.1175	-0.0699
NODE	140	-0.1175	-0.0589
NODE	141	-0.1175	-0.0479
NODE	142	-0.1175	-0.0369
NODE	143	-0.1175	-0.0258
NODE	144	-0.1175	-0.0148
NODE	145	-0.1175	-0.0050
NODE	146	-0.1175	0.0000
NODE	147	-0.1175	0.0050
NODE	148	-0.1175	0.0148
NODE	149	-0.1175	0.0258
NODE	150	-0.1175	0.0369
NODE	151	-0.1175	0.0479
NODE	152	-0.1175	0.0589
NODE	153	-0.1175	0.0699
NODE	154	-0.1175	0.0809
NODE	155	-0.1175	0.0919
NODE	156	-0.1175	0.1030
NODE	157	-0.1175	0.1140
NODE	158	-0.1175	0.1250
NODE	159	-0.1250	-0.1250

```
    NODE      160        -0.1250        -0.1140
    NODE      161        -0.1250        -0.1030
    NODE      162        -0.1250        -0.0919
    NODE      163        -0.1250        -0.0809
    NODE      164        -0.1250        -0.0699
    NODE      165        -0.1250        -0.0589
    NODE      166        -0.1250        -0.0479
    NODE      167        -0.1250        -0.0369
    NODE      168        -0.1250        -0.0258
    NODE      169        -0.1250        -0.0148
    NODE      170        -0.1250        -0.0050
    NODE      171        -0.1250         0.0000
    NODE      172        -0.1250         0.0050
    NODE      173        -0.1250         0.0148
    NODE      174        -0.1250         0.0258
    NODE      175        -0.1250         0.0369
    NODE      176        -0.1250         0.0479
    NODE      177        -0.1250         0.0589
    NODE      178        -0.1250         0.0699
    NODE      179        -0.1250         0.0809
    NODE      180        -0.1250         0.0919
    NODE      181        -0.1250         0.1030
    NODE      182        -0.1250         0.1140
    NODE      183        -0.1250         0.1250
    NODELINE          0        0
         YC_ZC        0        0
    FIXATIONS
        END_FIX
NODOFSOLID
  ELEM      1     1    26    27     2     1    0.
  ELEM      1     1    26    27     2     1    0.
  ELEM      2     2    27    28     3     1    0.
  ELEM      3     3    28    29     4     1    0.
  ELEM      4     4    29    30     5     1    0.
  ELEM      5     5    30    31     6     1    0.
  ELEM      6     6    31    32     7     1    0.
  ELEM      7     7    32    33     8     1    0.
  ELEM      8     8    33    34     9     1    0.
  ELEM      9     9    34    35    10     1    0.
  ELEM     10    10    35    36    11     1    0.
  ELEM     11    11    36    37    12     1    0.
  ELEM     12    12    37    38    13     1    0.
  ELEM     13    13    38    39    14     1    0.
  ELEM     14    14    39    40    15     1    0.
  ELEM     15    15    40    41    16     1    0.
```

ELEM	16	16	41	42	17	1	0.
ELEM	17	17	42	43	18	1	0.
ELEM	18	18	43	44	19	1	0.
ELEM	19	19	44	45	20	1	0.
ELEM	20	20	45	46	21	1	0.
ELEM	21	21	46	47	22	1	0.
ELEM	22	22	47	48	23	1	0.
ELEM	23	23	48	49	24	1	0.
ELEM	24	24	49	50	25	1	0.
ELEM	25	26	51	52	27	1	0.
ELEM	26	27	52	53	28	1	0.
ELEM	27	28	53	54	29	1	0.
ELEM	28	29	54	55	30	1	0.
ELEM	29	30	55	56	31	1	0.
ELEM	30	31	56	57	32	1	0.
ELEM	31	32	57	58	33	1	0.
ELEM	32	33	58	59	34	1	0.
ELEM	33	34	59	60	35	1	0.
ELEM	34	35	60	61	36	1	0.
ELEM	35	36	61	62	37	1	0.
ELEM	36	37	62	63	38	1	0.
ELEM	37	38	63	64	39	1	0.
ELEM	38	39	64	65	40	1	0.
ELEM	39	40	65	66	41	1	0.
ELEM	40	41	66	67	42	1	0.
ELEM	41	42	67	68	43	1	0.
ELEM	42	43	68	69	44	1	0.
ELEM	43	44	69	70	45	1	0.
ELEM	44	45	70	71	46	1	0.
ELEM	45	46	71	72	47	1	0.
ELEM	46	47	72	73	48	1	0.
ELEM	47	48	73	74	49	1	0.
ELEM	48	49	74	75	50	1	0.
ELEM	49	61	76	62	0	1	0.
ELEM	50	62	76	77	63	1	0.
ELEM	51	63	77	78	64	1	0.
ELEM	52	64	78	65	0	1	0.
ELEM	53	76	79	80	77	1	0.
ELEM	54	77	80	81	78	1	0.
ELEM	55	79	82	83	80	1	0.
ELEM	56	80	83	84	81	1	0.
ELEM	57	82	85	86	83	1	0.
ELEM	58	83	86	87	84	1	0.
ELEM	59	85	88	89	86	1	0.
ELEM	60	86	89	90	87	1	0.

ELEM	61	88	91	92	89	1	0.
ELEM	62	89	92	93	90	1	0.
ELEM	63	91	94	95	92	1	0.
ELEM	64	92	95	96	93	1	0.
ELEM	65	94	97	98	95	1	0.
ELEM	66	95	98	99	96	1	0.
ELEM	67	97	100	101	98	1	0.
ELEM	68	98	101	102	99	1	0.
ELEM	69	100	103	104	101	1	0.
ELEM	70	101	104	105	102	1	0.
ELEM	71	103	106	107	104	1	0.
ELEM	72	104	107	108	105	1	0.
ELEM	73	106	119	120	0	1	0.
ELEM	74	106	120	121	107	1	0.
ELEM	75	107	121	122	108	1	0.
ELEM	76	108	122	123	0	1	0.
ELEM	77	109	134	135	110	1	0.
ELEM	78	110	135	136	111	1	0.
ELEM	79	111	136	137	112	1	0.
ELEM	80	112	137	138	113	1	0.
ELEM	81	113	138	139	114	1	0.
ELEM	82	114	139	140	115	1	0.
ELEM	83	115	140	141	116	1	0.
ELEM	84	116	141	142	117	1	0.
ELEM	85	117	142	143	118	1	0.
ELEM	86	118	143	144	119	1	0.
ELEM	87	119	144	145	120	1	0.
ELEM	88	120	145	146	121	1	0.
ELEM	89	121	146	147	122	1	0.
ELEM	90	122	147	148	123	1	0.
ELEM	91	123	148	149	124	1	0.
ELEM	92	124	149	150	125	1	0.
ELEM	93	125	150	151	126	1	0.
ELEM	94	126	151	152	127	1	0.
ELEM	95	127	152	153	128	1	0.
ELEM	96	128	153	154	129	1	0.
ELEM	97	129	154	155	130	1	0.
ELEM	98	130	155	156	131	1	0.
ELEM	99	131	156	157	132	1	0.
ELEM	100	132	157	158	133	1	0.
ELEM	101	134	159	160	135	1	0.
ELEM	102	135	160	161	136	1	0.
ELEM	103	136	161	162	137	1	0.
ELEM	104	137	162	163	138	1	0.
ELEM	105	138	163	164	139	1	0.

```
 ELEM     106   139   164   165   140   1   0.
 ELEM     107   140   165   166   141   1   0.
 ELEM     108   141   166   167   142   1   0.
 ELEM     109   142   167   168   143   1   0.
 ELEM     110   143   168   169   144   1   0.
 ELEM     111   144   169   170   145   1   0.
 ELEM     112   145   170   171   146   1   0.
 ELEM     113   146   171   172   147   1   0.
 ELEM     114   147   172   173   148   1   0.
 ELEM     115   148   173   174   149   1   0.
 ELEM     116   149   174   175   150   1   0.
 ELEM     117   150   175   176   151   1   0.
 ELEM     118   151   176   177   152   1   0.
 ELEM     119   152   177   178   153   1   0.
 ELEM     120   153   178   179   154   1   0.
 ELEM     121   154   179   180   155   1   0.
 ELEM     122   155   180   181   156   1   0.
 ELEM     123   156   181   182   157   1   0.
 ELEM     124   157   182   183   158   1   0.
FRONTIER
 F    1   FIRE.FCT        NO        NO        NO
 F   25   FIRE.FCT        NO        NO        NO
 F   25         NO  FIRE.FCT        NO        NO
 F   26         NO  FIRE.FCT        NO        NO
 F   27         NO  FIRE.FCT        NO        NO
 F   28         NO  FIRE.FCT        NO        NO
 F   29         NO  FIRE.FCT        NO        NO
 F   30         NO  FIRE.FCT        NO        NO
 F   31         NO  FIRE.FCT        NO        NO
 F   32         NO  FIRE.FCT        NO        NO
 F   33         NO  FIRE.FCT        NO        NO
 F   34         NO  FIRE.FCT        NO        NO
 F   49   FIRE.FCT        NO        NO        NO
 F   53   FIRE.FCT        NO        NO        NO
 F   55   FIRE.FCT        NO        NO        NO
 F   57   FIRE.FCT        NO        NO        NO
 F   59   FIRE.FCT        NO        NO        NO
 F   61   FIRE.FCT        NO        NO        NO
 F   63   FIRE.FCT        NO        NO        NO
 F   65   FIRE.FCT        NO        NO        NO
 F   67   FIRE.FCT        NO        NO        NO
 F   69   FIRE.FCT        NO        NO        NO
 F   71   FIRE.FCT        NO        NO        NO
 F   73   FIRE.FCT        NO        NO        NO
 F   77         NO        NO        NO  FIRE.FCT
```

F	78	NO	NO	NO	FIRE.FCT
F	79	NO	NO	NO	FIRE.FCT
F	80	NO	NO	NO	FIRE.FCT
F	81	NO	NO	NO	FIRE.FCT
F	82	NO	NO	NO	FIRE.FCT
F	83	NO	NO	NO	FIRE.FCT
F	84	NO	NO	NO	FIRE.FCT
F	85	NO	NO	NO	FIRE.FCT
F	86	NO	NO	NO	FIRE.FCT
F	77	FIRE.FCT	NO	NO	NO
F	101	FIRE.FCT	NO	NO	NO
F	101	NO	FIRE.FCT	NO	NO
F	102	NO	FIRE.FCT	NO	NO
F	103	NO	FIRE.FCT	NO	NO
F	104	NO	FIRE.FCT	NO	NO
F	105	NO	FIRE.FCT	NO	NO
F	106	NO	FIRE.FCT	NO	NO
F	107	NO	FIRE.FCT	NO	NO
F	108	NO	FIRE.FCT	NO	NO
F	109	NO	FIRE.FCT	NO	NO
F	110	NO	FIRE.FCT	NO	NO
F	111	NO	FIRE.FCT	NO	NO
F	112	NO	FIRE.FCT	NO	NO
F	113	NO	FIRE.FCT	NO	NO
F	114	NO	FIRE.FCT	NO	NO
F	115	NO	FIRE.FCT	NO	NO
F	116	NO	FIRE.FCT	NO	NO
F	117	NO	FIRE.FCT	NO	NO
F	118	NO	FIRE.FCT	NO	NO
F	119	NO	FIRE.FCT	NO	NO
F	120	NO	FIRE.FCT	NO	NO
F	121	NO	FIRE.FCT	NO	NO
F	122	NO	FIRE.FCT	NO	NO
F	123	NO	FIRE.FCT	NO	NO
F	124	NO	FIRE.FCT	NO	NO
F	24	NO	NO	FIRE.FCT	NO
F	48	NO	NO	FIRE.FCT	NO
F	48	NO	FIRE.FCT	NO	NO
F	47	NO	FIRE.FCT	NO	NO
F	46	NO	FIRE.FCT	NO	NO
F	45	NO	FIRE.FCT	NO	NO
F	44	NO	FIRE.FCT	NO	NO
F	43	NO	FIRE.FCT	NO	NO
F	42	NO	FIRE.FCT	NO	NO
F	41	NO	FIRE.FCT	NO	NO

```
  F      40       NO        FIRE.FCT          NO           NO
  F      39       NO        FIRE.FCT          NO           NO
  F      52       NO        FIRE.FCT          NO           NO
  F      54       NO              NO    FIRE.FCT           NO
  F      56       NO              NO    FIRE.FCT           NO
  F      58       NO              NO    FIRE.FCT           NO
  F      60       NO              NO    FIRE.FCT           NO
  F      62       NO              NO    FIRE.FCT           NO
  F      64       NO              NO    FIRE.FCT           NO
  F      66       NO              NO    FIRE.FCT           NO
  F      68       NO              NO    FIRE.FCT           NO
  F      70       NO              NO    FIRE.FCT           NO
  F      72       NO              NO    FIRE.FCT           NO
  F      76       NO              NO    FIRE.FCT           NO
  F      91       NO              NO          NO     FIRE.FCT
  F      92       NO              NO          NO     FIRE.FCT
  F      93       NO              NO          NO     FIRE.FCT
  F      94       NO              NO          NO     FIRE.FCT
  F      95       NO              NO          NO     FIRE.FCT
  F      96       NO              NO          NO     FIRE.FCT
  F      97       NO              NO          NO     FIRE.FCT
  F      98       NO              NO          NO     FIRE.FCT
  F      99       NO              NO          NO     FIRE.FCT
  F     100       NO              NO          NO     FIRE.FCT
  F     100       NO              NO    FIRE.FCT           NO
  F     124       NO              NO    FIRE.FCT           NO
END_FRONT
  SYMMETRY
     END_SYM
PRECISION            1.E-3
MATERIALS
  STEELEC3
                       25.        9.      0.50
       TIME
                       12.     2400.
     END_TIME
IMPRESSION
  TIMEPRINT
                       60.     2400.
END_TIMEPR
```

Profile: Type: B30x30-0
Fire curve: FISO - Exposed Faces:0 0 0 0
Protected with: (None) 0 mm 0 mm 2 mm;

```
        NNODE           183
        NDIM                    2
    NDOFMAX                     1
        FROM              1     TO          183 STEP        1 NDOF      1
    END_NDOF
    TEMPERAT
        TETA            0.9
    TINITIAL           20.0
    MAKE.TEM
        NORENUM
B30x30-0.tem
            NMAT          1
        ELEMENTS
            SOLID       124
              NG          2
            NVOID         0
 END_ELEM
  NODES
     NODE          1          0.1500        -0.1500
     NODE          2          0.1500        -0.1371
     NODE          3          0.1500        -0.1241
     NODE          4          0.1500        -0.1112
     NODE          5          0.1500        -0.0982
     NODE          6          0.1500        -0.0853
     NODE          7          0.1500        -0.0724
     NODE          8          0.1500        -0.0594
     NODE          9          0.1500        -0.0465
     NODE         10          0.1500        -0.0335
     NODE         11          0.1500        -0.0206
     NODE         12          0.1500        -0.0075
     NODE         13          0.1500         0.0000
     NODE         14          0.1500         0.0075
     NODE         15          0.1500         0.0206
     NODE         16          0.1500         0.0335
     NODE         17          0.1500         0.0465
     NODE         18          0.1500         0.0594
     NODE         19          0.1500         0.0724
     NODE         20          0.1500         0.0853
     NODE         21          0.1500         0.0982
     NODE         22          0.1500         0.1112
     NODE         23          0.1500         0.1241
     NODE         24          0.1500         0.1371
     NODE         25          0.1500         0.1500
     NODE         26          0.1400        -0.1500
     NODE         27          0.1400        -0.1371
```

```
NODE    28    0.1400    -0.1241
NODE    29    0.1400    -0.1112
NODE    30    0.1400    -0.0982
NODE    31    0.1400    -0.0853
NODE    32    0.1400    -0.0724
NODE    33    0.1400    -0.0594
NODE    34    0.1400    -0.0465
NODE    35    0.1400    -0.0335
NODE    36    0.1400    -0.0206
NODE    37    0.1400    -0.0075
NODE    38    0.1400     0.0000
NODE    39    0.1400     0.0075
NODE    40    0.1400     0.0206
NODE    41    0.1400     0.0335
NODE    42    0.1400     0.0465
NODE    43    0.1400     0.0594
NODE    44    0.1400     0.0724
NODE    45    0.1400     0.0853
NODE    46    0.1400     0.0982
NODE    47    0.1400     0.1112
NODE    48    0.1400     0.1241
NODE    49    0.1400     0.1371
NODE    50    0.1400     0.1500
NODE    51    0.1300    -0.1500
NODE    52    0.1300    -0.1371
NODE    53    0.1300    -0.1241
NODE    54    0.1300    -0.1112
NODE    55    0.1300    -0.0982
NODE    56    0.1300    -0.0853
NODE    57    0.1300    -0.0724
NODE    58    0.1300    -0.0594
NODE    59    0.1300    -0.0465
NODE    60    0.1300    -0.0335
NODE    61    0.1300    -0.0206
NODE    62    0.1300    -0.0075
NODE    63    0.1300     0.0000
NODE    64    0.1300     0.0075
NODE    65    0.1300     0.0206
NODE    66    0.1300     0.0335
NODE    67    0.1300     0.0465
NODE    68    0.1300     0.0594
NODE    69    0.1300     0.0724
NODE    70    0.1300     0.0853
NODE    71    0.1300     0.0982
NODE    72    0.1300     0.1112
```

NODE	73	0.1300	0.1241
NODE	74	0.1300	0.1371
NODE	75	0.1300	0.1500
NODE	76	0.1169	-0.0075
NODE	77	0.1169	0.0000
NODE	78	0.1169	0.0075
NODE	79	0.0935	-0.0075
NODE	80	0.0935	0.0000
NODE	81	0.0935	0.0075
NODE	82	0.0701	-0.0075
NODE	83	0.0701	0.0000
NODE	84	0.0701	0.0075
NODE	85	0.0468	-0.0075
NODE	86	0.0468	0.0000
NODE	87	0.0468	0.0075
NODE	88	0.0234	-0.0075
NODE	89	0.0234	0.0000
NODE	90	0.0234	0.0075
NODE	91	0.0000	-0.0075
NODE	92	0.0000	0.0000
NODE	93	0.0000	0.0075
NODE	94	-0.0234	-0.0075
NODE	95	-0.0234	0.0000
NODE	96	-0.0234	0.0075
NODE	97	-0.0468	-0.0075
NODE	98	-0.0468	0.0000
NODE	99	-0.0468	0.0075
NODE	100	-0.0701	-0.0075
NODE	101	-0.0701	0.0000
NODE	102	-0.0701	0.0075
NODE	103	-0.0935	-0.0075
NODE	104	-0.0935	0.0000
NODE	105	-0.0935	0.0075
NODE	106	-0.1169	-0.0075
NODE	107	-0.1169	0.0000
NODE	108	-0.1169	0.0075
NODE	109	-0.1300	-0.1500
NODE	110	-0.1300	-0.1371
NODE	111	-0.1300	-0.1241
NODE	112	-0.1300	-0.1112
NODE	113	-0.1300	-0.0982
NODE	114	-0.1300	-0.0853
NODE	115	-0.1300	-0.0724
NODE	116	-0.1300	-0.0594
NODE	117	-0.1300	-0.0465

NODE	118	-0.1300	-0.0335
NODE	119	-0.1300	-0.0206
NODE	120	-0.1300	-0.0075
NODE	121	-0.1300	0.0000
NODE	122	-0.1300	0.0075
NODE	123	-0.1300	0.0206
NODE	124	-0.1300	0.0335
NODE	125	-0.1300	0.0465
NODE	126	-0.1300	0.0594
NODE	127	-0.1300	0.0724
NODE	128	-0.1300	0.0853
NODE	129	-0.1300	0.0982
NODE	130	-0.1300	0.1112
NODE	131	-0.1300	0.1241
NODE	132	-0.1300	0.1371
NODE	133	-0.1300	0.1500
NODE	134	-0.1400	-0.1500
NODE	135	-0.1400	-0.1371
NODE	136	-0.1400	-0.1241
NODE	137	-0.1400	-0.1112
NODE	138	-0.1400	-0.0982
NODE	139	-0.1400	-0.0853
NODE	140	-0.1400	-0.0724
NODE	141	-0.1400	-0.0594
NODE	142	-0.1400	-0.0465
NODE	143	-0.1400	-0.0335
NODE	144	-0.1400	-0.0206
NODE	145	-0.1400	-0.0075
NODE	146	-0.1400	0.0000
NODE	147	-0.1400	0.0075
NODE	148	-0.1400	0.0206
NODE	149	-0.1400	0.0335
NODE	150	-0.1400	0.0465
NODE	151	-0.1400	0.0594
NODE	152	-0.1400	0.0724
NODE	153	-0.1400	0.0853
NODE	154	-0.1400	0.0982
NODE	155	-0.1400	0.1112
NODE	156	-0.1400	0.1241
NODE	157	-0.1400	0.1371
NODE	158	-0.1400	0.1500
NODE	159	-0.1500	-0.1500
NODE	160	-0.1500	-0.1371
NODE	161	-0.1500	-0.1241
NODE	162	-0.1500	-0.1112

```
NODE        163     -0.1500     -0.0982
NODE        164     -0.1500     -0.0853
NODE        165     -0.1500     -0.0724
NODE        166     -0.1500     -0.0594
NODE        167     -0.1500     -0.0465
NODE        168     -0.1500     -0.0335
NODE        169     -0.1500     -0.0206
NODE        170     -0.1500     -0.0075
NODE        171     -0.1500      0.0000
NODE        172     -0.1500      0.0075
NODE        173     -0.1500      0.0206
NODE        174     -0.1500      0.0335
NODE        175     -0.1500      0.0465
NODE        176     -0.1500      0.0594
NODE        177     -0.1500      0.0724
NODE        178     -0.1500      0.0853
NODE        179     -0.1500      0.0982
NODE        180     -0.1500      0.1112
NODE        181     -0.1500      0.1241
NODE        182     -0.1500      0.1371
NODE        183     -0.1500      0.1500
NODELINE      0           0
YC_ZC         0           0
FIXATIONS
END_FIX
NODOFSOLID
ELEM     1     1    26    27     2     1    0.
ELEM     2     2    27    28     3     1    0.
ELEM     3     3    28    29     4     1    0.
ELEM     4     4    29    30     5     1    0.
ELEM     5     5    30    31     6     1    0.
ELEM     6     6    31    32     7     1    0.
ELEM     7     7    32    33     8     1    0.
ELEM     8     8    33    34     9     1    0.
ELEM     9     9    34    35    10     1    0.
ELEM    10    10    35    36    11     1    0.
ELEM    11    11    36    37    12     1    0.
ELEM    12    12    37    38    13     1    0.
ELEM    13    13    38    39    14     1    0.
ELEM    14    14    39    40    15     1    0.
ELEM    15    15    40    41    16     1    0.
ELEM    16    16    41    42    17     1    0.
ELEM    17    17    42    43    18     1    0.
ELEM    18    18    43    44    19     1    0.
ELEM    19    19    44    45    20     1    0.
```

```
ELEM 20 20 45 46 21 1 0.
ELEM 21 21 46 47 22 1 0.
ELEM 22 22 47 48 23 1 0.
ELEM 23 23 48 49 24 1 0.
ELEM 24 24 49 50 25 1 0.
ELEM 25 26 51 52 27 1 0.
ELEM 26 27 52 53 28 1 0.
ELEM 27 28 53 54 29 1 0.
ELEM 28 29 54 55 30 1 0.
ELEM 29 30 55 56 31 1 0.
ELEM 30 31 56 57 32 1 0.
ELEM 31 32 57 58 33 1 0.
ELEM 32 33 58 59 34 1 0.
ELEM 33 34 59 60 35 1 0.
ELEM 34 35 60 61 36 1 0.
ELEM 35 36 61 62 37 1 0.
ELEM 36 37 62 63 38 1 0.
ELEM 37 38 63 64 39 1 0.
ELEM 38 39 64 65 40 1 0.
ELEM 39 40 65 66 41 1 0.
ELEM 40 41 66 67 42 1 0.
ELEM 41 42 67 68 43 1 0.
ELEM 42 43 68 69 44 1 0.
ELEM 43 44 69 70 45 1 0.
ELEM 44 45 70 71 46 1 0.
ELEM 45 46 71 72 47 1 0.
ELEM 46 47 72 73 48 1 0.
ELEM 47 48 73 74 49 1 0.
ELEM 48 49 74 75 50 1 0.
ELEM 49 61 76 62 0 1 0.
ELEM 50 62 76 77 63 1 0.
ELEM 51 63 77 78 64 1 0.
ELEM 52 64 78 65 0 1 0.
ELEM 53 76 79 80 77 1 0.
ELEM 54 77 80 81 78 1 0.
ELEM 55 79 82 83 80 1 0.
ELEM 56 80 83 84 81 1 0.
ELEM 57 82 85 86 83 1 0.
ELEM 58 83 86 87 84 1 0.
ELEM 59 85 88 89 86 1 0.
ELEM 60 86 89 90 87 1 0.
ELEM 61 88 91 92 89 1 0.
ELEM 62 89 92 93 90 1 0.
ELEM 63 91 94 95 92 1 0.
ELEM 64 92 95 96 93 1 0.
```

ELEM	65	94	97	98	95	1	0.
ELEM	66	95	98	99	96	1	0.
ELEM	67	97	100	101	98	1	0.
ELEM	68	98	101	102	99	1	0.
ELEM	69	100	103	104	101	1	0.
ELEM	70	101	104	105	102	1	0.
ELEM	71	103	106	107	104	1	0.
ELEM	72	104	107	108	105	1	0.
ELEM	73	106	119	120	0	1	0.
ELEM	74	106	120	121	107	1	0.
ELEM	75	107	121	122	108	1	0.
ELEM	76	108	122	123	0	1	0.
ELEM	77	109	134	135	110	1	0.
ELEM	78	110	135	136	111	1	0.
ELEM	79	111	136	137	112	1	0.
ELEM	80	112	137	138	113	1	0.
ELEM	81	113	138	139	114	1	0.
ELEM	82	114	139	140	115	1	0.
ELEM	83	115	140	141	116	1	0.
ELEM	84	116	141	142	117	1	0.
ELEM	85	117	142	143	118	1	0.
ELEM	86	118	143	144	119	1	0.
ELEM	87	119	144	145	120	1	0.
ELEM	88	120	145	146	121	1	0.
ELEM	89	121	146	147	122	1	0.
ELEM	90	122	147	148	123	1	0.
ELEM	91	123	148	149	124	1	0.
ELEM	92	124	149	150	125	1	0.
ELEM	93	125	150	151	126	1	0.
ELEM	94	126	151	152	127	1	0.
ELEM	95	127	152	153	128	1	0.
ELEM	96	128	153	154	129	1	0.
ELEM	97	129	154	155	130	1	0.
ELEM	98	130	155	156	131	1	0.
ELEM	99	131	156	157	132	1	0.
ELEM	100	132	157	158	133	1	0.
ELEM	101	134	159	160	135	1	0.
ELEM	102	135	160	161	136	1	0.
ELEM	103	136	161	162	137	1	0.
ELEM	104	137	162	163	138	1	0.
ELEM	105	138	163	164	139	1	0.
ELEM	106	139	164	165	140	1	0.
ELEM	107	140	165	166	141	1	0.
ELEM	108	141	166	167	142	1	0.
ELEM	109	142	167	168	143	1	0.

```
  ELEM     110   143   168   169   144    1   0.
  ELEM     111   144   169   170   145    1   0.
  ELEM     112   145   170   171   146    1   0.
  ELEM     113   146   171   172   147    1   0.
  ELEM     114   147   172   173   148    1   0.
  ELEM     115   148   173   174   149    1   0.
  ELEM     116   149   174   175   150    1   0.
  ELEM     117   150   175   176   151    1   0.
  ELEM     118   151   176   177   152    1   0.
  ELEM     119   152   177   178   153    1   0.
  ELEM     120   153   178   179   154    1   0.
  ELEM     121   154   179   180   155    1   0.
  ELEM     122   155   180   181   156    1   0.
  ELEM     123   156   181   182   157    1   0.
  ELEM     124   157   182   183   158    1   0.
  FRONTIER
END_FRONT
  SYMMETRY
    END_SYM
  PRECISION        1.E-3
  MATERIALS
    STEELEC3
                   25.    9.   0.50
      TIME
                   12.  2400.
    END_TIME
IMPRESSION
    TIMEPRINT
                   60.  2400.
END_TIMEPR
```

Profile: B30x30-3s
Fire curve: FIRE4.fct - Exposed Faces:1 1 1 0
Protected with: (None) 0 mm 0 mm 0 mm;

```
  NNODE          183
    NDIM           2
  NDOFMAX          1
    FROM           1   TO     183 STEP   1  NDOF   1
  END_NDOF
  TEMPERAT
    TETA         0.9
  TINITIAL      20.0
  MAKE.TEM
    NORENUM
B30x30-3S.tem
```

```
        NMAT            1
    ELEMENTS
        SOLID         124
        NG              2
    NVOID               0
END_ELEM
   NODES
   NODE        1          0.1500          -0.1500
   NODE        2          0.1500          -0.1371
   NODE        3          0.1500          -0.1241
   NODE        4          0.1500          -0.1112
   NODE        5          0.1500          -0.0982
   NODE        6          0.1500          -0.0853
   NODE        7          0.1500          -0.0724
   NODE        8          0.1500          -0.0594
   NODE        9          0.1500          -0.0465
   NODE       10          0.1500          -0.0335
   NODE       11          0.1500          -0.0206
   NODE       12          0.1500          -0.0075
   NODE       13          0.1500           0.0000
   NODE       14          0.1500           0.0075
   NODE       15          0.1500           0.0206
   NODE       16          0.1500           0.0335
   NODE       17          0.1500           0.0465
   NODE       18          0.1500           0.0594
   NODE       19          0.1500           0.0724
   NODE       20          0.1500           0.0853
   NODE       21          0.1500           0.0982
   NODE       22          0.1500           0.1112
   NODE       23          0.1500           0.1241
   NODE       24          0.1500           0.1371
   NODE       25          0.1500           0.1500
   NODE       26          0.1400          -0.1500
   NODE       27          0.1400          -0.1371
   NODE       28          0.1400          -0.1241
   NODE       29          0.1400          -0.1112
   NODE       30          0.1400          -0.0982
   NODE       31          0.1400          -0.0853
   NODE       32          0.1400          -0.0724
   NODE       33          0.1400          -0.0594
   NODE       34          0.1400          -0.0465
   NODE       35          0.1400          -0.0335
   NODE       36          0.1400          -0.0206
   NODE       37          0.1400          -0.0075
   NODE       38          0.1400           0.0000
```

NODE	39	0.1400	0.0075
NODE	40	0.1400	0.0206
NODE	41	0.1400	0.0335
NODE	42	0.1400	0.0465
NODE	43	0.1400	0.0594
NODE	44	0.1400	0.0724
NODE	45	0.1400	0.0853
NODE	46	0.1400	0.0982
NODE	47	0.1400	0.1112
NODE	48	0.1400	0.1241
NODE	49	0.1400	0.1371
NODE	50	0.1400	0.1500
NODE	51	0.1300	-0.1500
NODE	52	0.1300	-0.1371
NODE	53	0.1300	-0.1241
NODE	54	0.1300	-0.1112
NODE	55	0.1300	-0.0982
NODE	56	0.1300	-0.0853
NODE	57	0.1300	-0.0724
NODE	58	0.1300	-0.0594
NODE	59	0.1300	-0.0465
NODE	60	0.1300	-0.0335
NODE	61	0.1300	-0.0206
NODE	62	0.1300	-0.0075
NODE	63	0.1300	0.0000
NODE	64	0.1300	0.0075
NODE	65	0.1300	0.0206
NODE	66	0.1300	0.0335
NODE	67	0.1300	0.0465
NODE	68	0.1300	0.0594
NODE	69	0.1300	0.0724
NODE	70	0.1300	0.0853
NODE	71	0.1300	0.0982
NODE	72	0.1300	0.1112
NODE	73	0.1300	0.1241
NODE	74	0.1300	0.1371
NODE	75	0.1300	0.1500
NODE	76	0.1169	-0.0075
NODE	77	0.1169	0.0000
NODE	78	0.1169	0.0075
NODE	79	0.0935	-0.0075
NODE	80	0.0935	0.0000
NODE	81	0.0935	0.0075
NODE	82	0.0701	-0.0075
NODE	83	0.0701	0.0000

NODE	84	0.0701	0.0075
NODE	85	0.0468	-0.0075
NODE	86	0.0468	0.0000
NODE	87	0.0468	0.0075
NODE	88	0.0234	-0.0075
NODE	89	0.0234	0.0000
NODE	90	0.0234	0.0075
NODE	91	0.0000	-0.0075
NODE	92	0.0000	0.0000
NODE	93	0.0000	0.0075
NODE	94	-0.0234	-0.0075
NODE	95	-0.0234	0.0000
NODE	96	-0.0234	0.0075
NODE	97	-0.0468	-0.0075
NODE	98	-0.0468	0.0000
NODE	99	-0.0468	0.0075
NODE	100	-0.0701	-0.0075
NODE	101	-0.0701	0.0000
NODE	102	-0.0701	0.0075
NODE	103	-0.0935	-0.0075
NODE	104	-0.0935	0.0000
NODE	105	-0.0935	0.0075
NODE	106	-0.1169	-0.0075
NODE	107	-0.1169	0.0000
NODE	108	-0.1169	0.0075
NODE	109	-0.1300	-0.1500
NODE	110	-0.1300	-0.1371
NODE	111	-0.1300	-0.1241
NODE	112	-0.1300	-0.1112
NODE	113	-0.1300	-0.0982
NODE	114	-0.1300	-0.0853
NODE	115	-0.1300	-0.0724
NODE	116	-0.1300	-0.0594
NODE	117	-0.1300	-0.0465
NODE	118	-0.1300	-0.0335
NODE	119	-0.1300	-0.0206
NODE	120	-0.1300	-0.0075
NODE	121	-0.1300	0.0000
NODE	122	-0.1300	0.0075
NODE	123	-0.1300	0.0206
NODE	124	-0.1300	0.0335
NODE	125	-0.1300	0.0465
NODE	126	-0.1300	0.0594
NODE	127	-0.1300	0.0724
NODE	128	-0.1300	0.0853

NODE	129	-0.1300	0.0982
NODE	130	-0.1300	0.1112
NODE	131	-0.1300	0.1241
NODE	132	-0.1300	0.1371
NODE	133	-0.1300	0.1500
NODE	134	-0.1400	-0.1500
NODE	135	-0.1400	-0.1371
NODE	136	-0.1400	-0.1241
NODE	137	-0.1400	-0.1112
NODE	138	-0.1400	-0.0982
NODE	139	-0.1400	-0.0853
NODE	140	-0.1400	-0.0724
NODE	141	-0.1400	-0.0594
NODE	142	-0.1400	-0.0465
NODE	143	-0.1400	-0.0335
NODE	144	-0.1400	-0.0206
NODE	145	-0.1400	-0.0075
NODE	146	-0.1400	0.0000
NODE	147	-0.1400	0.0075
NODE	148	-0.1400	0.0206
NODE	149	-0.1400	0.0335
NODE	150	-0.1400	0.0465
NODE	151	-0.1400	0.0594
NODE	152	-0.1400	0.0724
NODE	153	-0.1400	0.0853
NODE	154	-0.1400	0.0982
NODE	155	-0.1400	0.1112
NODE	156	-0.1400	0.1241
NODE	157	-0.1400	0.1371
NODE	158	-0.1400	0.1500
NODE	159	-0.1500	-0.1500
NODE	160	-0.1500	-0.1371
NODE	161	-0.1500	-0.1241
NODE	162	-0.1500	-0.1112
NODE	163	-0.1500	-0.0982
NODE	164	-0.1500	-0.0853
NODE	165	-0.1500	-0.0724
NODE	166	-0.1500	-0.0594
NODE	167	-0.1500	-0.0465
NODE	168	-0.1500	-0.0335
NODE	169	-0.1500	-0.0206
NODE	170	-0.1500	-0.0075
NODE	171	-0.1500	0.0000
NODE	172	-0.1500	0.0075
NODE	173	-0.1500	0.0206

```
NODE        174        -0.1500      0.0335
NODE        175        -0.1500      0.0465
NODE        176        -0.1500      0.0594
NODE        177        -0.1500      0.0724
NODE        178        -0.1500      0.0853
NODE        179        -0.1500      0.0982
NODE        180        -0.1500      0.1112
NODE        181        -0.1500      0.1241
NODE        182        -0.1500      0.1371
NODE        183        -0.1500      0.1500
NODELINE      0              0
YC_ZC         0              0
FIXATIONS
END_FIX
NODOFSOLID
ELEM    1    1    26    27    2    1    0.
ELEM    2    2    27    28    3    1    0.
ELEM    3    3    28    29    4    1    0.
ELEM    4    4    29    30    5    1    0.
ELEM    5    5    30    31    6    1    0.
ELEM    6    6    31    32    7    1    0.
ELEM    7    7    32    33    8    1    0.
ELEM    8    8    33    34    9    1    0.
ELEM    9    9    34    35   10    1    0.
ELEM   10   10    35    36   11    1    0.
ELEM   11   11    36    37   12    1    0.
ELEM   12   12    37    38   13    1    0.
ELEM   13   13    38    39   14    1    0.
ELEM   14   14    39    40   15    1    0.
ELEM   15   15    40    41   16    1    0.
ELEM   16   16    41    42   17    1    0.
ELEM   17   17    42    43   18    1    0.
ELEM   18   18    43    44   19    1    0.
ELEM   19   19    44    45   20    1    0.
ELEM   20   20    45    46   21    1    0.
ELEM   21   21    46    47   22    1    0.
ELEM   22   22    47    48   23    1    0.
ELEM   23   23    48    49   24    1    0.
ELEM   24   24    49    50   25    1    0.
ELEM   25   26    51    52   27    1    0.
ELEM   26   27    52    53   28    1    0.
ELEM   27   28    53    54   29    1    0.
ELEM   28   29    54    55   30    1    0.
ELEM   29   30    55    56   31    1    0.
ELEM   30   31    56    57   32    1    0.
```

ELEM	31	32	57	58	33	1	0.
ELEM	32	33	58	59	34	1	0.
ELEM	33	34	59	60	35	1	0.
ELEM	34	35	60	61	36	1	0.
ELEM	35	36	61	62	37	1	0.
ELEM	36	37	62	63	38	1	0.
ELEM	37	38	63	64	39	1	0.
ELEM	38	39	64	65	40	1	0.
ELEM	39	40	65	66	41	1	0.
ELEM	40	41	66	67	42	1	0.
ELEM	41	42	67	68	43	1	0.
ELEM	42	43	68	69	44	1	0.
ELEM	43	44	69	70	45	1	0.
ELEM	44	45	70	71	46	1	0.
ELEM	45	46	71	72	47	1	0.
ELEM	46	47	72	73	48	1	0.
ELEM	47	48	73	74	49	1	0.
ELEM	48	49	74	75	50	1	0.
ELEM	49	61	76	62	0	1	0.
ELEM	50	62	76	77	63	1	0.
ELEM	51	63	77	78	64	1	0.
ELEM	52	64	78	65	0	1	0.
ELEM	53	76	79	80	77	1	0.
ELEM	54	77	80	81	78	1	0.
ELEM	55	79	82	83	80	1	0.
ELEM	56	80	83	84	81	1	0.
ELEM	57	82	85	86	83	1	0.
ELEM	58	83	86	87	84	1	0.
ELEM	59	85	88	89	86	1	0.
ELEM	60	86	89	90	87	1	0.
ELEM	61	88	91	92	89	1	0.
ELEM	62	89	92	93	90	1	0.
ELEM	63	91	94	95	92	1	0.
ELEM	64	92	95	96	93	1	0.
ELEM	65	94	97	98	95	1	0.
ELEM	66	95	98	99	96	1	0.
ELEM	67	97	100	101	98	1	0.
ELEM	68	98	101	102	99	1	0.
ELEM	69	100	103	104	101	1	0.
ELEM	70	101	104	105	102	1	0.
ELEM	71	103	106	107	104	1	0.
ELEM	72	104	107	108	105	1	0.
ELEM	73	106	119	120	0	1	0.
ELEM	74	106	120	121	107	1	0.
ELEM	75	107	121	122	108	1	0.

ELEM	76	108	122	123	0	1	0.
ELEM	77	109	134	135	110	1	0.
ELEM	78	110	135	136	111	1	0.
ELEM	79	111	136	137	112	1	0.
ELEM	80	112	137	138	113	1	0.
ELEM	81	113	138	139	114	1	0.
ELEM	82	114	139	140	115	1	0.
ELEM	83	115	140	141	116	1	0.
ELEM	84	116	141	142	117	1	0.
ELEM	85	117	142	143	118	1	0.
ELEM	86	118	143	144	119	1	0.
ELEM	87	119	144	145	120	1	0.
ELEM	88	120	145	146	121	1	0.
ELEM	89	121	146	147	122	1	0.
ELEM	90	122	147	148	123	1	0.
ELEM	91	123	148	149	124	1	0.
ELEM	92	124	149	150	125	1	0.
ELEM	93	125	150	151	126	1	0.
ELEM	94	126	151	152	127	1	0
ELEM	95	127	152	153	128	1	0
ELEM	96	128	153	154	129	1	0.
ELEM	97	129	154	155	130	1	0.
ELEM	98	130	155	156	131	1	0.
ELEM	99	131	156	157	132	1	0.
ELEM	100	132	157	158	133	1	0.
ELEM	101	134	159	160	135	1	0.
ELEM	102	135	160	161	136	1	0.
ELEM	103	136	161	162	137	1	0.
ELEM	104	137	162	163	138	1	0.
ELEM	105	138	163	164	139	1	0.
ELEM	106	139	164	165	140	1	0.
ELEM	107	140	165	166	141	1	0.
ELEM	108	141	166	167	142	1	0.
ELEM	109	142	167	168	143	1	0.
ELEM	110	143	168	169	144	1	0.
ELEM	111	144	169	170	145	1	0.
ELEM	112	145	170	171	146	1	0.
ELEM	113	146	171	172	147	1	0.
ELEM	114	147	172	173	148	1	0.
ELEM	115	148	173	174	149	1	0.
ELEM	116	149	174	175	150	1	0.
ELEM	117	150	175	176	151	1	0.
ELEM	118	151	176	177	152	1	0.
ELEM	119	152	177	178	153	1	0.
ELEM	120	153	178	179	154	1	0.

```
 ELEM     121    154    179        180    155    1    0.
 ELEM     122    155    180        181    156    1    0.
 ELEM     123    156    181        182    157    1    0.
 ELEM     124    157    182        183    158    1    0.
FRONTIER
 F     1  FIRE4.fct         NO        NO          NO
 F    25  FIRE4.fct         NO        NO          NO
 F    25         NO  FIRE4.fct        NO          NO
 F    26         NO  FIRE4.fct        NO          NO
 F    27         NO  FIRE4.fct        NO          NO
 F    28         NO  FIRE4.fct        NO          NO
 F    29         NO  FIRE4.fct        NO          NO
 F    30         NO  FIRE4.fct        NO          NO
 F    31         NO  FIRE4.fct        NO          NO
 F    32         NO  FIRE4.fct        NO          NO
 F    33         NO  FIRE4.fct        NO          NO
 F    34         NO  FIRE4.fct        NO          NO
 F    49  FIRE4.fct         NO        NO          NO
 F    53  FIRE4.fct         NO        NO          NO
 F    55  FIRE4.fct         NO        NO          NO
 F    57  FIRE4.fct         NO        NO          NO
 F    59  FIRE4.fct         NO        NO          NO
 F    61  FIRE4.fct         NO        NO          NO
 F    63  FIRE4.fct         NO        NO          NO
 F    65  FIRE4.fct         NO        NO          NO
 F    67  FIRE4.fct         NO        NO          NO
 F    69  FIRE4.fct         NO        NO          NO
 F    71  FIRE4.fct         NO        NO          NO
 F    73  FIRE4.fct         NO        NO          NO
 F    77         NO         NO        NO   FIRE4.fct
 F    78         NO         NO        NO   FIRE4.fct
 F    79         NO         NO        NO   FIRE4.fct
 F    80         NO         NO        NO   FIRE4.fct
 F    81         NO         NO        NO   FIRE4.fct
 F    82         NO         NO        NO   FIRE4.fct
 F    83         NO         NO        NO   FIRE4.fct
 F    84         NO         NO        NO   FIRE4.fct
 F    85         NO         NO        NO   FIRE4.fct
 F    86         NO         NO        NO   FIRE4.fct
 F    77  FIRE4.fct         NO        NO          NO
 F   101  FIRE4.fct         NO        NO          NO
 F   101         NO  FIRE4.fct        NO          NO
 F   102         NO  FIRE4.fct        NO          NO
 F   103         NO  FIRE4.fct        NO          NO
 F   104         NO  FIRE4.fct        NO          NO
```

F	105	NO	FIRE4.fct	NO	NO
F	106	NO	FIRE4.fct	NO	NO
F	107	NO	FIRE4.fct	NO	NO
F	108	NO	FIRE4.fct	NO	NO
F	109	NO	FIRE4.fct	NO	NO
F	110	NO	FIRE4.fct	NO	NO
F	111	NO	FIRE4.fct	NO	NO
F	112	NO	FIRE4.fct	NO	NO
F	113	NO	FIRE4.fct	NO	NO
F	114	NO	FIRE4.fct	NO	NO
F	115	NO	FIRE4.fct	NO	NO
F	116	NO	FIRE4.fct	NO	NO
F	117	NO	FIRE4.fct	NO	NO
F	118	NO	FIRE4.fct	NO	NO
F	119	NO	FIRE4.fct	NO	NO
F	120	NO	FIRE4.fct	NO	NO
F	121	NO	FIRE4.fct	NO	NO
F	122	NO	FIRE4.fct	NO	NO
F	123	NO	FIRE4.fct	NO	NO
F	124	NO	FIRE4.fct	NO	NO
F	24	NO	NO	FIRE4.fct	NO
F	48	NO	NO	FIRE4.fct	NO
F	48	NO	FIRE4.fct	NO	NO
F	47	NO	FIRE4.fct	NO	NO
F	46	NO	FIRE4.fct	NO	NO
F	45	NO	FIRE4.fct	NO	NO
F	44	NO	FIRE4.fct	NO	NO
F	43	NO	FIRE4.fct	NO	NO
F	42	NO	FIRE4.fct	NO	NO
F	41	NO	FIRE4.fct	NO	NO
F	40	NO	FIRE4.fct	NO	NO
F	39	NO	FIRE4.fct	NO	NO
F	52	NO	FIRE4.fct	NO	NO
F	54	NO	NO	FIRE4.fct	NO
F	56	NO	NO	FIRE4.fct	NO
F	58	NO	NO	FIRE4.fct	NO
F	60	NO	NO	FIRE4.fct	NO
F	62	NO	NO	FIRE4.fct	NO
F	64	NO	NO	FIRE4.fct	NO
F	66	NO	NO	FIRE4.fct	NO
F	68	NO	NO	FIRE4.fct	NO
F	70	NO	NO	FIRE4.fct	NO
F	72	NO	NO	FIRE4.fct	NO
F	76	NO	NO	FIRE4.fct	NO
F	91	NO	NO	NO	FIRE4.fct

```
  F    92      NO          NO        NO       FIRE4.fct
  F    93      NO          NO        NO       FIRE4.fct
  F    94      NO          NO        NO       FIRE4.fct
  F    95      NO          NO        NO       FIRE4.fct
  F    96      NO          NO        NO       FIRE4.fct
  F    97      NO          NO        NO       FIRE4.fct
  F    98      NO          NO        NO       FIRE4.fct
  F    99      NO          NO        NO       FIRE4.fct
  F   100      NO          NO        NO       FIRE4.fct
  F   100      NO          NO     FIRE4.fct      NO
  F   124      NO          NO     FIRE4.fct      NO
END_FRONT
  SYMMETRY
    END_SYM
  PRECISION    1.E-3
  MATERIALS
    STEELEC3
               25.     9.     0.50
     TIME
               12.    7200.
    END_TIME
  IMPRESSION
    TIMEPRINT
               60.    7200.
END_TIMEPR
```

References

1. Scawthorn, C.R., O'Rourke, T.D. and Blackburn, F.T. The 1906 San Francisco earthquake and fire—Enduring lessons for fire protection and water supply. Earthq Spectra. 2006; 22: 135-58.
2. Scawthorn, C.R. Fire following earthquake. The shake out scenario. California: USGS-science for Changing World; 2008.
3. Chadwick, H.D. San Francisco Mission District burning in the aftermath of the San Francisco Earthquake of 1906. US Gov War Department. Office of the Chief Signal Officer; 1906.
4. James, C.D. The 1923 Tokyo Earthquake and Fire. University of California Berkeley. Retrieved. 2011; 16.
5. Unknown. The 1923 Great Kanto earthquake. Yokohama Central Library, Japan; 1923.
6. Dowrick, D.J. Damage and intensities in the magnitude 7.8 1931 Hawke's Bay, New Zealand, earthquake. Bulletin of the New Zealand National Society for Earthquake Engineering. 1998; 31: 139-63.
7. McSaveney, E. Historic earthquakes – The 1931 Hawke's Bay earthquake. Hawke's Bay Museum & Art Gallery; 1931.
8. Scawthorn, C. Fire following earthquake: estimates of the conflagration risk to insured property in greater Los Angeles and San Francisco: The Council; 1987.
9. Kawasumi, H. General report on the Niigata Earthquake of 1964: Electrical Engineering College Press; 1968.
10. Jennings, P.C. Engineering features of the San Fernando earthquake of February 9, 1971; 1971.
11. Kates, R.W., Haas, J.E., Amaral, D.J., Olson, R.A., Ramos, R. and Olson, R. Human impact of the Managua earthquake. Science. 1973; 182: 981-90.
12. Bakun, W., Clark, M., Cockerham, R., Ellsworth, W., Lindh, A., Prescott, W. et al. The 1984 Morgan Hill, California, earthquake. Science. 1984; 225: 288-91.
13. Dynes, R.R., Quarantelli, E.L. and Wenger, D. Individual and organizational response to the 1985 earthquake in Mexico City, Mexico. Disaster Research Center; 1990.
14. Metinides, E. The 1985 Mexico City earthquake. Mother Jones and the Foundation for National Progress; 1985.
15. Taly, N. The Whittier Narrows, California Earthquake of October 1, 1987—Performance of Buildings at California State University, Los Angeles. Earthq Spectra. 1988; 4: 277-317.

16. Bruneau, M. Preliminary report of structural damage from the Loma Prieta (San Francisco) earthquake of 1989 and pertinence to Canadian structural engineering practice. Canadian Journal of Civil Engineering. 1990; 17: 198-208.
17. http://www.mcclatchyreprints.com/. Loma Prieta earthquake. PARS International Corp; 1989.
18. Butcher, G., Beetham, R., Millar, P. and Tanaka, H. The Hokkaido-Nansei-Oki earthquake: final report of the NZNSEE reconnaissance team. Bulletin of the New Zealand National Society for Earthquake Engineering. 1994; 27: 2-54.
19. Shimbun, A. Hokkaido Nansei-oki earthquake. The Asahi Shimbun Company; 1993.
20. The January 17, 1994 Northridge, California earthquake: an EQE summary report. EQE International; 1994.
21. Lubas, K. Northridge earthquake. Los Angeles Times; 1994.
22. Schiff, A.J. Hyogoken-Nanbu (Kobe) Earthquake of January 17, 1995: Lifeline Performance. ASCE; 1999.
23. Scawthorn, C., Eidinger, J.M. and Schiff, A. Fire Following Earthquake: American Society of Civil Engineers 2005.
24. Hays, J. Kobe earthquake. Fact and Details; 1995.
25. Girgin, S. The natech events during the 17 August 1999 Kocaeli earthquake: aftermath and lessons learned. Natural Hazards and Earth System. Science. 2011; 11: 1129-40.
26. Ozmen, B. Isoseismic map, human casualty and building damage statistics of the Izmit earthquake of August 17, 1999. Third Japan-Turkey Workshop on Earthquake Engineering. 2000; p. 21-5.
27. Yamanaka, Y. and Kikuchi, M. Source process of the recurrent Tokachi-oki earthquake on September 26, 2003, inferred from teleseismic body waves. Earth, Planets and Space. 2003; 55: e21-e4.
28. Zama, S., Nishi, H., Yamada, M. and Hatayama, K. Damage of oil storage tanks caused by liquid sloshing in the 2003 Tokachi Oki earthquake and revision of design spectra in the long-period range. Proceedings of the 14th World Conference on Earthquake Engineering. 2008.
29. Hokkaido Refinery post-earthquake fire. Fire and Disaster Management Agency, Japan; 2003.
30. Mimura, N., Yasuhara, K., Kawagoe, S., Yokoki, H. and Kazama, S. Damage from the Great East Japan Earthquake and Tsunami – a quick report. Mitigation and Adaptation Strategies for Global Change. 2011; 16: 803-18.
31. Krausmann, E. and Cruz, A.M. Impact of the 11 March 2011, Great East Japan earthquake and tsunami on the chemical industry. Natural Hazards. 2013; 67: 811-28.
32. Nanto, D.K. Japan's 2011 Earthquake and Tsunami: Economic Effects and Implications for the United States: DIANE Publishing; 2011.
33. Tōhoku earthquake. www.chinasmack.com; 2011.
34. Sim, D. Northern Chile earthquake. IBTimes Co., Ltd; 2014.
35. Hamada, M. On fire spreading velocity in disasters. Sagami Shobo, Tokyo 1951.
36. Horiuchi, S., Kobayashi, M. and Nakai, S. Study on the Emergency Escape in City Area. Transactions of the Architectural Institute of Japan. 1974; 223: 45-71.
37. Mizuno, H. and Horiuchi, S. Study on the prediction about the numbers of the outbreak of fires caused by the earthquakes. Transactions of the Architectural Institute of Japan. 1976; 250: 81-90.

38. Scawthorn, C., Yamada, Y. and Iemura, H. A model for urban post-earthquake fire hazard. Disasters. 1981; 5: 125-32.
39. HAZUS. Technical manual. Federal Emergency Management Agency, Washington (DC). 1997.
40. Scawthorn, C. Enhancements in HAZUS-MH, Fire Following Earthquake Task 3: Updated Ignition Equation. SPA Project. 2009; 07-1.
41. Cousins, W. and Smith W. Estimated losses due to post-earthquake fire in three New Zealand cities. Proceedings, New Zealand Society of Earthquake Engineering Conference 2004.
42. Ren, A.Z. and Xie, Y. The simulation of post-earthquake fire-prone area based on GIS. Journal of Fire Sciences. 2004; 22: 421-39.
43. Davidson, R.A. Modeling postearthquake fire ignitions using generalized linear (mixed) models. Journal of Infrastructure Systems. 2009; 15: 351-60.
44. Zolfaghari, M.R., Peyghaleh, E. and Nasirzadeh, G. Fire following earthquake, intra-structure ignition modeling. Journal of Fire Sciences. 2009; 27: 45-79.
45. Yildiz, S.S. and Karaman, H. Post-earthquake ignition vulnerability assessment of **Küçükçekmece** District. Natural hazards and earth system sciences discussions. 2013; 1: 2005-40.
46. Himoto, K. and Nakamura, T. An Analysis of the Post-earthquake Fire Safety of Historic Buildings in Kyoto, Japan. Fire Technology. 2014; 50: 1107-25.
47. Lee, S., Davidson, R., Ohnishi, N. and Scawthorn, C. Fire Following Earthquake-Reviewing the State-of-the-Art of Modeling. Earthq Spectra. 2008; 24: 933-67.
48. Cousins, J., Heron, D., Mazzoni, S., Thomas, G. and Lloydd, D. Estimating risks from fire following earthquake: Institute of Geological & Nuclear Sciences Limited; 2002.
49. Drysdale, D. An Introduction to Fire Dynamics (3rd Edition). Chichester, England: John Wiley & Sons; 2011.
50. Audouin, L., Kolb, G., Torero, J. and Most, J. Average centreline temperatures of a buoyant pool fire obtained by image processing of video recordings. Fire Safety Journal. 1995; 24: 167-87.
51. Fan, S-g, Shu, G-P, She, G-J, Liew, J.R. Computational method and numerical simulation of temperature field for large space steel structures in fire. Advanced Steel Construction. 2014; 10: 151-78.
52. Zhang, G-w, Zhu, G-q, Yin, F. A Whole Process Prediction Method for Temperature Field of Fire Smoke in Large Spaces. Procedia Engineering. 2014; 71: 310-5.
53. Green, N.B. Earthquake Resistant: Building Design and Construction: Van Nostrand Reinhold Company; 1981.
54. Foliente, G., Leicester, R. and Pham, L. Development of the CIB proactive program on performance based building codes and standards. BCE Doc. 1998; 98: 232.
55. Sexton, M. and Barrett, P. Performance-based building and innovation: balancing client and industry needs. Building Research & Information. 2005; 33: 142-48.
56. Bukowski, R.W. Progress toward a performance-based codes system for the United States: National Institute of Standards and Technology; 1997.
57. Hadjisophocleous, G.V., Benichou, N. and Tamim, A.S. Literature Review of Performance-Based Fire Codes and Design Environment. Journal of Fire Protection Engineering. 1998; 9: 12-40.

58. Wood, K.E. The Effect of Performance-based Codes and Performance-based Design on the Office of the Illinois State Fire Marshal. Illinois The National Fire Academy; 2000.
59. Richardson, J.K. Changing the regulatory system to accept fire safety engineering methods. Journal of Fire Protection Engineering. 1993; 5: 135-40.
60. Meacham, B., Bowen, R., Traw, J. and Moore A. Performance-based building regulation: current situation and future needs. Building Research & Information. 2005; 33: 91-106.
61. Meacham, B. Using risk as a basis for establishing tolerable performance: an approach for building regulation. Proceedings of the special workshop on risk acceptance and risk communication, Stanford University, Stanford; 2007.
62. Averill, J.D. Performance-based codes: economics, documentation, and design: Worcester Polytechnic Institute; 1998.
63. International Code Council Performance Code. California: ICC; 2006.
64. Okada, T., Hiraishi, H., Ohashi, Y., Fujitani, H., Aoki, T., Akiyama, H. et al. A new framework for performance-based design of building structures. Proceedings of the 12th World Conference on Earthquake Engineering; 2000.
65. Xue, Q., Chen, C-C. Performance-based seismic design of structures: a direct displacement-based approach. Engineering Structures. 2003; 25: 1803-13.
66. Ghobarah, A. Performance-based design in earthquake engineering: state of development. Engineering Structures. 2001; 23: 878-84.
67. Fujitani, H., Tani, A., Aoki, Y. and Takahashi, I. Performance levels of building structures against the earthquake (concept of performnce-based design standing on questionnaires). Proceedings of 12th World Conference on Earthquake Engineering, New Zealand, CD-Rom, Paper; 2000.
68. Priestley, M. Performance based seismic design. Bulletin of the New Zealand Society for Earthquake Engineering. 2000; 33: 325-46.
69. Naeim, F. Performance Based Seismic Design of Tall Buildings. Earthquake Engineering in Europe: Springer; 2010; p. 147-69.
70. FEMA450. Recommended provisions for seismic regulations for new buildings and other structures. Part 1. Washington, DC: National Institute of Building Sciences; 2003.
71. Reinhorn, A., Mander, J., Bracci, J. and Kunnath S. A post-earthquake damage evaluation strategy for RC buildings. Proceedings of the Fourth US National Conference on Earthquake Engineering. 1990; p. 1047-56.
72. Bagheri, M. and Miri, M. Performance-based design in earthquake engineering. 5th National Congress on Civil Engineering. Ferdowsi University of Mashhad, Mashhad, Iran; 2010.
73. Priestley, M., Calvi, G. and Kowalsky, M. Direct displacement-based seismic design of structures. NZSEE Conference; 1993.
74. FEMA356. Prestandard and commentary for the seismic rehabilitation of buildings. Rehabilitation Requirements. Washington, DC: American Society of Civil Engineers; 2000.
75. JPDPA. Standard for Seismic Capacity Assessment of Existing Reinforced Concrete Buildings. Tokyo, Japan: Japanese Building Disaster Prevention Association; 1990.
76. Thermou, G. and Elnashai, A. Performance Parameters and Criteria for Assessment and Rehabilitation. Seismic Performance Evaluation and Retrofit of Structures (SPEAR), European Earthquake Engineering Research Network Report, Imperial College, UK; 2002.

77. NOAA/NGDC EVL, U.S. Geological Survey. Severe concrete splicing after the 1985 Mexico City earthquake 1985.
78. Behnam, B., Ronagh, H.R. and Lim, P.J. Numerical evaluation of the post-earthquake fire resistance of CFRP-strengthened reinforced concrete joints based on experimental observations. European Journal of Environmental and Civil Engineering. 2015; 1-19.
79. Buckingham, E. On Physically Similar Systems; Illustrations of the Use of Dimensional Equations. Physical Review. 1914; 4: 345-76.
80. Park, R. and Paulay, T. Reinforced concrete structures. New York: John Wiley & Sons; 1975.
81. Kamath, P., Sharma, U.K., Kumar, V., Bhargava, P., Usmani, A., Singh, B. et al. Full-scale fire test on an earthquake-damaged reinforced concrete frame. Fire Safety Journal. 2015; 73: 1-19.
82. Venture, S.J. Recommended Postearthquake Evaluation and Repair Criteria for Welded Steel Moment-Frame Buildings, Report No. FEMA-352. Federal Emergency Management Agency, Washington, DC; 2000.
83. Tomecek, D. and Milke, J. A study of the effect of partial loss of protection on the fire resistance of steel columns. Fire Technology. 1993; 29: 3-21.
84. Ryder, N.L., Wolin, S.D. and Milke, J.A. An Investigation of the Reduction in Fire Resistance of Steel Columns Caused by Loss of Spray-Applied Fire Protection. Journal of Fire Protection Engineering. 2002; 12: 31-44.
85. ASTM. Standard test methods for determining effects of large hydrocarbon pool fires on structural members and assemblies. ASTM E1529-06. America: American Society for Testing and Materials; 2006.
86. Kwon, K., Pessiki, S. and Lee, B-J. An Analytical Study of the Fire Load Behavior of Steel Building Columns with Damaged Spray-Applied Fire Resistive Material. ATLSS Report No 06-25. Bethlehem, Pennsylvania: Lehigh University; 2006; p. 132.
87. Wang, W-Y. and Li, G-Q. Fire-resistance study of restrained steel columns with partial damage to fire protection. Fire Safety Journal. 2009; 44: 1088-94.
88. Braxtan, N.J.L. Post-earthquake fire performance of steel moment frame building columns 3389946. United States, Pennsylvania: Lehigh University; 2010.
89. Braxtan, N.L. and Pessiki, S.P. Post Earthquake Fire Performance of Sprayed Fire-Resistive Material on Steel Moment Frames. J Struct Eng-ASCE. 2011; 137: 946-53.
90. Arablouei, A. and Kodur, V. A fracture mechanics-based approach for quantifying delamination of spray-applied fire-resistive insulation from steel moment-resisting frame subjected to seismic loading. Engineering Fracture Mechanics. 2014; 121–22: 67-86.
91. Keller, W.J. and Pessiki, S. Effect of Earthquake-Induced Damage on the Sidesway Response of Steel Moment-Frame Buildings during Fire Exposure. Earthq Spectra. 2015; 31: 273-92.
92. Behnam, B. and Abolghasemi, S. Investigating the Effect of Earthquake-induced Drift on the Response of Fireproofed Steel Structures. In: Keshvari, R., editor. The First National Conference on Fire and Urban Safety. Shahid Beheshti University, Tehran, Iran: Tehran Fire Rescue Organiztion; 2016; p. 172-80.
93. FEMA310. Handbook for the Seismic Evaluation of Buildigns – A Pre-standard. Washington DC: American Society of Civil Engineering; 1998.

94. IBC. International building code. International Code Council, Inc (formerly BOCA, ICBO and SBCCI) 2006; p. 60478-5795.
95. Soni, D.P. and Mistry, B.B. Qualitative review of seismic response of vertically irregular building frames. ISET Journal of Earthquake Technology, Technical Note. 2006; 43: 121-32.
96. Lavan, O. and De Stefano, M. Seismic Behaviour and Design of Irregular and Complex Civil Structures. Istanbul, Turkey: Springer; 2013.
97. Behnam, B. Structural Response of Vertically Irregular Tall Moment-resisting Steel Frames under Pre- and Post-earthquake Fire. The Structural Design of Tall and Special Buildings. 2015: n/a-n/a.
98. Le-Trung, K., Lee, K., Lee, J. and Lee, D.H. Evaluation of seismic behaviour of steel special moment frame buildings with vertical irregularities. The Structural Design of Tall and Special Buildings. 2012; 21: 215-32.
99. Elnashai, A.S. and Di Sarno, L. Fundamentals of Earthquake Engineering: From Source to Fragility. West Sussex, United Kingdom: John Wiley & Sons; 2015.
100. Behnam, B. Retrofitting management for residential buildings research. Tehran, Iran: Tehran Polytechnic; 2006.
101. EN1991-1-2. Eurocode 1: Actions on Structures - Part 1-2: General actions – Actions on structures exposed to fire. Brussels, Belgium CEN; 2002.
102. Wang, Y., Burgess, I., Wald, F. and Gillie, M. Performance-Based Fire Engineering of Structures. 1 ed. Hoboken: Taylor and Francis; 2012.
103. Hashemi Rezvani, F., Behnam, B., Ronagh, H.R. and Jeffers, A.E. Effect of Travelling Fire on Structural Response of a Generic Steel Fireprotected Moment Resisting Frame. In: Torero, J.L., editor. Second International Conference on Performance-based and Lifecycle Structural Engineering (PLSE 2015). Brisbane, Australia: Arinex Pty. Limited; 2015.
104. Harada, K. Performance Based Codes and Performance Based Fire Safety Design. Fire Science and Technology. 1999; 19: 1-10.
105. Barnfield, J., Cooke, G., Deakin, G., Hannah, M., Jones, T., Law, M. et al. British Standard Code of Practice for the Application of Fire Safety Engineering Principles to Fire Safety in Buildings: Warrington Fire Research Consultants; 1993.
106. Bukowski, R.W., Clarke, F., Hall, J.R. and Stiefel, S. Fire risk assessment method: description of methodology: Center for Fire Research; 1990.
107. Hertzberg, T., Sundström, B. and van Hees, P. Design fires for enclosures. SP Rapport. 2003; 02.
108. Buchanan, A.H. Association NZFP. Fire engineering design guide: University of Canterbury, Centre for Advanced Engineering; 1994.
109. Kodur, V. and Dwaikat, M. Performance-based Fire Safety Design of Reinforced Concrete Beams. Journal of Fire Protection Engineering. 2007; 17: 293-320.
110. Institution BS. Fire Tests on Building Materials and Structures; 1987.
111. Hadjisophocleous, G.V. and Benichou, N. Performance criteria used in fire safety design. Automation in Construction. 1999; 8: 489-501.
112. Malhotra, H.L. Fire Safety in Buildings. England: Department of the Environment, Building Research Establishment; 1986.
113. Ramachandran, G. Fire safety management and risk assessment. Facilities. 1999; 17: 363-77.
114. Guanquan, C. and Jinhua, S. Quantitative Assessment of Building Fire Risk to Life Safety. Risk analysis. 2008; 28: 615-25.

115. Chu, G. and Sun, J. Decision analysis on fire safety design based on evaluating building fire risk to life. Safety Science. 2008; 46: 1125-36.
116. Kobes, M., Helsloot, I., de Vries, B. and Post, J.G. Building safety and human behaviour in fire: A literature review. Fire Safety Journal. 2010; 45: 1-11.
117. Omidvari, M., Mansouri, N. and Nouri, J. A pattern of fire risk assessment and emergency management in educational center laboratories. Safety Science. 2015; 73: 34-42.
118. Sun, X-q and Luo, M-c. Fire Risk Assessment for Super High-rise Buildings. Procedia Engineering. 2014; 71: 492-501.
119. Hanea, D. and Ale, B. Risk of human fatality in building fires: A decision tool using Bayesian networks. Fire Safety Journal. 2009; 44: 704-10.
120. Charters, D., McGrail, D., Fajemirokun, N., Wang, Y., Townsend, N. and Holborn, P. Preliminary analysis of the number of occupants, fire growth, detection times and pre-movement times for probabilistic risk assessment. Proceedings of the 7th International Symposium on Fire Safety Science. Worcester, MA, USA; 2002; p. 357-68.
121. He, Y., Wang, J., Wu, Z., Hu, L., Xiong, Y. and Fan, W. Smoke venting and fire safety in an industrial warehouse. Fire Safety Journal. 2002; 37: 191-215.
122. Tanaka, T. Risk-based selection of design fires to ensure an acceptable level of evacuation safety. Fire Saf Sci 2008; p. 49-61.
123. Lee, G.C., Chen, S., Li, G. and Tong, M. On Performance Based Analysis of Buildings for Multi-Hazard Mitigation. Steel Structures. 2004; 4: 231-38.
124. LeGrone, P.D. An Analysis of Fire Sprinkler System Failures During The Northridge Earthquake and Comparison with The Seismic Design Standard for These Systems. 13 WCEE: 13th World Conference on Earthquake Engineering Conference Proceedings; 2004.
125. Chen, S., Lee, G.C. and Shinozuka, M. Hazard mitigation for earthquake and subsequent fire. Annual meeting: Networking of young earthquake engineering researchers and professionals. Honolulu, Hawaii: Multidisciplinary Centre for Earthquake Engineering Research, Buffalo, N.Y.; 2004.
126. Robertson, J. and Mehaffey, J. Accounting for fire following earthquake in the development of performance based building codes. 12th World Conference on Earthquake Engineering. Auckland, New Zealand: New Zealand Society for Earthquake Engineering; 2000.
127. Handmer, J.W., Reed, C. and Percovich, O. Disaster loss assessment guidelines. Canberra, Australia: Department of Emergency Services; 2002.
128. Bird, J.F., Bommer, J.J., Crowley, H. and Pinho, R. Modelling liquefaction-induced building damage in earthquake loss estimation. Soil Dynamics and Earthquake Engineering. 2006; 26: 15-30.
129. Pliefke, T., Sperbeck, S., Urban, M., Peil, U. and Budelmann, H. A standardized methodology for managing disaster risk – An attempt to remove ambiguity. Proceedings of the 5th IPW Ghent, Belgium; 2007.
130. Whitman, R.V., Reed, J.W. and Hong, S.T. Earthquake damage probability matrices. 5th World Conference on Earthquake Engineering. Rome; 1973; p. 2531-40.
131. Choudhury, G.S. and Jones, N. Identification and prioritization of data for collection in post-earthquake surveys. Natural Hazards. 1995; 12: 119-38.
132. Calvi, G., Pinho, R., Magenes, G., Bommer, J., Restrepo-Vélez, L. and Crowley, H. Development of seismic vulnerability assessment methodologies over the past 30 years. ISET J Earthq Technol. 2006; 43: 75-104.

133. Dumova-Jovanoska, E. Fragility curves for reinforced concrete structures in Skopje (Macedonia) region. Soil Dynamics and Earthquake Engineering. 2000; 19: 455-66.
134. Kappos, A.J., Stylianidis, K.C. and Pitilakis, K. Development of Seismic Risk Scenarios Based on a Hybrid Method of Vulnerability Assessment. Natural Hazards. 1998; 17: 177-92.
135. D'Ayala, D. and Speranza, E. An integrated procedure for the assessment of seismic vulnerability of historic buildings. 12th European Conference on Earthquake Engineering. London, U.K.; 2002; p. 561.
136. Cosenza, E., Manfredi, G., Polese, M. and Verderame, G.M. A multilevel approach to the capacity assessment of existing RC buildings. Journal of Earthquake Engineering. 2005; 9: 1-22.
137. FEMA174. Establishing Programs and Priorities for the Seismic Rehabilitation of Buildings Washington, D.C.; 1989.
138. FEMA228. Benefit-Cost Model for the Seismic Rehabilitation of Buildings. Washington, D.C.: Federal Emergency Management Agency; 1992.
139. Dunne, R. and Sonnenfeld, P. Estimation of homeless caseload for disaster assistance due to an earthquake: Pasadena. Calif, Southern California Earthquake Preparedness Project; 1991.
140. Rice, D.P. and Cooper, B.S. The economic value of human life. American Journal of Public Health and the Nations Health. 1967; 57: 1954-66.
141. Sturm, S. and Guinier, L. The Law School Matrix: Reforming Legal Education in a Culture of Competition and Conformity. V and L Rev. 2007; 60: 515.
142. Landefeld, J.S. and Seskin, E.P. The Economic Value of Life: Linking Theory to Practice. American Journal of Public Health. 1982; 72: 555-66.
143. Ferritto, J.M. Economic analysis procedure for earthquake hazard mitigation. California: DTIC Document; 1997.
144. Hallegatte, S. The Indirect Cost of Natural Disasters and an Economic Definition of Macroeconomic Resilience. SDRFI Impact Appraisal Project; 2014.
145. Tierney, K.J. Business impacts of the Northridge earthquake. Journal of Contingencies and Crisis Management. 1997; 5: 87-97.
146. Xin, J. and Huang, C. Fire risk analysis of residential buildings based on scenario clusters and its application in fire risk management. Fire Safety Journal. 2013; 62, Part A: 72-8.
147. Denner, L. and Diaz, T. Knowledge management in the public sector: an online presence as a tool for capture and sharing. ECLAC Subregional Headquarters for the Caribbean, Port of Spain. The Caribbean. No. LC/CAR/L. 351; 2011.
148. Coburn, A. and Spencer, R. Earthquake Protection. Chichester, U.K.: John Willey & Sons; 2002.
149. Bommer, J., Spence, R., Erdik, M., Tabuchi, S., Aydinoglu, N., Booth, E. et al. Development of an earthquake loss model for Turkish catastrophe insurance. Journal of Seismology. 2002; 6: 431-46.
150. Fleurbaey, M. Beyond GDP: The quest for a measure of social welfare. Journal of Economic Literature. 2009: 1029-75.
151. Charvériat, C. Natural disasters in Latin America and the Caribbean: An overview of risk. 2000.
152. Loayza, N., Olaberria, E., Rigolini, J. and Christiaensen, L. Natural disasters and growth-going beyond the averages. World Bank Policy Research Working Paper Series, Vol. 2009.

153. Lindell, M.K. and Prater, C.S. Assessing community impacts of natural disasters. Natural Hazards Review. 2003; 4: 176-85.
154. Hallegatte, S. and Przyluski, V. The economics of natural disasters: concepts and methods. World Bank Policy Research Working Paper Series, Vol. 2010.
155. Freedy, J.R., Shaw, D.L., Jarrell, M.P. and Masters, C.R. Towards an understanding of the psychological impact of natural disasters: An application of the conservation resources stress model. Journal of Traumatic Stress. 1992; 5: 441-54.
156. Glendon, A.I. and Clarke, S. Human Safety and Risk Management: A Psychological Perspective: CRC Press; 2015.
157. Vrijling, J.K., van Hengel, W. and Houben, R.J. A framework for risk evaluation. Journal of Hazardous Materials. 1995; 43: 245-61.
158. Pukeliene, V. Quality of Life: Factors Determining its Measurement Complexity. Inžinerinė ekonomika. 2011; 22: 147.
159. Denoël, J.F. Fire Safety and Concrete Structures. Brussels, Belgium: Federation of Belgian Cement Industry; 2007.
160. Karlsson, B. and Quintiere, J.G. Energy Release Rates. Enclosure Fire Dynamics: CRC Press; 1999.
161. Hanus, F. Analysis of simple connections in steel structures subjected to natural fires Research. Belgium: University of Liège; 2010.
162. Bailey, C.G., Burgess, I.W. and Plank, R.J. Analyses of the effects of cooling and fire spread on steel-framed buildings. Fire Safety Journal. 1996; 26: 273-93.
163. El-Rimawi, J.A., Burgess, I.W. and Plank, R.J. The treatment of strain reversal in structural members during the cooling phase of a fire. Journal of Constructional Steel Research. 1996; 37: 115-35.
164. Iu, C.K., Chan, S.L. and Zha, X.X. Nonlinear pre-fire and post-fire analysis of steel frames. Engineering Structures. 2005; 27: 1689-702.
165. Wang, P., Li, G. and Guo, S. Effects of the cooling phase of a fire on steel structures. Fire Safety Journal. 2008; 43: 451-58.
166. Yang, H., Han, L-H, Wang, Y-C. Effects of heating and loading histories on post-fire cooling behaviour of concrete-filled steel tubular columns. Journal of Constructional Steel Research. 2008; 64: 556-70.
167. Salah Dimia, M., Guenfoud, M., Gernay, T. and Franssen, J-M. Collapse of concrete columns during and after the cooling phase of a fire. Journal of Fire Protection Engineering. 2011; 21: 245-63.
168. Du, E-F., Shu, G-P. and Mao, X-Y. Analytical behavior of eccentrically loaded concrete encased steel columns subjected to standard fire including cooling phase. Int J Steel Struct. 2013; 13: 129-40.
169. Gernay, T. and Franssen, J.M. A performance indicator for structures under natural fire. Engineering Structures. 2015; 100: 94-103.
170. Wermiel, S.E. California Concrete, 1876-1906: Jackson, Percy, and the Beginnings of Reinforced Concrete Construction in the United States. Proceedings of the Third International Congress on Construction History; 2009.
171. Babrauskas, V. and Williamson, R. The historical basis of fire resistance testing – Part I. Fire Technology. 1978; 14: 184-94.
172. Babrauskas, V. and Williamson R. The historical basis of fire resistance testing – Part II. Fire Technology. 1978; 14: 304-16.
173. Cooper, L.Y. and Steckler, K.D. Methodology for developing and implementing alternative temperature-time curves for testing the fire resistance of barriers

for nuclear power plant applications. Nuclear Regulatory Commission, Washington, DC (United States). Div. of Systems Safety and Analysis; 1996.
174. Ingberg, S. Tests of the severity of building fires. NFPA Quarterly. 1928; 22: 43-61.
175. Gross, D. Fire research at NBS: the first 75 years. 3rd Int'l Symp Int'l Assoc Fire Safety Science2006. p. 119-33.
176. Ingberg, S. Fire Resistance Classifications of Building Constructions. NBS Buildings Materials and Structures BMS. 1942; 92.
177. Harmathy, T.Z., Sultan, M.A. and MacLaurin, J.W. Comparison of severity of exposure in ASTM E 119 and ISO 834 fire resistance tests. ASTM Journal of Testing and Evaluation. 1987; 15: 8.
178. Law, M. A review of formulae for T-equivalent. Proceedings of Fifth International Symposium on Fire Safety Science, Melbourne Australia; Mar 1997; p. 3-7.
179. Kawagoe, K. and Sekine, T. Estimation of fire temperature-time curve in rooms: Building Research Institute, Ministry of Construction, Japanese Government; 1963.
180. Hogg, J.M. Ministry of Technology and Fire Offices' Committee Joint Fire Research Organization: The Effect of Some Climatological Variations on the Incidence and Spread of Fires in Buildings in England and Wales from 1951 to 1961. Journal of the Royal Statistical Society Series C (Applied Statistics). 1965; 14: 140-61.
181. Law, M. A relationship between fire grading and building design and contents. Fire Safety Science. 1971; 877: 1-47.
182. Magnusson, S.E. and Thelandersson, S. Temperature-time curves of complete process of fire development. Bulletin of Division of Structural Mechanics and Concrete Construction, Bulletin 161970.
183. Pettersson, O. The Connection Between a Real Fire Exposure and the Heating Conditions According to Standard Fire Resistance Tests-With Special Application to Steel Structures. Bulletin of Division of Structural Mechanics and Concrete Construction, Bulletin 39; 1975.
184. Harmathy, T.Z. and Mehaffey, J.R. Post-Flashover compartment fires. Fire and Materials. 1983; 7: 49-61.
185. Wickström, U. Temperature Calculation of Insulated Steel Columns Exposed to Natural Fire. Fire Safety Journal. 1981; 4: 219-25.
186. Lennon, T. Designers' Guide to EN 1991-1-2, EN 1993-1-2 and EN 1994-1-2: Handbook for the Fire Design of Steel, Composite and Concrete Structures to the Eurocodes: Thomas Telford; 2007.
187. Harmathy, T. and Sultan, M. Correlation between the severities of the ASTM E119 and ISO 834 fire exposures. Fire Safety Journal. 1988; 13: 163-68.
188. Eurocode 1, Part 1-2, BS EN 1991-1-2. General actions – Actions on structures exposed to fire. CEN, Brussels: European Standard EN 1991-1-2; 1991.
189. Zehfuss, J. and Hosser, D. A parametric natural fire model for the structural fire design of multi-storey buildings. Fire Safety Journal. 2007; 42: 115-26.
190. Franssen, J-M. and Real, P.V. Fire design of steel structures: Eurocode 1: actions on structures, part 1-2: General actions – Actions on structures exposed to fire – Eurocode 3: Design of steel structures, part 1-2: General rules – Structural fire design. Berlin: European Convention for Constructional Steel Work; 2010.
191. Zalok, E. Validation of Methodologies to Determine Fire Load for Use in Structural Fire Protection. The Fire Protection Research Foundation. Ottawa,

Canada: Department of Civil and Environmental Engineering, Carleton University; 2011.

192. ASTME119-01. Standard Methods of Fire Test of Building Construction and Materials. West Conshohocken, PA2001.
193. Manzello, S., Park, S.H., Mizukami, T. and Bentz, D. Measurment of thermal properties of gypsum board at elevated temperatures. In: Tan, K.H., Kodur, V., Tan, T.H., editors. Fifth International Conference Structures in Fire SiF'08. Nanyang Technological University, Singapore: Research Publishing Services; 2008; p. 656-65.
194. Cadorin, J. A tool to design steel elements submitted to compartment fires—OZone V2. Part 2: Methodology and application. Fire Safety Journal. 2003; 38: 429-51.
195. Cadorin, J. A tool to design steel elements submitted to compartment fires—OZone V2. Part 1: pre- and post-flashover compartment fire model. Fire Safety Journal. 2003; 38: 395-427.
196. McCaffrey, B. Flame height. DiNenno, PJ, SFPE Handbook of Fire Protection Engineering National Fire Protection Association, Quincy, MA. 1995.
197. Zukoski, E., Cetegen, B. and Kubota, T. Visible structure of buoyant diffusion flames. Symposium (International) on Combustion: Elsevier; 1985; p. 361-66.
198. Heskestad, G. Fire plumes, flame height, and air entrainment. The SFPE handbook of fire protection engineering. 1988.
199. Hasemi, Y., Yokobayashi, S., Wakamatsu, T. and Ptchelintsev, A. Firesafety of building components exposed to a localized fire: scope and experiments on ceiling/beam system exposed to a localized fire. Proceedings of ASIAFLAM. 1995: 351-61.
200. Stratton, B.J. Determining Flame Height And Flame Pulsation Frequency And Estimating Heat Release Rate From 3D Flame Reconstruction. Christchurch, New Zealand: University of Canterbury; 2005.
201. Cox, G. and Kumar, S. Modeling enclosure fires using CFD. SFPE Handbook for Fire Protection Engineering, 3rd ed, National Fire Protection Association, Quincy, MA. 2002.
202. Kebriai, A. and Bashirnejad, K. Fire and smoke simulation of high-rise buildings using FDS. In: Keshvari, R., editor. The First National Conference on Fire and Urban Safety. Shahid Beheshti University, Tehran, Iran: Tehran Fire Rescue Organiztion; 2016; p. 112-20.
203. Giraldo, M.P., Lacasta, A., Avellaneda, J. and Burgos, C. Computer-simulation study on fire behaviour in the ventilated cavity of ventilated façade systems. MATEC Web of Conferences: EDP Sciences; 2013. p. 03002.
204. Hasemi, Y. Thermal modeling of upward wall flame spread. Fire Safety Science: Proceedings of the First International Symposium, International Association for Fire Safety Science; 1986; pp. 87-96.
205. Delichatsios, M. and Delichatsios, M. Upward flame spread and critical conditions for PE/PVC cables in a tray configuration. Fire Safety Science. 1994; 4: 433-44.
206. Saito, K., Quintiere, J. and Williams, F. Upward turbulent flame spread. Fire Safety Science. Proceedings of the First International Symposium; 1986; pp. 75-86.
207. Keski-Rahkonen, O., Mangs, J. and Turtola, A. Ignition of and fire spread on cables and electronic components: Technical Research Centre of Finland; 1999.

208. Behnam, B. and Ronagh, H.R. A Study on the Effect of Sequential Post-Earthquake Fire on the Performance of Reinforced Concrete Structures. International Journal of Structural Integrity 2014; 5: 141-66.
209. Gales, J. Travelling Fires and the St. Lawrence Burns Project. Fire Technology. 2014; 50: 1535-43.
210. Clifton, G. Fire Models for Large Firecells. Report R4-83. New Zealand: HERA; 1996.
211. Cooke, G.M.F. Tests to Determine the Behaviour of Fully Developed Natural Fires in a Large Compartment. Fire Note 4, Fire Research Station. Watford, UK: Building Research Establishment; 1998.
212. Kirby, B.R., Wainman, D.E., Tomlinson, L.N., Kay, T.R. and Peacock, B.N. Natural Fires in Large Scale Compartments. International Journal on Engineering Performance-Based Fire Codes. 1999; 1: 43-58.
213. Abecassis-Empis, C., Reszka, P., Steinhaus, T., Cowlard, A., Biteau, H., Welch, S. et al. Characterisation of Dalmarnock fire Test One. Experimental Thermal and Fluid Science. 2008; 32: 1334-43.
214. Welch, S., Jowsey, A., Deeny, S., Morgan, R. and Torero, J.L. BRE large compartment fire tests—Characterising post-flashover fires for model validation. Fire Safety Journal. 2007; 42: 548-67.
215. Ellobody, E. and Bailey, C.G. Structural performance of a post-tensioned concrete floor during horizontally travelling fires. Engineering Structures. 2011; 33: 1908-17.
216. Gillie, M., Stratford, T. and Chen, J-F. Behaviour of a concrete structure in a real compartment fire. Proceedings of the ICE – Structures and Buildings; 2012; p. 421-33.
217. Stern-Gottfried, J. Travelling Fires for Structural Design. Edinburgh, Scotland The University of Edinburgh; 2011.
218. Stern-Gottfried, J. and Rein, G. Travelling fires for structural design – Part I: Literature review. Fire Safety Journal. 2012; 54: 74-85.
219. Stern-Gottfried, J. and Rein, G. Travelling fires for structural design – Part II: Design methodology. Fire Safety Journal. 2012; 54: 96-112.
220. Rackauskaite, E., Hamel, C., Law, A. and Rein, G. Improved Formulation of Travelling Fires and Application to Concrete and Steel Structures. Structures. 2015; 3: 250-60.
221. Audouin, L., Kolb, G., Torero, J.L. and Most, J.M. Average centreline temperatures of a buoyant pool fire obtained by image processing of video recordings. Fire Safety Journal. 1995; 24: 167-87.
222. Harmathy, T.Z. A new look at compartment fires, part I. Fire Technology. 1972; 8: 196-217.
223. Harmathy, T.Z. A new look at compartment fires, part II. Fire Technology. 1972; 8: 326-51.
224. Alpert, R.L. Calculation of response time of ceiling-mounted fire detectors. Fire Technology. 1972; 8: 181-95.
225. Law, A., Stern-Gottfried, J., Gillie, M. and Rein, G. The influence of travelling fires on a concrete frame. Engineering Structures. 2011; 33: 1635-42.
226. Horová, K., Wald, F. and Bouchair, A. Travelling Fire in Full-Scale Experimental Building. In: Jármai, K., Farkas, J., editors. Design, Fabrication and Economy of Metal Structures: Springer Berlin Heidelberg; 2013; pp. 371-76.

227. Cheng, X-d., Zhou, Y., Yang, H. and Li, K-y. Numerical Study on Temperature Distribution of Structural Components Exposed to Travelling Fire. Procedia Engineering. 2014; 71: 166-72.
228. Behnam, B. and Hashemi, Rezvani, F. Structural Evaluation of Tall Steel Moment-Resisting Structures in Simulated Horizontally Traveling Postearthquake Fire. ASCE's Journal of Performance of Constructed Facilities. 2014; 0: 04014207.
229. Behnam, B. On the Effect of Travelling Fire on the Stability of Seismic-damaged Large Reinforced Concrete Structures. Accepted (In Press) International Journal of Civil Engineering; 2016.
230. Wang, Y.C. Steel and composite structures: behaviour and design for fire safety. CRC Press; 2003.
231. Thomas, L.C. Fundamentals of heat transfer. 1980.
232. Torero, J.L. Assessing the true performance of structures in fire. Performance-based and Life-cycle Structural Engineering Hong Kong: the University of Hong Kong; 2012.
233. Sandstrom J. Temperature calculations in fire exposed structures with the use of adiabatic surface temperatures. Stockholm, Sweden: Lulea University of Technology; 2008.
234. Wong, M-B. and Ghojel, J. Sensitivity analysis of heat transfer formulations for insulated structural steel components. Fire Safety Journal. 2003; 38: 187-201.
235. Kwasniewski, A. Analyses of structures under fire. Warsaw, Poland: Warsaw University of Technology; 2011.
236. Lamont, S. The behaviour of multi-storey composite steel framed structures in response to compartment fires U164473. Scotland: The University of Edinburgh (United Kingdom); 2002.
237. Marchant, E.W. Fire and buildings. Building Science. 1973; 8: 94.
238. Cooke, G.M.F. An introduction to the mechanical properties of structural steel at elevated temperatures. Fire Safety Journal. 1988; 13: 45-54.
239. Rösler, J., Bäker, M., Harders, H. Creep. Mechanical Behaviour of Engineering Materials: Springer Berlin Heidelberg; 2007; p. 383-406.
240. Youssef, M.A. and Moftah, M. General stress-strain relationship for concrete at elevated temperatures. Engineering Structures. 2007; 29: 2618-34.
241. Bamonte, P., Gambarova, P.G. and Meda, A. Today's concretes exposed to fire—test results and sectional analysis. Structural Concrete. 2008; 9: 19-29.
242. Harmathy, T.Z. Fire Safety Desgin and Concrete: Longman Sceintific and Technical; 1993.
243. Chen, J. and Young, B. Stress-strain curves for stainless steel at elevated temperatures. Engineering Structures. 2006; 28: 229-39.
244. Petzold, A., Röhrs, M. and Neville, A.M. Concrete for high temperatures: translated from the German by A.B. Phillips and F.M. Turner, Maclaren 1970, 235 pp. £6. Building Science. 1971; 6: 33-4.
245. Bažant, Z.P. Creep of Concrete. In: Editors-in-Chief: KHJB, Robert, W.C., Merton, C.F., Bernard, I., Edward, J.K., Subhash, M. et al., editors. Encyclopedia of Materials: Science and Technology (Second Edition). Oxford: Elsevier; 2001; p. 1797-800.
246. Bocca, P.G. and Antonaci, P. Experimental study for the evaluation of creep in concrete through thermal measurements. Cement and Concrete Research. 2005; 35: 1776-83.

247. Franssen, J.M. User's manual for SAFIR 2011 a computer program for analysis of structures subjected to fire. Liege, Belgium: University of Liege, Belgium; 2011.
248. Luc, T. and Niels, P.H. Fire design of concrete structures – structural behaviour and assessment. Germany: International Federation for Structural Concrete (fib); 2008.
249. Malhotra, H.L. Design of Fire-Resisting Structures. First ed. East kilbride, Scotland: Thomson Litho Ltd.; 1982.
250. Schneeberger, H. A contribution to the stress-strain relationship of concrete. Materiaux et constructions. 1992; 25: 145-48.
251. Purkiss, J.A. Fire Safety Engineering: Design of Structures. 2 ed. Burlington, MA, USA: Elsevier; 1996.
252. Bamonte, P. Today's concretes exposed to fire-test results and sectional analysis. Structural Concrete. 2008; 9.
253. Omer, A. Effects of elevated temperatures on properties of concrete. Fire Safety Journal. 2007; 42: 516-22.
254. Minson A. Eurocode 2-3. Concrete Structures. 2006; 40: 30-1.
255. Smith, F.P. Concrete Spalling – Controlled Fire Tests and Review. Journal: Forensic Science Society. 1991; 31: 67-75.
256. Kodur, V.K.R., Wang, T.C. and Cheng, F.P. Predicting the fire resistance behaviour of high strength concrete columns. Cement and Concrete Composites. 2004; 26: 141-53.
257. Debicki, G., Haniche, R. and Delhomme, F. An experimental method for assessing the spalling sensitivity of concrete mixture submitted to high temperature. Cement and Concrete Composites. 2012; 34: 958-63.
258. Deeny, S., Stratford, T., Dhakal, R., Moss, P. and Buchanan, A. Spalling of concrete, implications for structural performance in fire. International Conference of Structural Fire Engineering Prague: Pražská technika, Czech Technical University; 2009; p. 202-7.
259. Raut, N. and Kodur, V. Response of Reinforced Concrete Columns under Fire-Induced Biaxial Bending. ACI structural journal. 2011; 108: 610-9, 26-27.
260. Khoury, G.A. and Anderberg, Y. Concrete spalling review Sweden Swedish National Road Administration; 2000.
261. Kodur, V.K.R. Guidelines for Fire Resistance Design of High-strength Concrete Columns. Journal of Fire Protection Engineering. 2005; 15: 93-106.
262. Jansson, R. and Boström, L. The Influence of Pressure in the Pore System on Fire Spalling of Concrete. Fire Technology. 2010; 46: 217-30.
263. Chandra, S. and Berntsson, L. Lightweight Aggregate Concrete – Science, Technology, and Applications. William Andrew Publishing/Noyes; 2002; p. 294.
264. Wong, Y-L., Fu, Y-F., Poon, C-S. and Tang, C-A. Spalling of concrete cover of fiber-reinforced polymer reinforced concrete under thermal loads. Materials and Structures. 2006; 39: 991-9.
265. Jansson, R. Experimental Investigation on Concrete Spalling in Fire. In: Pietro G. Gambarova, Roberto Felicetti, Alberto Meda, Riva P, editors. Fire Design of Concrete Structures: What now? What next? Milan, Italy: Milan University of Technology; 2005; p. 109-13.
266. Majorana, C.E., Salomoni, V.A., Mazzucco, G. and Khoury, G.A. An approach for modelling concrete spalling in finite strains. Mathematics and Computers in Simulation. 2010; 80: 1694-712.

267. Venkatesh, Kodur and Monther Dwaikat. Fire-induced spalling in reinforced concrete beams. Proceedings of the Institution of Civil Engineers Structures and Buildings. 2012; 165: 347.
268. Hertz, K.D. Limits of spalling of fire-exposed concrete. Fire Safety Journal. 2003; 38: 103-16.
269. Hertz, K.D. and Sørensen, L.S. Test method for spalling of fire exposed concrete. Fire Safety Journal. 2005; 40: 466-76.
270. Ali, F.A., Nadjai, A., Glackin, P., Silcock, G.W.H. and Abu-tair, A. Structural Performance of High Strength Concrete Columns in Fire. Seventh International Association of Fire Safety Science Symposium. Worcester, MA, USA; 2002; p. 1001-12.
271. Ali, F., Ali, N. and Abu-Tair, A. Explosive spalling of normal strength concrete slabs subjected to severe fire. Materials and Structures. 2011; 44: 943-56.
272. Huang, Z. Modelling of reinforced concrete structures in fire. Proceedings of the Institution of Civil Engineers Engineering and Computational Mechanics. 2010; 163: 43.
273. Jansson, R. Measurement of Concrete Thermal Properties at High Temperature. In: Pietro G. Gambarova, Roberto Felicetti, Alberto Meda, Riva P, editors. Fire Design of Concrete Structures: What now? What next? Milan, Italy: Milan University of Technology; 2005; p. 101-7.
274. Bamonte, P. Structural Behavior and Failure Modes of R/C at High Temperature: R/C Sections and 2-D Members. In: Pietro G. Gambarova, Roberto Felicetti, Alberto Meda, Riva P, editors. Fire Design of Concrete Structures: What now? What next? Milan, Italy: Milan University of Technology; 2005; p. 159-74.
275. Chandrasekaran, S. and Roy A. Seismic Evaluation of Multi-Storey RC Frame Using Modal Pushover Analysis. Nonlinear Dyn. 2006; 43: 329-42.
276. Billings, S. and Fakhouri, S. Identification of systems containing linear dynamic and static nonlinear elements. Automatica. 1982; 18: 15-26.
277. Berrah, M. and Kausel, E. Response spectrum analysis of structures subjected to spatially varying motions. Earthquake Engineering & Structural Dynamics. 1992; 21: 461-70.
278. Wilson, E.L., Der Kiureghian, A. and Bayo, E. A replacement for the SRSS method in seismic analysis. Earthquake Engineering & Structural Dynamics. 1981; 9: 187-92.
279. Sinha, R., Igusa, T. CQC and SRSS methods for non□classically damped structures. Earthquake Engineering & Structural Dynamics. 1995; 24: 615-19.
280. Humar, J. and Mahgoub, M.A. Determination of seismic design forces by equivalent static load method. Canadian Journal of Civil Engineering. 2003; 30: 287-307.
281. Chopra, A.K. Dynamics of structures: theory and applications to earthquake engineering. 2nd ed. Upper Saddle River, NJ: Prentice Hall; 2001.
282. Saatcioglu, M. and Humar, J. Dynamic analysis of buildings for earthquake-resistant design. Canadian Journal of Civil Engineering. 2003; 30: 338-59.
283. Fardis, M. Guidelines for displacement-based design of buildings and bridge. Risk Mitigation for Earthquake and Landslides. Pavia, Italy: IUSS Press; 2007.
284. Krawinkler, H. and Seneviratna, G.D.P.K. Pros and cons of a pushover analysis of seismic performance evaluation. Engineering Structures. 1998; 20: 452-64.
285. Fajfar, P. A Nonlinear Analysis Method for Performance Based Seismic Design. Earthq Spectra. 2000; 16: 573-92.

286. Kalkan, E. and Chopra, A.K. Modal-pushover-based ground-motion scaling procedure.(Author abstract). Journal of Structural Engineering (New York, NY). 2011; 137: 298.
287. Azimi, H., Galal, K. and Pekau, O.A. Incremental Modified Pushover Analysis. The Structural Design of Tall and Special Buildings. 2009; 18: 839-59.
288. Han, S.W. and Chopra, A.K. Approximate incremental dynamic analysis using the modal pushover analysis procedure. Earthquake Engineering & Structural Dynamics. 2006; 35: 1853-73.
289. Bagchi, A. A simplified methd of evaluating the seismic performance of buildings. Earthq Eng Eng Vib. 2004; 3: 223-36.
290. Chopra, A.K. and Goel, R.K. A modal pushover analysis procedure to estimate seismic demands for unsymmetric-plan buildings. Earthquake Engineering & Structural Dynamics. 2004; 33: 903-27.
291. Mwafy, A.M. and Elnashai, A.S. Static pushover versus dynamic collapse analysis of RC buildings. Engineering Structures. 2001; 23: 407-24.
292. Shakeri, K., Tarbali, K. and Mohebbi, M. An adaptive modal pushover procedure for asymmetric-plan buildings. Engineering Structures. 2012; 36: 160-72.
293. Shakeri, K., Shayanfar, M.A. and Kabeyasawa, T. A story shear-based adaptive pushover procedure for estimating seismic demands of buildings. Engineering Structures. 2010; 32: 174-83.
294. Chopra, A.K., Goel, R.K. and Chintanapakdee, C. Evaluation of a modified MPA procedure assuming higher modes as elastic to estimate seismic demands. Earthq Spectra. 2004; 20: 757-78.
295. FEMA273. NEHRP Guidelines for the Seismic Rehabilitation of Buildings. Washington, DC: American Society of Civil Engineers; 1997.
296. Papanikolaou, V.K. and Elnashai, A.S. Evaluation of Conventional and Adaptive Pushover Analysis I: Methodology. Journal of Earthquake Engineering. 2005; 9: 923-41.
297. Mortezaei, A., Ronagh, H.R., Kheyroddin, A. and Amiri, G.G. Effectiveness of modified pushover analysis procedure for the estimation of seismic demands of buildings subjected to near□fault earthquakes having forward directivity. The Structural Design of Tall and Special Buildings. 2011; 20: 679-99.
298. Tirca, L. and Chen, L. The influence of lateral load patterns on the seismic design of zipper braced frames. Engineering Structures. 2012; 40: 536-55.
299. Kalkan, E. and Kunnath, S.K. Assessment of current nonlinear static procedures for seismic evaluation of buildings. Engineering Structures. 2007; 29: 305-16.
300. Council, B.S.S. NEHRP recommended provisions for the development of seismic regulations for new buildings: Part 1: Provisions. Earthquake Hazard Reductions Series; 2003.
301. Fajfar, P. Capacity spectrum method based on inelastic demand spectra. Earthquake Engineering and Structural Dynamics. 1999; 28: 979-93.
302. Council, B.S.S., Agency, U.S.F.E.M. and Council, A.T. NEHRP guidelines for the seismic rehabilitation of buildings. Washington, D.C.: Federal Emergency Management Agency; 2003.
303. Roh, H., Reinhorn, A.M. and Lee, J.S. Power spread plasticity model for inelastic analysis of reinforced concrete structures. Engineering Structures. 2012; 39: 148-61.
304. Karthik, M.M. and Mander, J.B. Stress-block parameters for unconfined and confined concrete based on a unified stress-strain model. (Technical Notes)

(Author abstract). Journal of Structural Engineering (New York, NY). 2011; 137: 270.
305. Firat, Alemdar Z. Plastic hinging behavior of reinforced concrete bridge columns 3403076. United States – Kansas: University of Kansas; 2010.
306. Inel, M. and Ozmen, H.B. Effects of plastic hinge properties in nonlinear analysis of reinforced concrete buildings. Engineering Structures. 2006; 28: 1494-502.
307. Kwak, H-G. and Kim, S-P. Nonlinear analysis of RC beams based on moment–curvature relation. Computers & Structures. 2002; 80: 615-28.
308. Warner, R.F. Concrete structures. South Melbourne, Vic.: Longman; 1998.
309. Bae, S. and Bayrak, O. Seismic Performance of Full-Scale Reinforced Concrete Columns. ACI Structural Journal. 2008; 105: 123-33.
310. Scott, B.D., Park, R. and Priestley, M.J.N. Stress-Strain Behavior of Concrete Confined by Overlapping Hoops at Low and High Strain Rates. Journal of the American Concrete Institute. 1982; 79: 13-27.
311. Teng-Hooi, T. and Ngoc-Ba, N. Flexural Behavior of Confined High-Strength Concrete Columns. ACI Structural Journal. 2005; 102: 198-205.
312. Kent, D.C. and Park R. Flexural members with confined concrete. Journal of the Structural Division. 1971; 97: 1969-90.
313. Mander, J.B., Priestley, M.J.N. and Park, R. Theoretical Stress▯Strain Model for Confined Concrete. Journal of Structural Engineering (New York, NY). 1988; 114: 1804-26.
314. Priestley, M.J.N., Calvi, G.M. and Seible, F. Seismic design and retrofit of bridges. New York: Wiley; 1996.
315. Bayrak, O. and Sheikh, S.A. Plastic hinge analysis. Journal of Structural Engineering (New York, NY). 2001; 127: 1092-100.
316. Bae, S. and Bayrak, O. Plastic Hinge Length of Reinforced Concrete Columns. ACI Structural Journal. 2008; 105: 290-300.
317. Franssen, J-M. SAFIR: A thermal/structural program for modeling structures under fire. Engineering Journal. 2005; 42: 143-50.
318. Lim, L., Buchanan, A., Moss, P. and Franssen, J-M. Numerical modelling of two-way reinforced concrete slabs in fire. Engineering Structures. 2004; 26: 1081-91.
319. Moss, P.J., Dhakal, R.P., Wang, G. and Buchanan, A.H. The fire behaviour of multi-bay, two-way reinforced concrete slabs. Engineering Structures. 2008; 30: 3566-73.
320. Chang, J.J. Computer simulation of hollowcore concrete flooring systems exposed to fire research. Christchurch, New Zealand: University of Canterbury; 2007.
321. Lee, T-H. and Mosalam, K.M. Probabilistic fiber element modeling of reinforced concrete structures. Computers & Structures. 2004; 82: 2285-99.
322. Irschik, H. Analogy between refined beam theories and the Bernoulli-Euler theory. International Journal of Solids and Structures. 1991; 28: 1105-12.
323. Gernay, T., Franssen, J.M. SAFIR MANUAL Materials SILCON_ETC and CALCON_ETC. University of Liege, Belgium: Environnement & Constructions Structural Engineering; 2011.
324. Crisfield, M.A. A fast incremental/iterative solution procedure that handles "snap-through". Computers & Structures. 1981; 13: 55-62.
325. Franssen, J.M., Kodur, V.K.R. and Mason, J. Elements of Theory for SAFIR2001 Free: A Computer Program for Analysis of Structures at Elevated Temperature Conditions. University of Liege, Belgium; 2002.

326. SAP2000-V14. Integrated finite element analysis and design of structures basic analysis reference manual. Berkeley, CA, USA; 2002.
327. Behnam, B. and Ronagh, H.R. Methodology for Investigating the Behavior of Reinforced Concrete Structures Subjected to Post Earthquake Fire. Advances in Concrete Construction, An International Journal. 2013; 1: 29-44.
328. Behnam, B. and Ronagh, H.R. Post-Earthquake Fire Performance-Based Behavior of Reinforced Concrete Structures. Earthquakes and Structures, An International Journal. 2013; 5: 379-94.
329. Beeby, A.W. and Scott, R.H. Insights into the cracking and tension stiffening behaviour of reinforced concrete tension members revealed by computer modelling. Magazine of Concrete Research. 2004; 56: 179-90.
330. Beeby, A.W. and Scottt, R.H. Cracking and deformation of axially reinforced members subjected to pure tension. Magazine of Concrete Research. 2005; 57: 611-21.
331. Kong, K.L., Beeby, A.W., Forth, J.P. and Scott, R.H. Cracking and tension zone behaviour in RC flexural members. Proceedings of the ICE – Structures and Buildings; 2007; p. 165-72.
332. Kong, S.M.K. A Study of Implementing Performance-Based Design for Fire Safety Provisions in Higher Education Institutes 3500878. Hong Kong: Hong Kong Polytechnic University (Hong Kong); 2011.
333. Shi, X., Tan, T., Tan, K. and Guo, Z. Influence of Concrete Cover on Fire Resistance of Reinforced Concrete Flexural Members. Journal of Structural Engineering. 2004; 130: 1225-32.
334. Vejmelková, E., Padevět, P. and Černý R. Effect of cracks on hygric and thermal characteristics of concrete. Bauphysik. 2008; 30: 438-44.
335. Ervine, A., Gillie, M., Stratford, T.J. and Pankaj, P. Thermal propagation through tensile cracks in reinforced concrete. Journal of Materials in Civil Engineering. 2012; 24: 516-22.
336. Wu, B., Xiong, W. and Wen, B. Thermal fields of cracked concrete members in fire. Fire Safety Journal. 2014.
337. ACI318. Building code requirements for structural concrete (ACI 318-08) and commentary. America: American Concrete Institute; 2008.
338. Harmathy, T.Z. Properties of building materials; 1988.
339. Ronagh, H.R. and Behnam, B. Investigating the Effect of Prior Damage on the Post-earthquake Fire Resistance of Reinforced Concrete Portal Frames. International Journal of Concrete Structures and Materials. 2012; 6: 209-20.
340. Usmani A. Stability of the World Trade Center Twin Towers Structural Frame in Multiple Floor Fires. Journal of Engineering Mechanics. 2005; 131: 654-7.
341. Flint, G. Fire Induced Collapse of Tall Buildings. Edinburgh: The University of Edinburgh; 2005.
342. Röben, C., Gillie, M. and Torero, J. Structural behaviour during a vertically travelling fire. Journal of Constructional Steel Research. 2010; 66: 191-97.
343. Quiel, S.E. and Garlock, M.E. Modeling high-rise steel framed buildings under fire. ASCE Struct Congr. 2008; 1-10.
344. Behnam, B. and Ronagh, H. Performance of reinforced concrete structures subjected to fire following earthquake. European Journal of Environmental and Civil Engineering. 2013; 17: 270-92.
345. Behnam, B. Investigating the Effect of Vertically Travelling Post-Earthquake Fire on Tall Reinforced Concrete Structures. Journal of Iranian Society of Civil Engineering. 2014; 16: 14-26.

346. Chung, K.F. Fire resistance design of composite slabs in building structures: from research to practice. The Structural Engineer. 2006; 84.
347. Fletcher, I.A., Borg, A., Hitchen, N. and Welch, S. Performance of concrete in fire: a review of the state of the art, with a case study of the windsor tower fire. The 4th International Workshop in Structures in Fire. Aveiro, Portugal; 2006; p. 779-90.
348. Behnam, B. and Ronagh, H.R. Post-Earthquake Fire performance-based behavior of unprotected moment resisting 2D steel frames. KSCE Journal of Civil Engineering. 2015; 19: 274-84.
349. ASCE. Minimum design loads for buildings and other structures SEI/ASCE 7-10. Washington (DC): American Society of Civil Engineers; 2013.
350. Behnam, B. and Ronagh, H.R. Behavior of moment-resisting tall steel structures exposed to a vertically traveling post-earthquake fire. The Structural Design of Tall and Special Buildings. 2014; 23: 1083–96.
351. Holicky, M., Alois Meterna, Gerhard Sedlacek and Schleich, J-B. Implementation of Eurocodes, Handbook 5, Design of buildings for the fire situation. Luxemboug: Leonardo da Vinci Pilot Project; 2005.
352. Dorfman, M.S. Introduction to risk management and insurance. 8th ed. Upper Saddle River, NJ: Pearson Prentice Hall; 2005.
353. Behnam, B., Skitmore, M. and Ronagh, H.R. Risk mitigation of post-earthquake fire in urban buildings. Journal of Risk Research. 2014; 1-20.
354. Coppola, D.P. Introduction to international disaster management. Boston: Butterworth-Heinemann; 2011.
355. Lin, Moe T. and Pathranarakul, P. An integrated approach to natural disaster management: public project management and its critical success factors. Disaster Prevention and Management: An International Journal. 2006; 15: 396-413.
356. Wisner, B., Blaikie, P., Cannon, T. and Davis, I. At risk: Natural hazards, people's vulnerability and disasters. 2004.
357. Nishino, T., Tanaka, T. and Hokugo, A. An evaluation method for the urban post-earthquake fire risk considering multiple scenarios of fire spread and evacuation. Fire Safety Journal. 2012; 54: 167-80.
358. Koike, T., Kanamori, T. and Sato, Y. An evacuation plan from a fire following earthquake in a congested city area. Journal of Earthquake and Tsunami. 2010; 04: 33-49.
359. Flint, C. and Brennan, M. Community emergency response teams: From disaster responders to community builders. Rural Realities. 2006; 1: 1-9.
360. Okumura, T., Suzuki, K., Fukuda, A., Kohama, A., Takasu, N., Ishimatsu, S. et al. The Tokyo Subway Sarin Attack: Disaster Management, Part 1: Community Emergency Response. Academic Emergency Medicine. 1998; 5: 613-17.
361. Simpson, D.M. Community Emergency Response Training (CERTs): A Recent History and Review. Natural Hazards Review. 2001; 2: 54-63.
362. Blanchard, B.S. and Fabrycky, W.J. Systems engineering and analysis. Prentice Hall Englewood Cliffs, New Jersey; 1990.
363. Wallenius, J. and Zionts, S. Multiple criteria decision making: from early history to the 21st century: World Scientific; 2011.
364. Linstone, H.A. and Turoff, M. The Delphi method: Techniques and applications: Addison-Wesley Reading, MA; 1975.
365. Saaty, T.L. Decision making with the analytic hierarchy process (AHP). International Journal of Services Sciences. 2008; 1: 83-98.

366. Behnam, B. and Ronagh, H.R. An Engineering Solution to Improve Post-Earthquake Fire Resistance in Important Reinforced Concrete Structures. Adv Struct Eng. 2014; 17: 993-1009.
367. Behnam, B. and Ronagh, H.R. A Post-Earthquake Fire Factor to Improve the Fire Resistance of Damaged Ordinary Reinforced Concrete Structures. Journal of Structural Fire Engineering. 2013; 4: 207-26.
368. Grace, N.F., Ragheb, W.F. and Abdel-Sayed, G. Ductile FRP Strengthening Systems. Concrete International. 2005; 31-6.
369. Bousselham, A. State of Research on Seismic Retrofit of RC Beam-Column Joints with Externally Bonded FRP. Journal of Composites for Construction. 2010; 14: 49-61.
370. Mahini, S.S. and Ronagh, H.R. Web-bonded FRPs for relocation of plastic hinges away from the column face in exterior RC joints. Composite Structures. 2011; 93: 2460-72.
371. Mahini, S.S. and Ronagh, H.R. Strength and ductility of FRP web-bonded RC beams for the assessment of retrofitted beam–column joints. Composite Structures. 2010; 92: 1325-32.
372. Le-Trung, K., Lee, K., Lee, J., Lee, D.H. and Woo, S. Experimental study of RC beam–column joints strengthened using CFRP composites. Composites Part B: Engineering. 2010; 41: 76-85.
373. Attari, N., Amziane, S. and Chemrouk, M. Efficiency of Beam–Column Joint Strengthened by FRP Laminates. Advanced Composite Materials. 2010; 19: 171-83.
374. Asaro, R.J., Lattimer, B. and Ramroth, W. Structural response of FRP composites during fire. Composite Structures. 2009; 87: 382-93.
375. Firmo, J.P., Correia, J.R. and França, P. Fire behaviour of reinforced concrete beams strengthened with CFRP laminates: Protection systems with insulation of the anchorage zones. Composites Part B: Engineering. 2012; 43: 1545-56.
376. Keller, T. and Bai, Y. Structural Performance of FRP Composites in Fire. Adv Struct Eng. 2010; 15: 793-804.
377. Gibson, A.G. and Mouritz, A.P. Fire properties of polymer composite materials. Dordrecht: Kluwer Academic Publishers; 2006.
378. Williams, B., Kodur, V., Green, M.F. and Bisby, L. Fire Endurance of Fiber-Reinforced Polymer Strengthened Concrete T-Beams. ACI Structural Journal. 2008; 105: 60-7.
379. Behnam, B., Lim, P.J. and Ronagh, H.R. Plastic Hinge Relocation in Reinforced Concrete Frames as a Method of Improving Post-earthquake Fire Resistance Structures. 2015.
380. Behnam, B. and Ronagh, H.R. Post-earthquake fire resistance of CFRP strengthened reinforced concrete structures. The Structural Design of Tall and Special Buildings. 2014; 23: 814–32.
381. Behnam, B. Risk Assessment of Post-earthquake Fire in Dense Urban Regions. In: Keshvari, R., editor. The First National Conference on Fire and Urban Safety. Shahid Beheshti University, Tehran, Iran: Tehran Fire Rescue Organiztion; 2016; p. 72-80.
382. Behnam, B. and Ronagh, H.R. Firewalls and post-earthquake fire resistance of reinforced-concrete frames. Proceedings of the Institution of Civil Engineers – Structures and Buildings. 2016; 169: 20-33.

Index

F

G

H

I

L

M

N

O

P

U

V

W